Hydrothermal Analysis in Engineering Using Control Volume Finite Element Method

Hydrothermal Analysis in Engineering Using Control Volume Finite Element Method

Mohsen Sheikholeslami Kandelousi
Department of Mechanical Engineering, Babol University of Technology

Davood Domairry Ganji
Department of Mechanical Engineering, Babol University of Technology

AMSTERDAM • BOSTON • HEIDELBERG • LONDON
NEW YORK • OXFORD • PARIS • SAN DIEGO
SAN FRANCISCO • SINGAPORE • SYDNEY • TOKYO
Academic Press is an imprint of Elsevier

Academic Press is an imprint of Elsevier
125, London Wall, EC2Y 5AS
525 B Street, Suite 1800, San Diego, CA 92101-4495, USA
225 Wyman Street, Waltham, MA 02451, USA
The Boulevard, Langford Lane, Kidlington, Oxford OX5 1GB, UK

British Library Cataloguing in Publication Data

A catalogue record for this book is available from the British Library

Library of Congress Cataloging-in-Publication Data

A catalog record for this book is available from the Library of Congress

ISBN: 978-0-12-802950-3

For information on all Academic Press publications
visit our website at http://store.elsevier.com/

Contents

Nomenclature

$\boldsymbol{A}$ amplitude
$\boldsymbol{C_p}$ specific heat at a constant pressure
$\boldsymbol{D_B}$ Brownian diffusion coefficient
$\boldsymbol{D_T}$ thermophoretic diffusion coefficient
$\boldsymbol{Gr_f}$ Grashof number
$\boldsymbol{Ha}$ Hartmann number $\left(HB_x\sqrt{\sigma_f/\mu_f}\right)$
$\boldsymbol{Le}$ Lewis number (α/D_B)
$\boldsymbol{N}$ number of undulations
$\boldsymbol{Nb}$ Brownian motion parameter $\left((\rho c)_p D_B(\phi_h-\phi_c)/(\rho c)_f\alpha\right)$
$\boldsymbol{Nt}$ thermophoretic parameter $\left((\rho c)_p D_T(T_h-T_c)/\left[(\rho c)_f\alpha T_c\right]\right)$
$\boldsymbol{Nu}$ Nusselt number
$\boldsymbol{Nr}$ Buoyancy ratio $\left(\left(\rho_p-\rho_0\right)(\phi_h-\phi_c)/\left[(1-\phi_c)\rho_{f_0}\beta L(T_h-T_c)\right]\right)$
$\boldsymbol{Pr}$ Prandtl number (υ_f/α_f)
$\boldsymbol{J}$ electric current
$\boldsymbol{T}$ fluid temperature
$\boldsymbol{u,v}$ velocity components in the x-direction and y-direction
$\boldsymbol{U,V}$ dimensionless velocity components in the X-direction and Y-direction
$\boldsymbol{x,y}$ space coordinates
$\boldsymbol{X,Y}$ dimensionless space coordinates
$\boldsymbol{r}$ nondimensional radial distance
$\boldsymbol{k}$ thermal conductivity
$\boldsymbol{L}$ gap between inner and outer boundary of the enclosure $L=r_{out}-r_{in}$
$\vec{g}$ gravitational acceleration vector
$\boldsymbol{Ra}$ Rayleigh number
$\boldsymbol{q''}$ heat flux
$\boldsymbol{Rd}$ radiation parameter

GREEK SYMBOLS

γ angle measured from right plane
ζ inclination angle
γ angle of turn of the semiannulus enclosure
ε eccentricity
α thermal diffusivity
σ electrical conductivity
ϕ volume fraction
μ dynamic viscosity
ω vorticity
Ω dimensionless vorticity
υ kinematic viscosity
ψ stream function

Ψ dimensionless stream function
Θ dimensionless temperature
ρ fluid density
β thermal expansion coefficient
μ_0 magnetic permeability of vacuum

SUBSCRIPTS

c cold
h hot
loc local
ave average
nf nanofluid
f base fluid
p solid particles
in inner
out outer

Preface

In this book we provide readers with the fundamentals of the control volume finite element method (CVFEM) for heat and fluid flow problems. The CVFEM comprises interesting characteristics of both the finite volume and finite element methods. It combines the flexibility of the finite element methods to discretize complex geometry with the conservative formulation of the finite volume methods, in which variables can be easily interpreted physically in terms of fluxes, forces, and sources. Most other available texts concentrate on solids problems. We applied this method for flow and heat transfer. Application of CVFEM in different interesting fields such as nanofluid flow and heat transfer, magnetohydrodynamics, ferrohydrodynamics, and porous media is considered. Several examples have been prepared in these fields of science. This text is suitable for senior undergraduate students, postgraduate students, engineers, and scientists.

The first chapter of the book deals with the essential fundamentals of the CVFEM. The necessary ingredients in numerical solutions are discussed. Chapter 2 deals with solving Navier-Stokes equations (in vorticity stream function form) and energy equations. In this chapter two basic important problems are solved via the CVFEM. The third chapter gives a complete account of the nanofluid hydrothermal behavior and application of CVFEM to solve such problems. All the relevant differential equations are derived from first principles. All three types of convection modes—forced, mixed, and natural convection—are discussed in detail. Examples and comparisons are provided to support the accuracy and flexibility of the CVEFM. Examples start with a single-phase model and then extend to a two-phase model.

The application of the CVEFM to heat and fluid flow in the presence of a magnetic field are discussed in detail in Chapter 4. Two kinds of magnetic fields are considered: a constant magnetic field and a spatially variable magnetic field. Several examples included in Chapter 4 give the reader a full account of the theory and practice associated with the CVFEM. Chapter 5 gives the procedures for solving equations of flow and heat transfer in porous media. Some examples provide readers with an opportunity to learn about the effect of active parameters. A sample FORTRAN code for advection-diffusion in lid-driven cavity geometry is presented in Appendix. Readers will be able to extend this code and solve all of the examples within this book.

Mohsen Sheikholeslami Kandelousi
Davood Domairry Ganji

CHAPTER

Control volume finite element method (CVFEM)

1

1.1 INTRODUCTION

Fluid flow has several applications in engineering and nature. Mathematically, real flows are governed by a set of nonlinear partial differential equations in complex geometry. So, suitable solutions can be obtained through numerical techniques such as the finite difference method, the finite volume method (FVM), and the finite element method (FEM). In the past decade the FEM has been developed for use in the area of computational fluid dynamics; this method has now become a powerful method to simulate complex geometry. However, the FVM is applied most in calculating fluid flows. The control volume finite element method (CVFEM) combines interesting characteristics from both the FVM and FEM. The CVFEM was presented by Baliga and Patankar [1,2] using linear triangular finite elements and by Raw and Schneider [3] using linear quadrilateral elements. Several authors have improved the CVFEM from then to now. Raw et al. [4] applied a nine-nodded element to solve heat conduction problems. Banaszek [5] compared the Galerkin and CVFEM methods in diffusion problems using six-nodded and nine-nodded elements. Campos Silva et al. [6] developed a computational program using nine-nodded finite elements based on a control volume formulation to simulate two-dimensional transient, incompressible, viscous fluid flows. Campos Silva and Moura [7] and Campos Silva [8] presented results for fluid flow problems. The CVFEM combines the flexibility of FEMs to discretize complex geometry with the conservative formulation of the FVMs, in which the variables can be easily interpreted physically in terms of fluxes, forces, and sources. Saabas and Baliga [9,10] referenced a list of several works related to FVMs and CVFEMs. Voller [11] presented the application of CVFEM for fluids and solids. Sheikholeslami et al. [12] studied the problem of natural convection between a circular enclosure and a sinusoidal cylinder. They concluded that streamlines, isotherms, and the number, size, and formation of cells inside the enclosure strongly depend on the Rayleigh number, values of amplitude, and the number of undulations of the enclosure.

1.2 DISCRETIZATION: GRID, MESH, AND CLOUD

In general there are three ways to place node points into a domain [11].

1.2.1 GRID

A basic approach assigns the location of nodes using a structured grid where, in a two-dimensional domain, the location of a node is uniquely specified by a row and a column index (Fig. 1.1a). Although such a structured approach can lead to convenient and efficient discrete equations, it lacks flexibility in accommodating complex geometries or allowing for the local concentration of nodes in solution regions of particular interest.

1.2.2 MESH

Geometric flexibility, usually at the expense of solution efficiency, can be added by using an unstructured mesh. Figure 1.1b shows an unstructured mesh of triangular elements. In two-dimensional domains triangular meshes are good selections because they can tessellate any planar surface. Note, however, that other choices of elements can be used in place of or in addition to triangular elements. The mesh can be used to determine the placement of the nodes. A common choice is to place the nodes at the vertices of the elements. In the case of triangles this allows for the

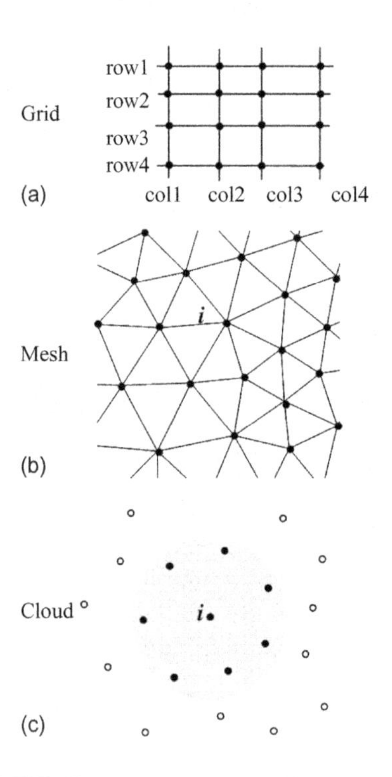

FIGURE 1.1

Different forms of discretization [11], including a grid (a), mesh (b), and cloud (c).

approximation of a dependent variable over the element, by linear interpolation between the vertex nodes. Higher-order approximations can be arrived by using more nodes (e.g., placing nodes at midpoints) or alternative elements (e.g., quadrilaterals). When considering an unstructured mesh recognizing the following is important:

1. The quality of the numerical solution obtained is critically dependent on the mesh. For example, avoiding highly acute angles is a key quality requirement for a mesh of triangular elements. The generation of appropriate meshes for a given domain is a complex topic worthy of a monograph in its own right. Fortunately, for two-dimensional problems in particular, there is a significant range of commercial and free software that can be used to generate quality meshes.
2. The term *unstructured* is used to indicate a lack of a global structure that relates the position of all the nodes in the domain. In practice, however, a local structure—the region of support—listing the nodes connected to a given node i is required. Establishing, storing, and using this local data structure is one of the critical ingredients in using an unstructured mesh.

1.2.3 CLOUD

The most flexible discretization is to simply populate the domain with node points that have no formal background grid or mesh connecting the nodes. Solution approaches based on this "meshless" form of discretization create local and structures, usually based on a "cloud" of neighboring nodes that fall within a given length scale of a given node i [13] (Fig. 1.1c).

1.3 ELEMENT AND INTERPOLATION SHAPE FUNCTIONS

A building block of discretization is the triangular element (Fig. 1.2). For linear triangular elements the node points are placed at the vertices. In Fig. 1.2, the nodes, moving in a counterclockwise direction, are labeled 1, 2, and 3. Values of the dependent variable ϕ are calculated and stored at these node points.

In this way, values at an arbitrary point (x, y) within the element can be approximated with linear interpolation

$$\phi \approx ax + by + c, \tag{1.1}$$

where the constant coefficients a, b, and c satisfy the nodal relationships

$$\phi_i = ax_i + by_i + c, \quad i = 1, 2, 3. \tag{1.2}$$

Equation (1.1) can be more conveniently written in terms of the shape function N_1, N_2, and N_3, where

$$N_i(x, y) = \begin{cases} 1 & \text{At node } i \\ 0 & \text{At all points on side opposite node } i \end{cases} \tag{1.3}$$

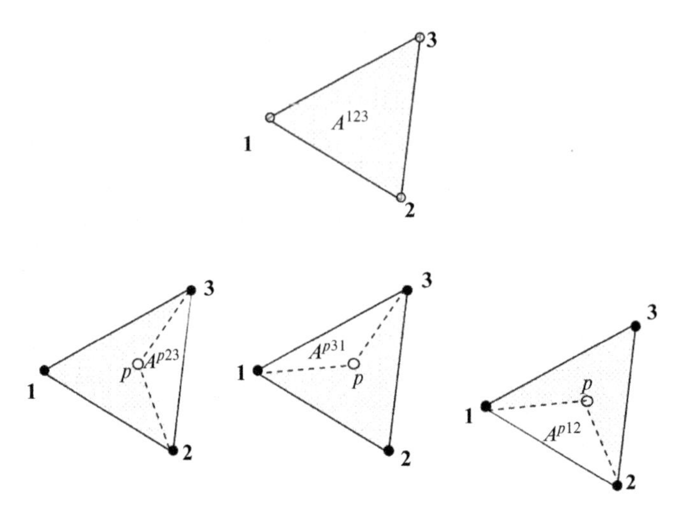

FIGURE 1.2

An element indicating the areas used in shape function definitions [11].

$$\sum_{i=1}^{3} N_i(x, y) = 1 \quad \text{At every point in the element} \tag{1.4}$$

such that, over the element the continuous unknown field can be expressed as the linear combination of the values at nodes $i = 1,2,3$:

$$\phi(x, y) \approx \sum_{i=1}^{3} N_i(x, y)\phi_i. \tag{1.5}$$

With linear triangular elements a straightforward geometric derivation for the shape functions can be obtained. With reference to Fig. 1.2, observe that the area of the element is given by

$$A^{123} = \frac{1}{2}\begin{vmatrix} 1 & x_1 & y_1 \\ 1 & x_1 & y_1 \\ 1 & x_1 & y_1 \end{vmatrix} = \frac{1}{2}[(x_2y_3 - x_3y_2) - x_1(y_3 - y_2) + y_1(x_3 - x_2)] \tag{1.6}$$

and the area of the subelements with vertices at points $(p, 2, 3)$, $(p, 3, 1)$, and $(p, 1, 2)$, where p is an arbitrary and variable point in the element, are given by

$$\begin{aligned} A^{p23} &= [(x_2y_3 - x_3y_2) - x_p(y_3 - y_2) + y_p(x_3 - x_2)] \\ A^{p31} &= [(x_3y_1 - x_1y_3) - x_p(y_1 - y_3) + y_p(x_1 - x_3)]. \\ A^{p12} &= [(x_1y_2 - x_2y_1) - x_p(y_2 - y_1) + y_p(x_2 - x_1)] \end{aligned} \tag{1.7}$$

Based on these definitions, it follows that the shape functions are given by

$$N_1 = A^{P23}/A^{123}, \quad N_2 = A^{P31}/A^{123}, \quad N_3 = A^{P12}/A^{123}. \tag{1.8}$$

Note that, when point p coincides with node i $(1, 2, \text{or } 3)$, the shape function $N_i = 1$, and when point p is anywhere on the element side opposite node i, the associated subelement area is zero, and, through Eq. (1.8), the shape function $N_i = 0$. Hence the shape functions defined by Eq. (1.8) satisfy the required condition in Eq. (1.3). Further, note that at any point p, the sum of the areas:

$$A^{P23} + A^{P31} + A^{P12} = A^{123} \tag{1.9}$$

is such that the shape functions at (x_p, y_p) sum to unity. Hence the shape functions defined by Eq. (1.8) also satisfy the conditions in Eq. (1.4). For future reference, it is worthwhile to note that the following constants are the derivatives of the shape functions in Eq. (1.8) over the element:

$$\begin{aligned}
N_{1x} &= \frac{\partial N_1}{\partial x} = \frac{(y_2 - y_3)}{2A^{123}}, \quad N_{1y} = \frac{\partial N_1}{\partial y} = \frac{(x_2 - x_3)}{2A^{123}} \\
N_{2x} &= \frac{\partial N_2}{\partial x} = \frac{(y_3 - y_2)}{2A^{123}}, \quad N_{2y} = \frac{\partial N_2}{\partial y} = \frac{(x_1 - x_3)}{2A^{123}} \\
N_{3x} &= \frac{\partial N_3}{\partial x} = \frac{(y_1 - y_2)}{2A^{123}}, \quad N_{3y} = \frac{\partial N_3}{\partial x} = \frac{(x_2 - x_1)}{2A^{123}}
\end{aligned} \tag{1.10}$$

1.4 REGION OF SUPPORT AND CONTROL VOLUME

The local structure on the mesh in Fig. 1.1b is defined in terms of the region of support—the list of nodes that share a common element with a given node i [11] (Fig. 1.3). In this region of support, as illustrated in Fig. 1.3, a control volume is created by joining the center of each element in the support to the midpoints of the

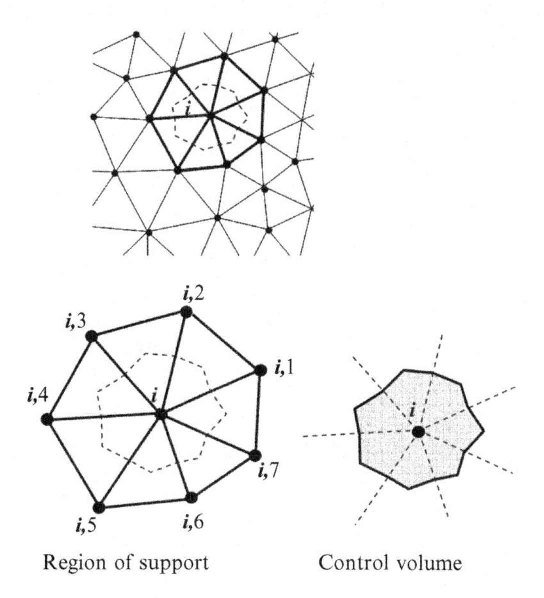

FIGURE 1.3

Region of support and control volume for node i in an unstructured mesh of linear triangular elements [11].

element sides that pass through node i. This creates a closed polygonal control volume with $2m$ sides (faces), where m is the number of elements in the support. Each element contributes one-third of its area to the control volume area, and the volumes from all the nodes tessellate the domain without overlapping.

1.5 DISCRETIZATION AND SOLUTION

1.5.1 STEADY-STATE ADVECTION-DIFFUSION WITH SOURCE TERMS

To illustrate a solution procedure using the CVFEM, one can consider the general form of advection-diffusion equation for node i in integral form:

$$-\int_V Q\mathrm{d}V - \int_A k\nabla\phi\cdot n\mathrm{d}A + \int_A (v\cdot n)\phi\mathrm{d}A = 0 \tag{1.11}$$

or point form

$$-\nabla\cdot(k\nabla\phi) + \nabla\cdot(v\phi) - Q = 0, \tag{1.12}$$

which can be represented by the system of CVFEM discrete equations as:

$$[a_i + Qc_i + Bc_i]\phi_i = \sum_{j=1}^{n_i} a_{i,j}\phi_{S_{i,j}} + Q_{B_i} + B_{B_i}. \tag{1.13}$$

In Eq. (1.13), the a's are the coefficients, the index (i, j) indicates the jth node in the support of node i, the index $S_{i,j}$ provides the node number of the jth node in the support, the Bs account for boundary conditions, and the Qs for source terms. For the selected triangular element shown in Fig. 1.4, this approximation (without considering the source term) leads to:

$$-\left(a_1^k + a_1^u\right)\phi_i + \left(a_2^k + a_2^u\right)\phi_{S_{i,3}} + \left(a_2^k + a_2^u\right)\phi_{S_{i,4}} = 0 \tag{1.14}$$

Using upwind method for advection coefficients identified by the superscript u, are given by

$$\begin{aligned} a_1^u &= \max\left[q_{f1}, 0\right] + \max\left[q_{f2}, 0\right] \\ a_2^u &= \max\left[-q_{f1}, 0\right] \\ a_3^u &= \max\left[-q_{f2}, 0\right] \end{aligned} \tag{1.15}$$

The diffusion coefficients, identified with the superscript k, are given by

$$\begin{aligned} a_1^k &= -k_{f1}N_{1x}\Delta\vec{y}_{f1} + k_{f1}N_{1y}\Delta\vec{x}_{f1} - k_{f2}N_{1x}\Delta\vec{y}_{f2} + k_{f2}N_{1y}\Delta\vec{x}_{f2} \\ a_2^k &= -k_{f1}N_{2x}\Delta\vec{y}_{f1} + k_{f1}N_{2y}\Delta\vec{x}_{f1} - k_{f2}N_{2x}\Delta\vec{y}_{f2} + k_{f2}N_{2y}\Delta\vec{x}_{f2}\,. \\ a_2^k &= -k_{f1}N_{3x}\Delta\vec{y}_{f1} + k_{f1}N_{3y}\Delta\vec{x}_{f1} - k_{f2}N_{3x}\Delta\vec{y}_{f2} + k_{f2}N_{3y}\Delta\vec{x}_{f2} \end{aligned} \tag{1.16}$$

In Eq. (1.15), the volume flow across faces 1 and 2 in the direction of the outward normal is

$$\begin{aligned} q_{f1} &= v\cdot nA|_{f1} = v_x^{f1}\Delta\vec{y}_{f1} - v_y^{f1}\Delta\vec{y}_{f1} \\ q_{f2} &= v\cdot nA|_{f2} = v_x^{f2}\Delta\vec{y}_{f2} - v_y^{f2}\Delta\vec{y}_{f2} \end{aligned} \tag{1.17}$$

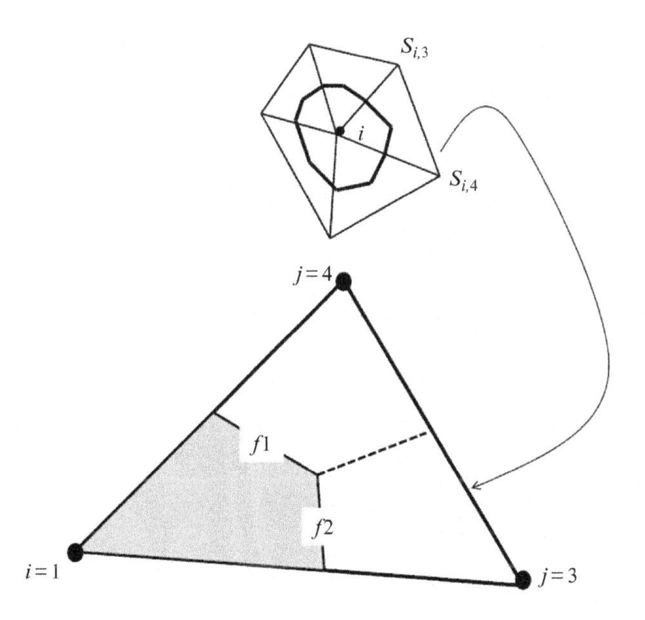

FIGURE 1.4

A sample triangular element and its corresponding control volume.

The value of the diffusivity at the midpoint of face 1 can be obtained as

$$k_{f1} = [N_1k_1 + N_2k_2 + N_3k_3]_{f1} = \frac{5}{12}k_1 + \frac{5}{12}k_2 + \frac{2}{12}k_3 \tag{1.18}$$

and at the midpoint of face 2 as

$$k_{f2} = [N_1k_1 + N_2k_2 + N_3k_3]_{f2} = \frac{5}{12}k_1 + \frac{2}{12}k_2 + \frac{5}{12}k_3. \tag{1.19}$$

The velocity component at the midpoint of face 1 is:

$$\begin{aligned} v_x^{f1} &= \frac{5}{12}v_{x_1} + \frac{5}{12}v_{x_2} + \frac{2}{12}v_{x_3} \\ v_y^{f1} &= \frac{5}{12}v_{y_1} + \frac{5}{12}v_{y_2} + \frac{2}{12}v_{y_3} \end{aligned}. \tag{1.20}$$

On face 2 the velocity component is:

$$\begin{aligned} v_x^{f2} &= \frac{5}{12}v_{x_1} + \frac{2}{12}v_{x_2} + \frac{5}{12}v_{x_3} \\ v_y^{f2} &= \frac{5}{12}v_{y_1} + \frac{2}{12}v_{y_2} + \frac{5}{12}v_{y_3} \end{aligned}. \tag{1.21}$$

These values can be used to update the ith support coefficients using the following equation:

$$\begin{aligned} a_i &= a_i + a_1^k \\ a_{i,3} &= a_{i,3} + a_2^k \\ a_{i,4} &= a_{i,4} + a_3^k \end{aligned}. \tag{1.22}$$

In Eq. (1.16), moving counterclockwise around node i, the signed distances are:

$$\Delta\vec{x}_{f1}=\frac{x_3}{3}-\frac{x_2}{6}-\frac{x_1}{6},\quad \Delta\vec{x}_{f2}=-\frac{x_2}{3}+\frac{x_3}{6}+\frac{x_1}{6}$$
$$\Delta\vec{y}_{f1}=\frac{y_3}{3}-\frac{y_2}{6}-\frac{y_1}{6},\quad \Delta\vec{y}_{f2}=-\frac{y_2}{3}+\frac{y_3}{6}+\frac{y_1}{6},\tag{1.23}$$

the derivatives of the shape functions are:

$$N_{1x}=\frac{\partial N_1}{\partial x}=\frac{(y_2-y_3)}{2V^{\text{ele}}},\quad N_{1y}=\frac{\partial N_1}{\partial y}=\frac{(x_3-x_2)}{2V^{\text{ele}}}$$
$$N_{2x}=\frac{\partial N_2}{\partial x}=\frac{(y_3-y_1)}{2V^{\text{ele}}},\quad N_{2y}=\frac{\partial N_1}{\partial y}=\frac{(x_1-x_3)}{2V^{\text{ele}}},\tag{1.24}$$
$$N_{3x}=\frac{\partial N_2}{\partial x}=\frac{(y_1-y_2)}{2V^{\text{ele}}},\quad N_{3y}=\frac{\partial N_3}{\partial y}=\frac{(x_2-x_1)}{2V^{\text{ele}}}$$

and the volume of the element is

$$V^{\text{ele}}=\frac{(x_2y_3-x_3y_2)+x_1(y_2-y_3)+y_1(x_3-x_2)}{2}.\tag{1.25}$$

The obtained algebraic equations from the discretization procedure using CVFEM are solved using the Gauss-Seidel method.

1.5.2 IMPLEMENTATION OF SOURCE TERMS AND BOUNDARY CONDITIONS

The boundary conditions for the present problem can be enforced using B_{B_i} and B_{C_i} as follows [14–16]:

$$\text{Insulated boundary:}\quad B_{B_i}=0\ \text{ and }\ B_{C_i}=0\tag{1.26}$$

$$\text{Fixed value boundary:}\quad B_{B_i}=\phi_{\text{value}}\times 10^{16}\ \text{ and }\ B_{C_i}=10^{16}\tag{1.27}$$

$$\text{Fixed flux boundary:}\quad B_{B_i}=A_k\times q''\ \text{ and }\ B_{C_i}=0\tag{1.28}$$

where ϕ_{value} is the prescribed value on the boundary and A_k is the length of the control volume surface on the boundary segment.

To provide a general treatment for boundary conditions, some preliminary calculation of the boundary face areas associated with each node j in a given boundary segment is required. Figure 1.5 shows a schematic of the kth ($k=3$) boundary, indicating the data structure. Assuming unit depth, the face area associated with any node j of the boundary segment highlighted in Fig. 1.5 is given by

$$A_{k,j}=\begin{cases}\text{Upper}_1\\ \text{Upper}_j+\text{Lower}_j, & 2\le j\le n_{B,k}-1\\ \text{Lower}_{n_{B,k}}\end{cases}\tag{1.29}$$

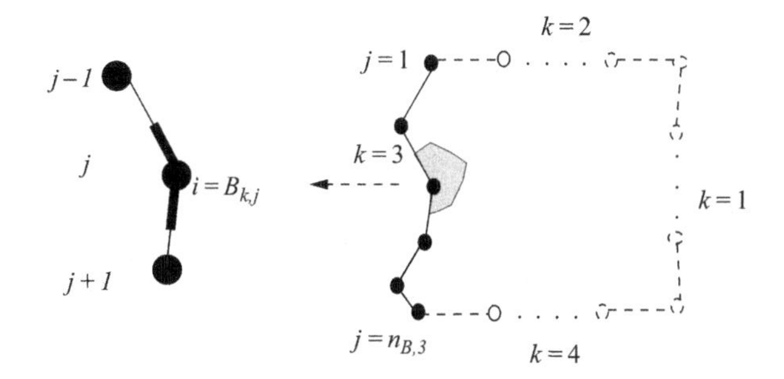

FIGURE 1.5

A domain boundary segment with three sequential points detailed [11].

where

$$
\begin{aligned}
\text{Upper}_j &= \frac{1}{2}\sqrt{(x_j - x_{j+1})^2 + (y_j - y_{j+1})^2} \\
\text{Lower}_j &= \frac{1}{2}\sqrt{(x_j - x_{j-1})^2 + (y_j - y_{j-1})^2}
\end{aligned} \tag{1.30}
$$

The first and last lines on the right-hand side of Eq. (1.29) account for the first and last node on the boundary segment, respectively. This treatment assumes that there are at least two boundary segments, i.e., at least two contiguous regions of the domain boundary where different boundary conditions are applied. In cases where only one boundary condition is applied (e.g., a constant value over the whole boundary) the condition of a two-segment boundary can be artificially imposed. The volume source terms can be applied to Eq. (1.13) as:

$$
\sum_{j=1}^{\text{elements}} \int_{V_j} Q\mathrm{d}V \approx Q_i V_i \tag{1.31}
$$

or, after linearizing the source term

$$
Q_i V_i = -Q_{C_i}\phi_i + Q_{B_i} \tag{1.32}
$$

1.5.3 UNSTEADY ADVECTION-DIFFUSION WITH SOURCE TERMS

The net flow rate of quantity into the control volume around node i can be approximated as

$$
Net_i = \int_{V_i} Q\mathrm{d}V + \sum_{j=1}^{n_i} \int_{A^j} \kappa\nabla\phi\cdot n\mathrm{d}A - \int_{A^j} (v\cdot n)\phi\mathrm{d}A \approx -[a_i + Q_{C_i} + B_{C_i}]\phi_i + \sum_{j=1}^{n_i} a_{i,j}\phi_{S_{i,j}} + Q_{B_i} + B_{B_i}. \tag{1.33}
$$

In a steady-state problem this flow rate is identically zero. In a transient problem, however, it results in a change in storage of the quantity in the control volume, i.e.,

$$\frac{\mathrm{d}}{\mathrm{d}t}\int_{V_i}\phi \mathrm{d}V = \mathrm{Net}_i. \tag{1.34}$$

Using nodal lumping for the volume integration and finite difference in time, this equation can be used to evaluate the nodal field values at time $t+\Delta t$ in terms of the nodal field values at time t,

$$V_i\phi_i^{\mathrm{new}} = V_i\phi_i + \Delta t\left[(1-\theta)\mathrm{Net}_i + \theta \mathrm{Net}_i^{\mathrm{new}}\right], \tag{1.35}$$

where Δt is a time step and the superscript "new" indicates evaluation at time $t+\Delta t$. The parameter $0 \le \theta \le 1$ is a user-defined weighting factor used to approximate the net flow into control volume i during the time interval $[t, t+\Delta t]$ in terms of the net flows at the beginning and end of the time step. Neglecting for now contributions from the boundaries and sources, the resulting discrete equations for three choices of θ are:

Fully implicit $\theta = 1$:

$$V_i\phi_i^{\mathrm{new}} = V_i\phi_i + \Delta t\left(\sum_{j=1}^{n_i} a_{i,j}\phi_{S_{i,j}}^{\mathrm{new}} - a_i\phi_i^{\mathrm{new}}\right) \tag{1.36}$$

The advantage of this choice is that it is unconditionally stable, i.e., for any choice of time step errors (induced or inherent) do not grow. The downside is that a system of equations needs to be solved to obtain the field values at the new time step $\phi_i^{\mathrm{new}}\,(i = 1, \ldots, n)$.

Crank-Nicolson $\theta = 0.5$:

$$V_i\phi_i^{\mathrm{new}} = V_i\phi_i + \frac{\Delta t}{2}\left(\sum_{j=1}^{n_i} a_{i,j}\phi_{S_{i,j}}^{\mathrm{new}} - a_i\phi_i^{\mathrm{new}}\right) + \frac{\Delta t}{2}\left(\sum_{j=1}^{n_i} a_{i,j}\phi_{S_{i,j}} - a_i\phi_i\right) \tag{1.37}$$

This unconditionally stable scheme also requires a system solution. Unlike the first-order-in-time, fully implicit method, however, the Crank-Nicolson is second-order-in-time (i.e., time errors scale with Δt^2)

Fully explicit $\theta = 0$:

$$V_i\phi_i^{\mathrm{new}} = V_i\phi_i + \Delta t\left(\sum_{j=1}^{n_i} a_{i,j}\phi_{S_{i,j}} - a_i\phi_i\right) \tag{1.38}$$

This choice can be used to directly update the new time values from the current time values without solving a system of equations. This, however, comes at the price of a restriction on the time step size to ensure stability. The solution of Eq. (1.38) is likely to become unstable (error will increase as the solution advances) if the net coefficient for ϕ_i becomes negative. This requires that the time step is chosen such that

$$\Delta t < \min\left(\frac{V_i}{a_i}\right), \quad (i = 1, \ldots, n) \tag{1.39}$$

If a fine grid is used this value could be prohibitively small, i.e., the advance of time could be too slow to reach a practical value using reasonable computational resources. In many cases, however, the drawback of using a small time step is offset by the ability to solve Eq. (1.29) without iteration. Driven by its simplicity and flexibility of modeling complex nonlinear terms, the explicit time integration approach is the preferred choice in this work. When sources and boundary condition treatments are added, the explicit scheme Eq. (1.29) can be written as

$$(V_i + B_{C_i})\phi_i^{\text{new}} = V_i\phi_i + \Delta t\left(\sum_{j=1}^{n_i} a_{i,j}\phi_{S_{i,j}} - a_i\phi_i + Q_{B_i} - Q_{C_i}\phi_i\right) + B_{B_i} \tag{1.40}$$

The addition of the source term in Eq. (1.40) could require a further reduction in the time step to retain positive coefficients and stability, i.e.,

$$\Delta t < \min\left(\frac{V_i}{a_i + Q_{C_i}}\right), \quad (i = 1, \ldots, n) \tag{1.41}$$

REFERENCES

[1] B.R. Baliga, S.V. Patankar, A new finite element formulation for convection diffusion problems, Numer. Heat Transfer 3 (1980) 393–409.

[2] B.R. Baliga, S.V. Patankar, A control volume finite-element method for two dimensional fluid flow and heat transfer, Numer. Heat Transfer 6 (1983) 245–261.

[3] M.J. Raw, G.E. Schneider, A skewed, positive influence coefficient up-winding procedure for control volume based finite element convection-diffusion computation, Numer. Heat Transfer 9 (1986) 1–26.

[4] M.J. Raw, G.E. Schneider, V. Hassani, A nine-noded quadratic control volume based finite element for heat conduction, J. Spacecraft 22 (1985) 523–529.

[5] J. Banaszek, Comparison of control volume and Galerkin finite element methods for diffusion type problems, Numer. Heat Transfer 16 (1989) 59–78.

[6] J.B. Campos Silva, E.D.R. Vieira, L.F.M. Moura, Control volume finite element and flow visualization methods applied for unsteady viscous flow past a circular cylinder, in: Proc. V Congresso de Engenhari a Mecanica do Norte-Nordeste, Fortaleza, CE, Brazil, vol. II, 1998, pp. 80–87.

[7] J.B. Campos Silva, L.F.M. Moura, Numerical simulation of fluid flow by the control volume finite element method, in: Proceedings (in CD-ROM) of the XIV Brazilian Congress of Mechanical Engineering (COBEM97), Bauru, SP, Brazil, 1997, pp. 1–8, (Paper Code 041).

[8] J.B. Campos Silva, Numerical Simulation of Fluid Flow by the Finite Element Method Based on Control Volumes, Ph.D. ThesisState University of Campinas, Campinas, SP, Brazil, 1998 (in Portuguese).

[9] H.J. Saabas, B.R. Baliga, Co-located equal order control volume finite element method for multidimensional incompressible fluid flow—part I: formulation, Numer. Heat Transfer 26 (1994) 381–407.

[10] H.J. Saabas, B.R. Baliga, Co-located equal order control volume finite element method for multidimensional incompressible fluid flow—part II: verification, Numer. Heat Transfer B 26 (1994) 409–424.

[11] V.R. Voller, Basic Control Volume Finite Element Methods for Fluids and Solids, World Scientific Publishing Co. Pte. Ltd. 5, Tohccxxvc, 2009.
[12] M. Sheikholeslami, M. Gorji-Bandpy, I. Pop, S. Soleimani, Numerical study of natural convection between a circular enclosure and a sinusoidal cylinder using control volume based finite element method, Int. J. Therm. Sci. 72 (2013) 147–158.
[13] D.W. Pepper, Meshless methods, in: W.J. Minkowycz, E.M. Sparrow, J.Y. Murthu (Eds.), Handbook of Numerical Heat Transfer, Wiley, Hoboken, 2006.
[14] M. Sheikholeslami, R. Ellahi, M. Hassan, S. Soleimani, A study of natural convection heat transfer in a nanofluid filled enclosure with elliptic inner cylinder, Int. J. Numer. Methods for Heat & Fluid Flow 24 (8) (2014) 1906–1927.
[15] M. Sheikholeslami Kandelousi, Effect of spatially variable magnetic field on ferrofluid flow and heat transfer considering constant heat flux boundary condition, Eur. Phys. J. Plus (2014) 129–248.
[16] M. Sheikholeslami, D.D. Ganji, M.M. Rashidi, Ferrofluid flow and heat transfer in a semi annulus enclosure in the presence of magnetic source considering thermal radiation, J. Taiwan Inst. Chem. Eng. (2014), in press. http://dx.doi.org/10.1016/j.jtice.2014.09.026.

CHAPTER

CVFEM stream function-vorticity solution

2

2.1 CVFEM STREAM FUNCTION-VORTICITY SOLUTION FOR A LID-DRIVEN CAVITY FLOW

2.1.1 DEFINITION OF THE PROBLEM AND GOVERNING EQUATION

A classic test problem in computational fluid mechanics is a lid-driven cavity flow. In this two-dimensional problem, the flow in a square conduit is induced by sliding the upper surface (the lid) at a constant x velocity U [1] (Fig. 2.1).

The nature of the flow in the cavity is controlled by a Reynolds number, defined as:

$$Re = \frac{UL}{\upsilon} \tag{2.1}$$

where L is the dimension of the cavity and υ is the kinematic viscosity.

In terms of the stream function Ψ and vorticity ω, the governing equations in integral form are

$$\int_S \omega v \cdot n - \upsilon \nabla \omega \cdot n \mathrm{d}S \tag{2.2}$$

and

$$-\int_A \omega \mathrm{d}A = \int_S \nabla \Psi \cdot n \mathrm{d}S \tag{2.3}$$

In terms of the velocity field v, the vorticity and stream function are defined by

$$\omega = \frac{\partial v_y}{\partial x} - \frac{\partial v_x}{\partial y} \tag{2.4}$$

and

$$v_x = \frac{\partial \Psi}{\partial y}, \quad v_y = -\frac{\partial \Psi}{\partial x} \tag{2.5}$$

The boundary condition on Eq. (2.3) is

$$\Psi = 0 \ \text{ on all solid boundaries.} \tag{2.6}$$

There is no explicit boundary condition for the vorticity equation (Eq. 2.2). In setting up the numerical solution of Eqs. (2.2) and (2.3), a boundary condition for the

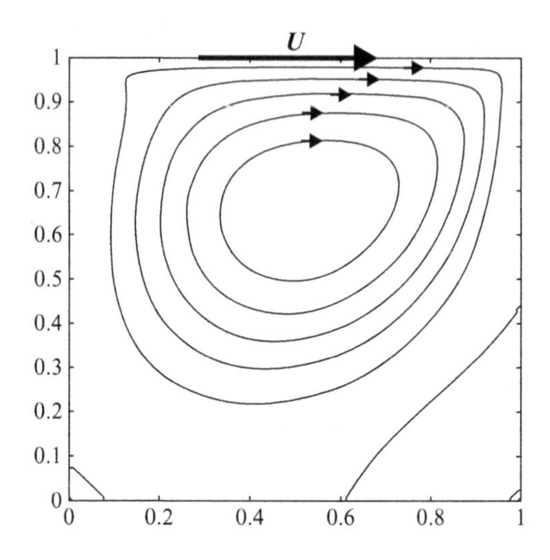

FIGURE 2.1

Lid-driven cavity flow with the streamline indicating flow [1].

solution of the discrete form of Eq. (2.2) is established by using the discrete form of Eq. (2.3) at boundary nodes coupled with known velocity conditions, that is,

$$\begin{aligned} &u_x = U \quad \text{on the lid,} \quad u_x = 0 \quad \text{on all other boundaries} \\ &u_y = 0 \quad \text{on all boundaries} \end{aligned} \tag{2.7}$$

2.1.2 THE CVFEM DISCRETIZATION OF THE STREAM FUNCTION EQUATION

In keeping with rest of this book, the improvement of discrete equations is based on linear triangular elements. The key components of this discretization—the support, control volume, the element, and the control volume faces $(f1, f2)$ in an element—are illustrated in Fig. 2.2 [1]. It is convenient to first consider the discretization of the stream function equation (Eq. 2.3). The form of this equation can be identified as steady-state diffusion with a volume source. The general form for the discrete equation for node i in the support shown in Fig. 2.2 is

$$\left[a_i^{\Psi} + Q_{C_i}^{\Psi} + B_{C_i}^{\Psi}\right]\Psi_i = \sum_{j=1}^{n_i} a_{i,j}^{\Psi}\Psi_{S_{i,j}} + Q_{B_i}^{\Psi} + B_{B_i}^{\Psi} \tag{2.8}$$

where the a's are the coefficients; the index $(i,\ j)$ indicates the jth node $(j = 1, 2, \ldots, n_i)$ in the support of node i; the index $S_{i,j}$ provides the global node number $(i = 1, 2, \ldots, n)$ of the jth node in the support; the B's account for boundary conditions; and the Q's represent source terms.

2.1.2.1 Diffusion contributions

Following the presentation in previous chapter, the contribution to the coefficients a in Eq. (2.8), obtained by considering the diffusion flux across the control volume faces $(f1, f2)$ for the particular element in Fig. 2.2, are

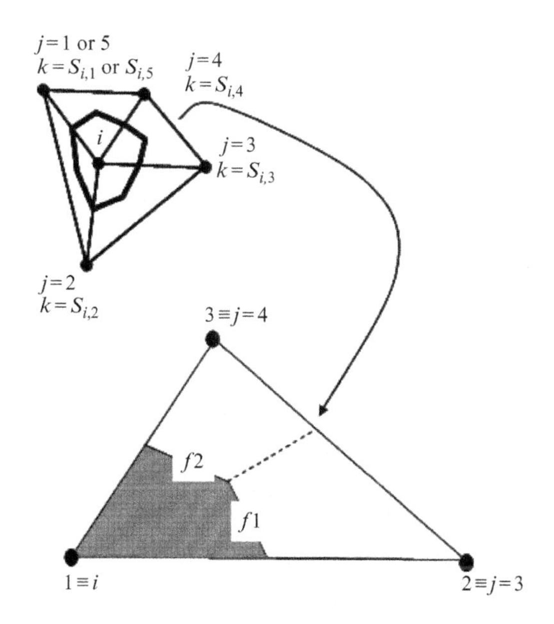

FIGURE 2.2

Arrangement of control volumes and elements [1].

$$\begin{aligned}
a_1{}^k &= -N_{1x}\Delta\vec{y}_{f1} + N_{1y}\Delta\vec{x}_{f1} - N_{1x}\Delta\vec{y}_{f2} + N_{1y}\Delta\vec{x}_{f2} \\
a_2{}^k &= N_{2x}\Delta\vec{y}_{f1} - N_{2y}\Delta\vec{x}_{f1} + N_{2x}\Delta\vec{y}_{f2} - N_{2y}\Delta\vec{x}_{f2} \\
a_3{}^k &= N_{3x}\Delta\vec{y}_{f1} - N_{3y}\Delta\vec{x}_{f1} + N_{3x}\Delta\vec{y}_{f2} - N_{3y}\Delta\vec{x}_{f2}
\end{aligned} \tag{2.9}$$

These values can be used to update the ith support coefficients through

$$\begin{aligned}
a_i^{\Psi} &= a_i^{\Psi} + a_1^{\kappa} \\
a_{i,3}^{\Psi} &= a_{i,3}^{\Psi} + a_2^{\kappa} \\
a_{i,4}^{\Psi} &= a_{i,4}^{\Psi} + a_4^{\kappa}
\end{aligned} \tag{2.10}$$

In Eq. (2.9), moving counterclockwise around node i, the signed distances are

$$\begin{aligned}
\Delta\vec{x}_{f1} &= \frac{x_3}{3} - \frac{x_2}{6} - \frac{x_1}{6}, \quad \Delta\vec{x}_{f2} = -\frac{x_2}{3} + \frac{x_3}{6} + \frac{x_1}{6} \\
\Delta\vec{y}_{f1} &= \frac{y_3}{3} - \frac{y_2}{6} - \frac{y_1}{6}, \quad \Delta\vec{y}_{f1} = -\frac{y_2}{3} + \frac{y_3}{6} + \frac{y_1}{6}
\end{aligned}, \tag{2.11}$$

the derivatives of the shape functions are

$$\begin{aligned}
N_{1x} &= \frac{\partial N_1}{\partial x} = \frac{(y_2 - y_3)}{2V^{\text{ele}}}, \quad N_{1y} = \frac{\partial N_1}{\partial y} = \frac{(x_3 - x_2)}{2V^{\text{ele}}} \\
N_{2x} &= \frac{\partial N_2}{\partial x} = \frac{(y_3 - y_1)}{2V^{\text{ele}}}, \quad N_{2y} = \frac{\partial N_2}{\partial y} = \frac{(x_1 - x_3)}{2V^{\text{ele}}}, \\
N_{3x} &= \frac{\partial N_3}{\partial x} = \frac{(y_1 - y_2)}{2V^{\text{ele}}}, \quad N_{3y} = \frac{\partial N_3}{\partial y} = \frac{(x_2 - x_1)}{2V^{\text{ele}}}
\end{aligned} \tag{2.12}$$

and the volume of the element is

$$V^{\text{ele}} = \frac{(x_2y_3 - x_3y_2) + x_1(y_2 - y_3) + y_1(x_3 - x_2)}{2} \tag{2.13}$$

2.1.2.2 Source terms

The left-hand side of Eq. (2.3) is identified as a volume source term, which can be represented in Eq. (2.8) by the setting

$$Q_{B_i}^{\Psi} = \omega_i V_i, \quad Q_{C_i}^{\Psi} = 0 \tag{2.14}$$

where V_i is the volume of the control volume, made up of contributions of $V^{\text{ele}}/3$ from each element in the support of node i. Coupling with the nodal vorticity field requires an iterative solution to Eq. (2.8). For any iteration, the source term in Eq. (2.14) will need to be reevaluated to reflect updates to the iterative values of the nodal vorticity field.

2.1.2.3 Boundary conditions

The boundary conditions of $\Psi = 0$ on all boundaries are set in Eq. (2.8) by setting, for each node i lying on the domain boundary,

$$B_{B_i}^{\Psi} = 0, \quad B_{C_i}^{\Psi} = 10^{15} \tag{2.15}$$

2.1.3 THE CVFEM DISCRETIZATION OF THE VORTICITY EQUATION

The vorticity equation (Eq. 2.2) has the form of a steady-state advection diffusion equation for a scalar. The general form for this equation for node i in the support shown in Fig. 2.2 is

$$\left[a_i^{\omega} + B_{C_i}^{\omega}\right]\omega_i = \sum_{j=1}^{n_i} a_{i,j}^{\omega}\omega_{S_{i,j}} + B_{B_i}^{\omega} \tag{2.16}$$

2.1.3.1 Diffusion contributions

Based on consideration of the diffusion flux across the control volume faces $(f1, f2)$ in Fig. 2.2, the contributions to the coefficients a in Eq. (2.16) are

$$\begin{aligned}
a_1^{\kappa} &= -\nu N_{1x}\Delta\vec{y}_{f1} + \nu N_{1y}\Delta\vec{x}_{f1} - \nu N_{1x}\Delta\vec{y}_{f2} + \nu N_{1y}\Delta\vec{x}_{f2} \\
a_2^{\kappa} &= \nu N_{2x}\Delta\vec{y}_{f1} - \nu N_{2y}\Delta\vec{x}_{f1} + \nu N_{2x}\Delta\vec{y}_{f2} - \nu N_{2y}\Delta\vec{x}_{f2} \\
a_3^{\kappa} &= \nu N_{3x}\Delta\vec{y}_{f1} - \nu N_{3y}\Delta\vec{x}_{f1} + \nu N_{3x}\Delta\vec{y}_{f2} - \nu N_{3y}\Delta\vec{x}_{f2}
\end{aligned} \tag{2.17}$$

These values can be used to update the ith support coefficients through

$$\begin{aligned}
a_i^{\omega} &= a_i^{\omega} + a_1^{\kappa} \\
a_{i,3}^{\omega} &= a_{i,3}^{\omega} + a_2^{\kappa} \\
a_{i,4}^{\omega} &= a_{i,4}^{\omega} + a_4^{\kappa}
\end{aligned} \tag{2.18}$$

2.1.3.2 Advection coefficients

Based on consideration of the advection flux across the control volume faces $(f1, f2)$ in Fig. 2.2, the contributions to the coefficients a in Eq. (2.8) using up winding are

$$\begin{aligned} a_1^u &= \max\left[q_{f1}, 0\right] + \max\left[q_{f2}, 0\right] \\ a_2^u &= \max\left[-q_{f1}, 0\right] \\ a_3^u &= \max\left[-q_{f2}, 0\right] \end{aligned} \tag{2.19}$$

These values can be used to update the ith support coefficients through

$$\begin{aligned} a_i^{\omega} &= a_i^{\omega} + a_1^k + a_1^u \\ a_{i,3}^{\omega} &= a_{i,j}^{\omega} + a_2^k + a_2^u \\ a_{i,4}^{\omega} &= a_{i,j}^{\omega} + a_3^k + a_3^u \end{aligned} \tag{2.20}$$

In Eq. (2.19), the volume flow rates across the control volume faces are calculated by

$$\begin{aligned} q_{f1} &= v \cdot nA|_{f1} = v_x^e \Delta \vec{y}_{f1} - v_y^e \Delta \vec{x}_{f1} \\ q_{f2} &= v \cdot nA|_{f2} = v_x^e \Delta \vec{y}_{f2} - v_y^e \Delta \vec{x}_{f2} \end{aligned} \tag{2.21}$$

where the velocities at the midpoint of the faces are approximated in terms of the constant element velocity, $v^e = (v_x^e, v_y^e)$. This value, in turn, is calculated directly from the nodal stream function field through the approximation of Eq. (2.5), for example, for the element in Fig. 2.2,

$$\begin{aligned} v_x^e &\approx N_{1y} \Psi_i + N_{2y} \Psi_{S_{i,3}} + N_{3y} \Psi_{S_{i,4}} \\ v_y^e &\approx -N_{1x} \Psi_i - N_{2x} \Psi_{S_{i,3}} - N_{3x} \Psi_{S_{i,4}} \end{aligned} \tag{2.22}$$

This coupling with the nodal stream function field requires an iterative solution to Eq. (2.16). At each iteration the advection coefficients in Eq. (2.19) need to be evaluated to reflect updates to the iterative values of the nodal stream function field.

2.1.3.3 Boundary conditions

Before Eq. (2.16) can be solved, boundary conditions need to be prescribed. Fixed-value boundary conditions are used. At each node i on a domain boundary the discrete form of the stream function Eq. (2.3) can be used to prescribe a value for the nodal vorticity. Because the stream function takes the known fixed value $\Psi = 0$ on all boundaries, use of this equation is allowed since it is not needed in the stream function solution. At a node i on the boundary, the control volume finite element discretization of Eq. (2.3), informed by Eq. (2.8), can be written as

$$\omega_i = \frac{1}{V_i}\left[a_i^{\Psi}\Psi_i - \sum_{j=1}^{n_i} a_{i,j}^{\Psi}\Psi_{S_{i,j}} - \sum_{\text{boundary}} A\nabla\Psi \cdot n\right] \tag{2.23}$$

where the coefficients a^{Ψ} are given by Eq. (2.10) and the definition of the source term in Eq. (2.14) has been used to isolate ω_i. The last term in Eq. (2.23) represents contributions from control volume faces that coincide with a boundary segment [1] (the double line in Fig. 2.3). In the sliding lid problem in Fig. 2.1, the only nonzero contribution from this term is for nodes on the sliding lid and has the form

$$\left[\sum_{\text{boundary}} A\nabla\Psi \cdot n\right] = \Delta U \tag{2.24}$$

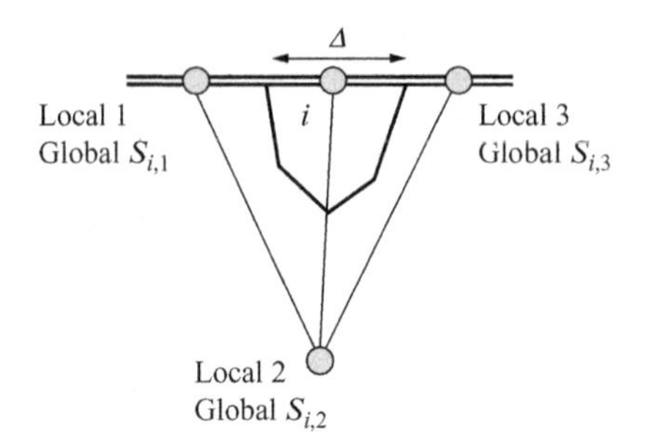

FIGURE 2.3

Node points on a sliding lid boundary [1].

where Δ is the length of the control volume surface on the boundary segment [1] (Fig. 2.3).

Hence, with current iterative values of the stream function known, Eqs. (2.23) and (2.24) can be used to estimate fixed nodal values of vorticity for boundary nodes i:

$$\omega_i = \frac{1}{V_i}\left[a_i^{\Psi}\Psi_i - \sum_{j=1}^{n_i} a_{i,j}^{\Psi}\Psi_{S_{i,j}} - \begin{cases} \Delta U & i \ \varepsilon \ \text{lid} \\ 0 & \text{otherwise} \end{cases}\right] \tag{2.25}$$

These values are forced into the solution of Eq. (2.16) by setting

$$B_{B_i}^{\omega} = \omega_i \times 10^{15}, \quad B_{C_i}^{\omega} = 10^{15} \tag{2.26}$$

2.1.4 CALCULATING THE NODAL VELOCITY FIELD

Following the calculation of the converged Ψ field, the nodal velocity field v can be calculated. Integrating Eq. (2.5) over the ith control volume results in

$$\int_{V_i} v_x \mathrm{d}V = \int_{V_i} \frac{\partial \Psi}{\partial y}\mathrm{d}V, \quad \int_{V_i} v_y \mathrm{d}V = -\int_{V_i} \frac{\partial \Psi}{\partial x}\mathrm{d}V, \tag{2.27}$$

or, using the divergence theorem,

$$\int_{V_i} v_x \mathrm{d}V = \oint_S \Psi n_y \mathrm{d}S, \quad \int_{V_i} v_y \mathrm{d}V = -\oint_S \Psi n_x \tag{2.28}$$

where the surface integral is over the control volume faces, and $n = (n_x \cdot n_y)$ is the outward-pointing normal on the control volume face. Using midpoint integration approximations,

$$v_{x_i} = \frac{1}{V_i}\sum_{\text{faces}} \Psi A n_y, \quad v_{y_i} = -\frac{1}{V_i}\sum_{\text{faces}} \Psi A n_x \tag{2.29}$$

where A is the area (length × unit depth) of a given control volume face, and, since boundary values of $\Psi = 0$, the summations are restricted to control faces that do not coincide with the domain boundaries. Recalling that the outward normal area product on a volume face is

$$A n_x = \Delta \vec{y}, \quad A n_y = -\Delta \vec{x} \tag{2.30}$$

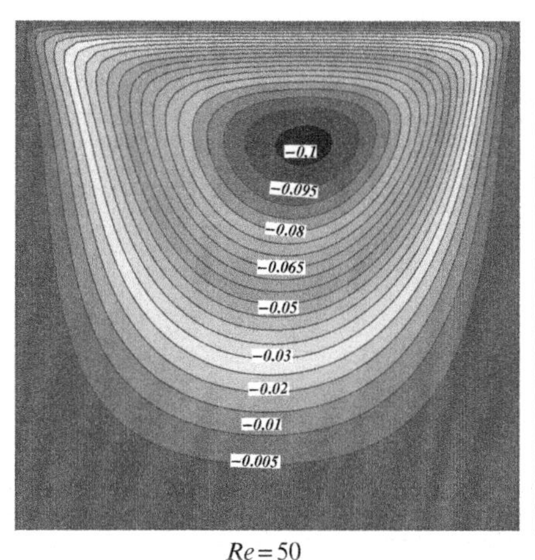

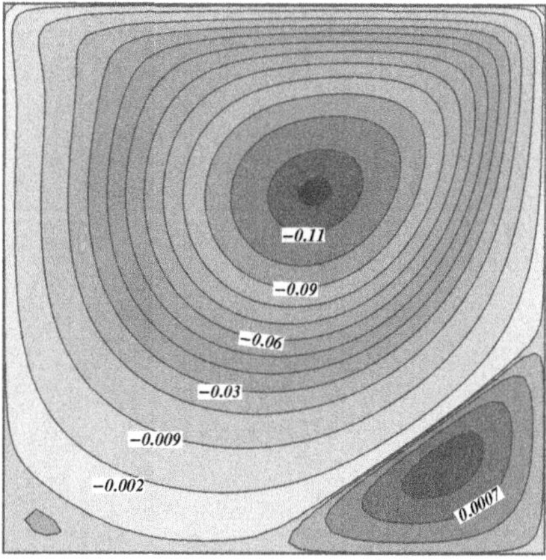

$Re=50$ $Re=500$

FIGURE 2.4

Streamlines for different Reynolds numbers.

(see Eq. 2.11 for definitions of $\Delta\vec{y}$ and $\Delta\vec{x}$), the contributions to control volume i from the right-hand side of Eq. (2.29), for the selected element from the control volume in Fig. 2.2, can be calculated as

$$\sum_{f1,f2} \Psi An_y = -\left(\frac{5}{12}\Psi_i + \frac{5}{12}\Psi_{S_{i,3}} + \frac{2}{12}\Psi_{S_{i,4}}\right)\Delta\vec{x}_{f1} - \left(\frac{5}{12}\Psi_i + \frac{2}{12}\Psi_{S_{i,3}} + \frac{5}{12}\Psi_{S_{i,4}}\right)\Delta\vec{x}_{f2}$$
$$\sum_{f1,f2} \Psi An_x = -\left(\frac{5}{12}\Psi_i + \frac{5}{12}\Psi_{S_{i,3}} + \frac{2}{12}\Psi_{S_{i,4}}\right)\Delta\vec{y}_{f1} - \left(\frac{5}{12}\Psi_i + \frac{2}{12}\Psi_{S_{i,3}} + \frac{5}{12}\Psi_{S_{i,4}}\right)\Delta\vec{y}_{f2} \quad (2.31)$$

With the nodal velocity field calculated from Eq. (2.29), the velocity at given point $x=(x, y)$ can be determined by determining the triangular element (local nodes 1, 2, and 3), which contains the point x, and then using the shape function interpolations

$$v_x(x) = N_1 v_{x1} + N_2 v_{x2} + N_3 v_{x3}$$
$$v_y(x) = N_1 v_{y1} + N_2 v_{y2} + N_3 v_{y3} \quad (2.32)$$

where the shape functions N_i $(i=1,2,3)$ are defined in Eqs. (2.7) and (2.8).

2.1.5 RESULTS

Figure 2.4 shows the streamline for different Reynolds numbers. A FORTRAN code for the control volume finite element method (CVFEM) solution of a steady-state advection diffusion equation is provided in Appendix.

2.2 CVFEM STREAM FUNCTION-VORTICITY SOLUTION FOR NATURAL CONVECTION

2.2.1 DEFINITION OF THE PROBLEM AND GOVERNING EQUATION

The physical model and the corresponding triangular elements used in the present CVFEM program are shown in Fig. 2.5 [2], where r_{out} is the radius of the outer cylinder. The surfaces of the inner and outer cylinders are maintained at constant

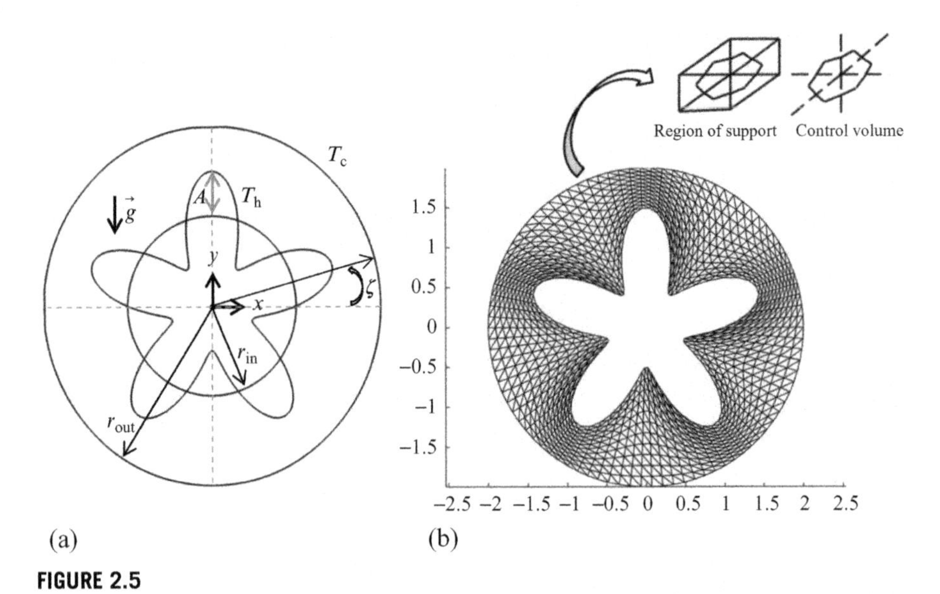

FIGURE 2.5

(a) Geometry and boundary conditions. (b) Computational mesh [2].

temperatures T_h and T_c, respectively, under the condition that $T_h > T_c$. The profile of the inner cylinder follows the pattern:

$$r = r_{in} + A\cos(N(\zeta - \zeta_0)) \tag{2.33}$$

where r_{in} is the base circle radius ($r_{in} = 1$), A and N are the amplitude and number of undulations, respectively, and ζ is the rotation angle.

Under the Boussinesq approximation, the continuity, momentum, and energy equations for the laminar and steady-state natural convection in the cylindrical enclosure can be written as follows:

$$\frac{\partial u}{\partial x} + \frac{\partial v}{\partial y} = 0 \tag{2.34}$$

$$u\frac{\partial u}{\partial x} + v\frac{\partial u}{\partial y} = -\frac{1}{\rho}\frac{\partial P}{\partial x} + \upsilon\left(\frac{\partial^2 u}{\partial x^2} + \frac{\partial^2 u}{\partial y^2}\right) \tag{2.35}$$

$$u\frac{\partial v}{\partial x} + v\frac{\partial v}{\partial y} = -\frac{1}{\rho}\frac{\partial P}{\partial y} + \upsilon\left(\frac{\partial^2 v}{\partial x^2} + \frac{\partial^2 v}{\partial y^2}\right) + g\beta(T - T_c) \tag{2.36}$$

$$u\frac{\partial T}{\partial x} + v\frac{\partial T}{\partial y} = \alpha\left(\frac{\partial^2 T}{\partial x^2} + \frac{\partial^2 T}{\partial y^2}\right) \tag{2.37}$$

where u and v are the velocity components along the axes x and y, T is the fluid temperature, and P is the pressure. The physical meaning of the other quantities is mentioned in Nomenclature section.

We introduce the following dimensionless variables defined as

$$X=\frac{x}{L},\quad Y=\frac{y}{L},\quad \Omega=\frac{\omega L^2}{\alpha},\quad \Psi=\frac{\psi}{\alpha},\quad \Theta=\frac{T-T_c}{T_h-T_c} \tag{2.38}$$

where $L=r_{out}-r_{in}=r_{in}=1$. Using these variables, Eqs. (2.34)–(2.36) can be written in the dimensionless form as

$$\frac{\partial\Psi}{\partial Y}\frac{\partial\Omega}{\partial X}-\frac{\partial\Psi}{\partial X}\frac{\partial\Omega}{\partial Y}=Pr\left(\frac{\partial^2\Omega}{\partial X^2}+\frac{\partial^2\Omega}{\partial Y^2}\right)+RaPr\left(\frac{\partial\Theta}{\partial X}\right) \tag{2.39}$$

$$\frac{\partial\Psi}{\partial Y}\frac{\partial\Theta}{\partial X}-\frac{\partial\Psi}{\partial X}\frac{\partial\Theta}{\partial Y}=\left(\frac{\partial^2\Theta}{\partial X^2}+\frac{\partial^2\Theta}{\partial Y^2}\right) \tag{2.40}$$

$$\frac{\partial^2\Psi}{\partial X^2}+\frac{\partial^2\Psi}{\partial Y^2}=-\Omega \tag{2.41}$$

where $Ra=g\beta(T_h-T_c)L^3/(\alpha\upsilon)$ is the Rayleigh number and $Pr=\upsilon/\alpha$ is the Prandtl number. As shown in Fig. 2.5, the boundary conditions of Eqs. (2.39)–(2.41) are

$$\begin{aligned}&\Theta=1.0 \text{ on the inner sinusoidal circular cylinder}\\&\Theta=0.0 \text{ on the outer circular enclosure}\\&\Psi=0.0 \text{ on all solid boundaries}\end{aligned} \tag{2.42}$$

The values of the vorticity Ω on the boundaries of the enclosure can be obtained using the definition of the stream function and the known velocity conditions during the iterative solution procedure. The local Nusselt number on the cold wall Nu_{loc} can be expressed as

$$Nu_{loc}=\frac{\partial\Theta}{\partial n} \tag{2.43}$$

where n is the normal distance to the surface of the outer cylinder. The average Nusselt number on the cold circular wall Nu_{ave} is evaluated as

$$Nu_{ave}=\frac{1}{2\pi}\int_0^{360^\circ} Nu_{loc}(\zeta)d\zeta \tag{2.44}$$

2.2.2 EFFECT OF ACTIVE PARAMETERS

Various values of Rayleigh numbers $Ra=10^3,\ 10^4,\ 10^5$ and 10^6, amplitude parameter $A=0.1,\ 0.3$, and 0.5, and the number of undulations $N=2$, 3, 5, and 6 at a constant Prandtl number, $Pr=0.71$ (air) are calculated. Streamlines and isotherm obtained by the present code are compared with those reported by Kim et al. [3] for different Rayleigh number values (Fig. 2.6). The shape, size, and location of vortices are the same in the results obtained by both Kim et al. [3] and the current code. These comparisons illustrate excellent agreement between the present calculations and previously published results. We are, therefore, confident that the present results using the CVFEM code are correct and accurate.

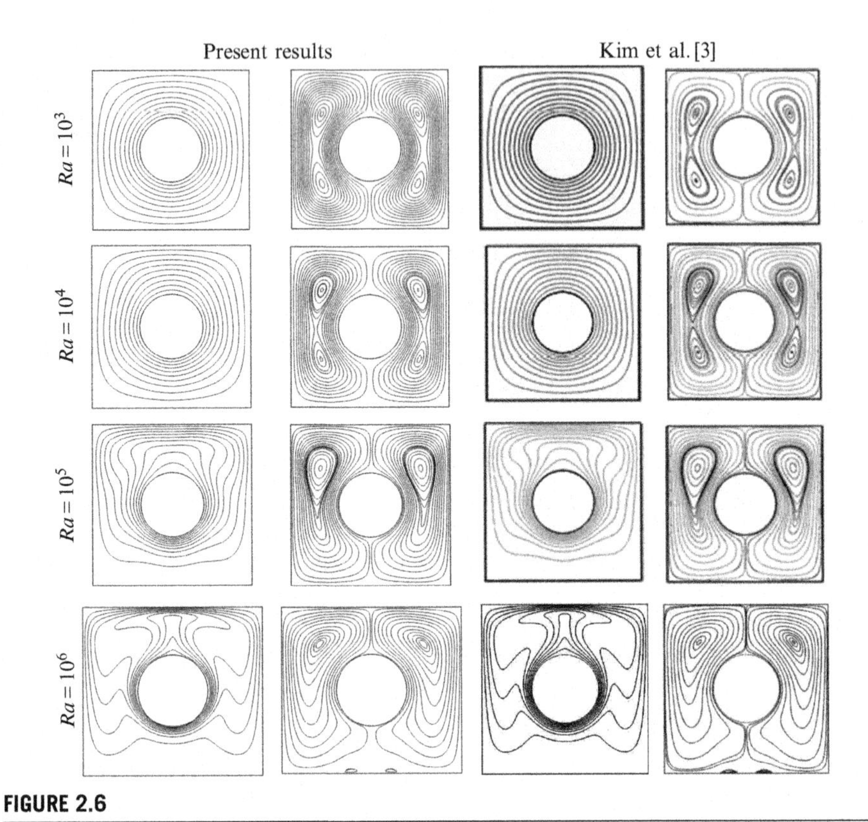

FIGURE 2.6

Comparison of the present results with those reported by Kim et al. [3] for different Rayleigh numbers when $Pr = 0.7$.

Figure 2.7 shows the isotherms and streamlines for different values of amplitude parameter and number of undulations at $Ra = 10^3$. For all cases at this Rayleigh number, the isotherms follow the shape of the inner and outer walls. Moreover, for lower Rayleigh numbers the absolute values of stream function are small, which indicates the domination of a conduction heat transfer mechanism. For $N = 3$, increasing the amplitude of undulation parameter A enhances the absolute value of stream function $|\Psi_{\max}|$. This occurs because with an increase in A the gap between the hot and cold walls decreases, which results in higher heat transfer.

Figure 2.7 also shows that as the amplitude of undulation increases from $A = 0.1$ to $A = 0.5$, the single vortex inside the enclosure divides into two vortices, with different strengths, located at the top and bottom of the enclosure. Moreover, the streamlines indicate that the upper vortex is stronger than the lower one, possibly because of the additional space available for circulation of fluid at the top of the enclosure and the existence of a hot surface beneath the cold outer wall in this region, which helps the flow circulate. As seen at the bottom of the enclosure, a small secondary vortex near the vertical centerline appears because of the existence of the

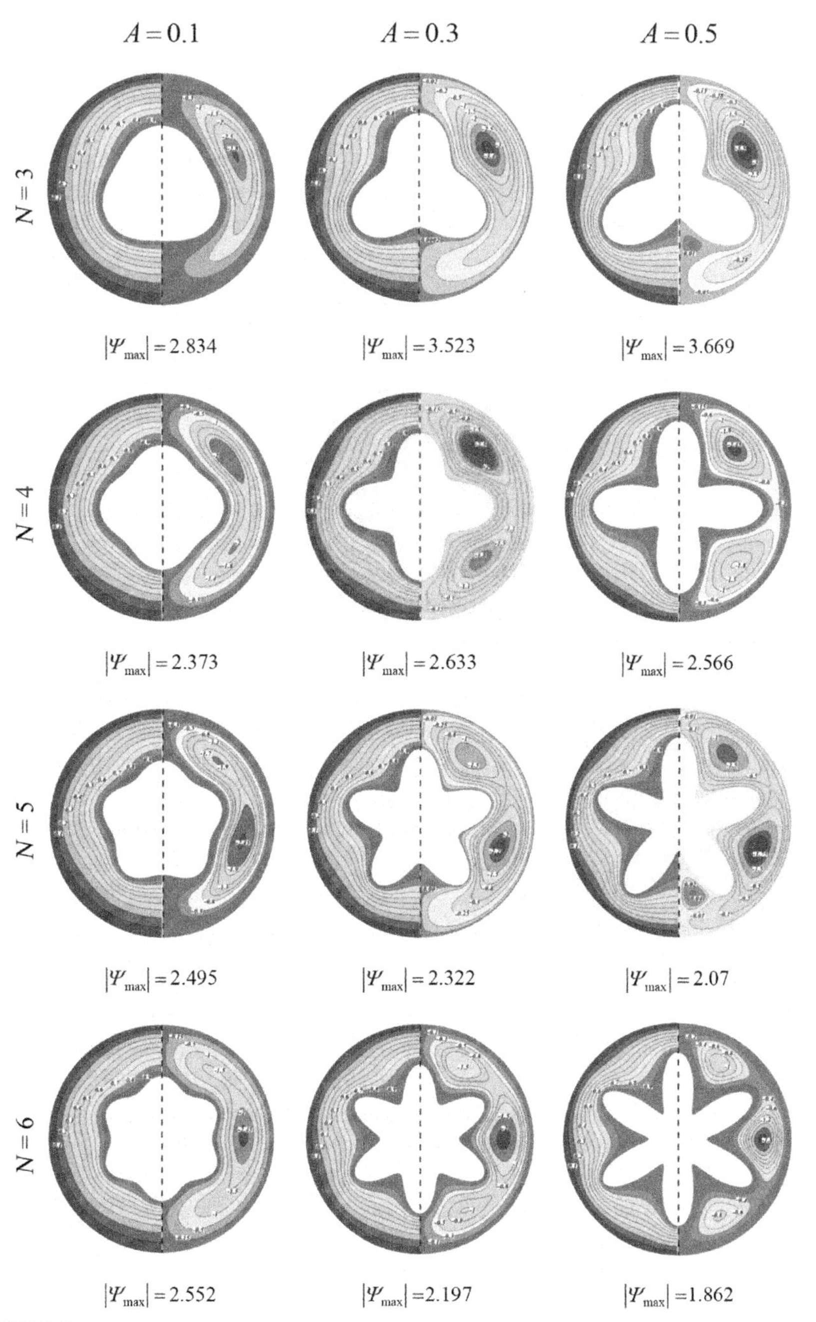

FIGURE 2.7

Isotherms (left) and streamlines (right) for different values of amplitude A and number of undulations N at $Ra = 10^3$.

crest in that region. A similar flow pattern is seen at this region for $N=5$. The values of streamlines show that, for $N=4$, increasing A from 0.1 to 0.3 enhances the values of $|\Psi_{max}|$, but as A increases further, its value decreases, which could be related to the effect of less available space for fluid circulation in this area.

The streamlines also show that there are two main vortices between the inner and outer cylinders for all values of A: an upper stronger vortex and a lower weaker one. Since the spaces available for circulation are equal for both vortices, it can be concluded that the existence of a hot surface beneath the cold one assists in the circulation of flow in the upper half of the enclosure, whereas the existence of a hot surface over the cold one in the lower half of the enclosure has a reverse effect on the flow circulation. At $N=4$ there is no small secondary vortex at the bottom half of the enclosure because the lower crest is parallel to the flow path. A similar pattern exists for $N=6$. In addition, the flow pattern, size, and formation of vortices are nearly similar for $N=5$ and 6. As we have seen, increasing A for these values of the undulation number divides the space between the inner and outer cylinders into three circulation regions. An increase in A decreases $|\Psi_{max}|$ considerably at $N=5$ and 6. Although increase in A expands the effective area of the hot surface, the fluid flow is dampened, especially in the gaps between undulated surfaces, so these areas are almost stagnant. The streamlines and isotherms for different values of N and A at $Ra=10^4$ are shown in Fig. 2.8. As can be seen, the isotherms are disturbed at this Ra, and thermal plumes gradually appear on the hot surface of inner cylinder, which indicates that the convective heat transfer is now comparable with the conduction mode of heat transfer at this Rayleigh number. In addition, the thickness of the thermal boundary layer between the crests of the inner cylinder becomes thinner compared with that at $Ra=10^3$, which indicates that the flow ventilates the region between each both crests; hence the thermal boundary layer is thinner at these areas.

At $Ra=10^4$, the variation of $|\Psi_{max}|$ with N and A is similar to that at $Ra=10^3$. For $N=3$, with the increase in A from 0.1 to 0.3, a thermal plume forms over the upper crest at $\varsigma=90°$ in the reverse direction because of the two secondary vortices that are newly generated at this area. With a larger increase in A, these secondary vortices disappear because the available space between the inner hot wall and the outer cold cylinder decreases. For $N=4$, the primary vortex divides into two smaller vortices: an upper strong one and a lower weak one. The flow pattern and number of vortices for $N=5$ and 6 is similar to those at $Ra=10^3$, expect the value of $|\Psi_{max}|$ is higher at $Ra=10^4$ because the convective heat transfer mechanism is more pronounced.

Figures 2.9 and 2.10 show the effects of A and N on the streamlines and isotherms at $Ra=10^5$ and 10^6, respectively. As a general observation, the isotherms are crowed near the crests, which demonstrates the diminution in the thermal boundary layer thickness at the bottom of the enclosure. At the top of the inner cylinder ($\varsigma=90°$), a strong plume appears, which strongly impinges upon the outer wall of the enclosure for all values of A and N.

The maximum value of the stream function has a direct relation to the value of A at $N=6$; it enhances up to the value $A=0.3$ and then decreases. For $N=3$ and $A=0.1$, a primary vortex exists at the top of the inner cylinder. When A increases,

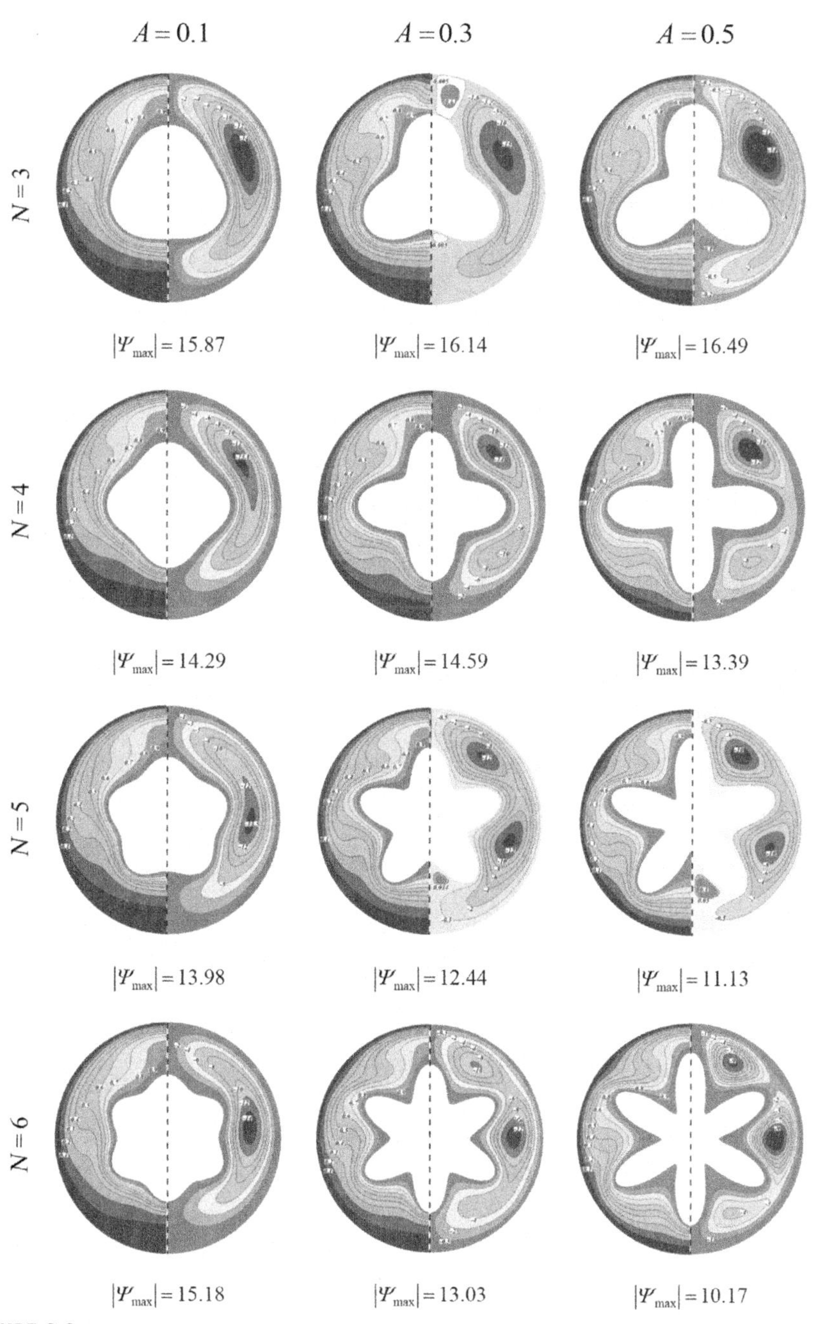

FIGURE 2.8

Isotherms (left) and streamlines (right) for different values of amplitude A and number of undulations N at $Ra = 10^4$.

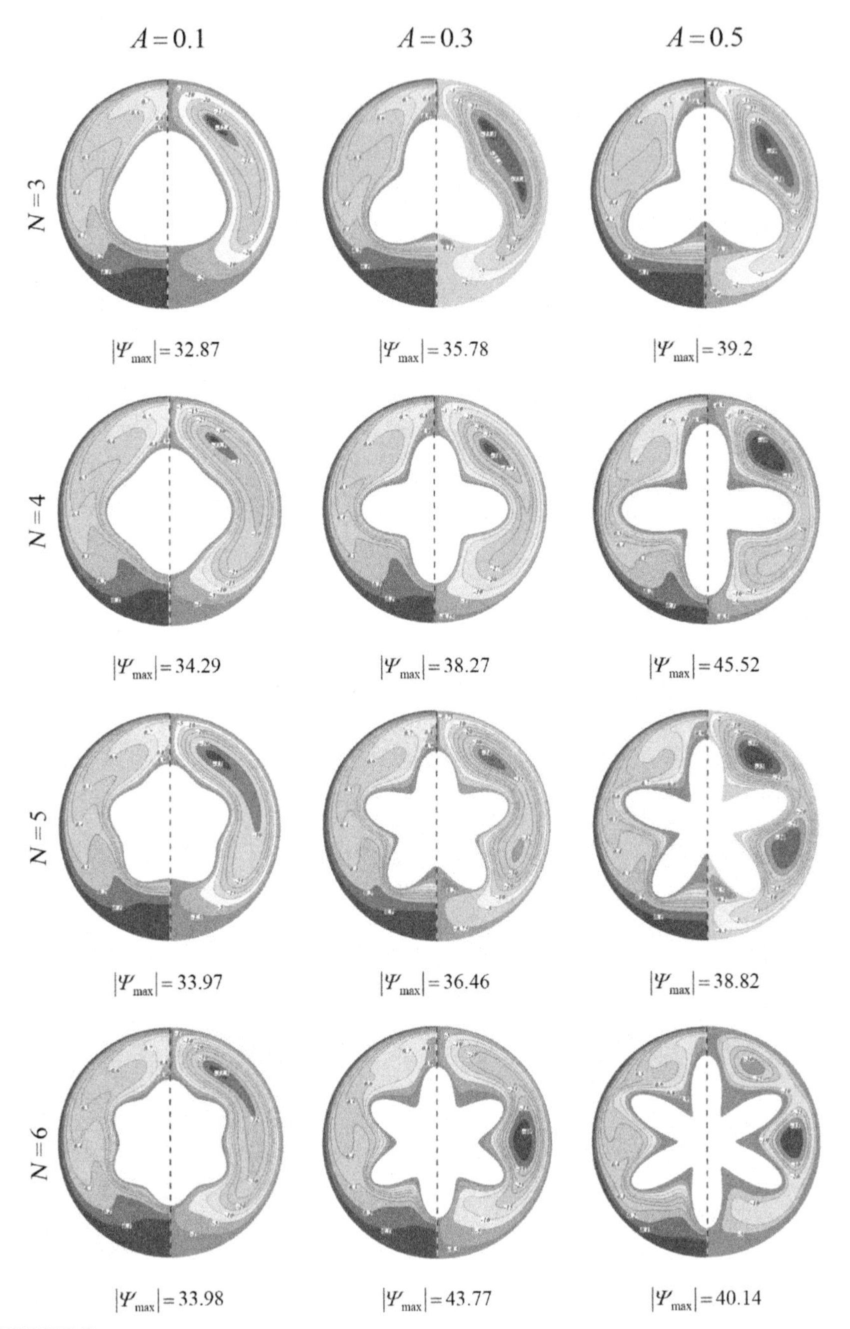

FIGURE 2.9

Isotherms (left) and streamlines (right) for different values of amplitude A and number of undulations N at $Ra = 10^5$.

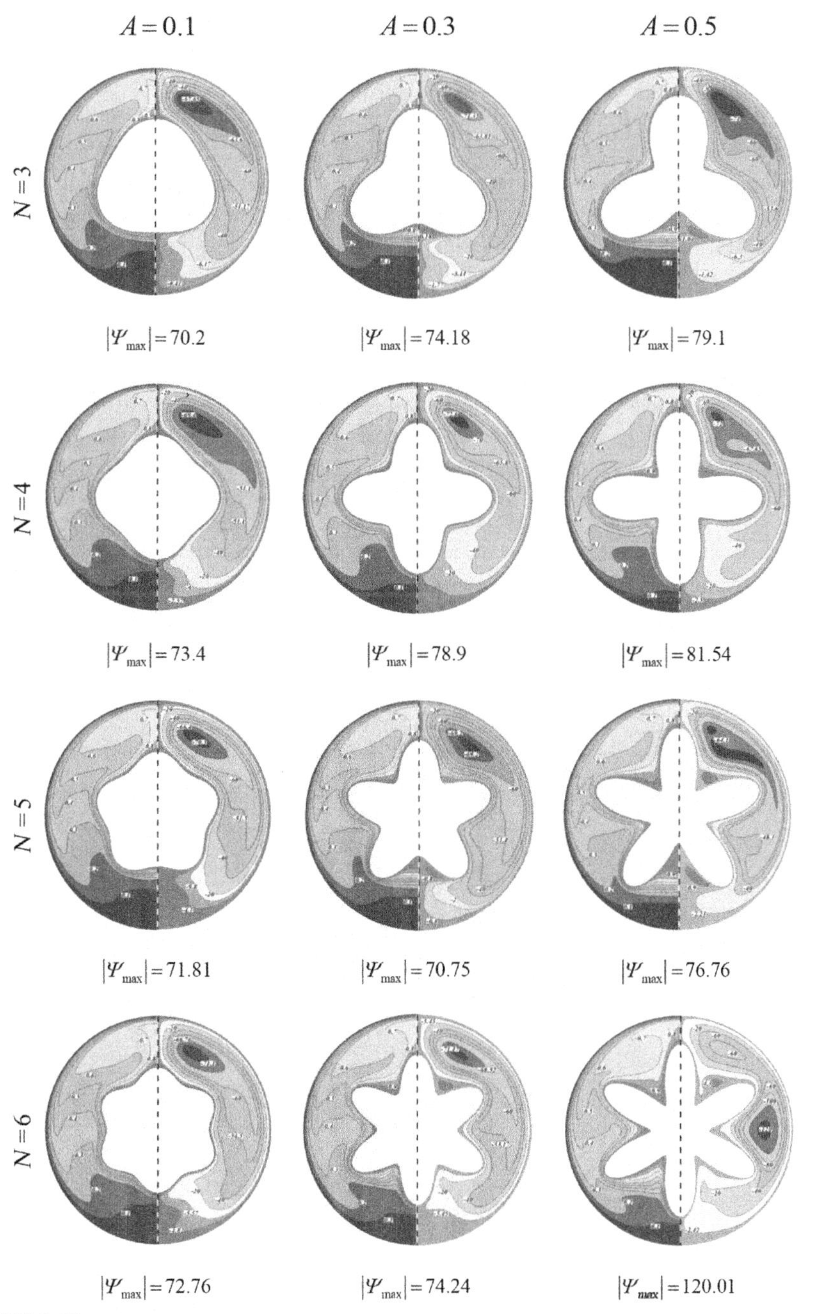

FIGURE 2.10

Isotherms (left) and streamlines (right) for different values of amplitude A and number of undulations N at $Ra = 10^6$.

this main vortex divides into two vortices with small cores; these are located at the upper half of the cylinder. Finally, for $A = 0.5$, these two vortices merge together and form a single vortex, which is stronger than the first one. For $N = 4$, increasing A from 0.1 to 0.3 and then to 0.5 divides the main cell into two cells with different strengths. For $N = 5$, 6 and $A = 0.1$, a single vortex exists inside the enclosure. Increasing A to 0.3 and then to 0.5 pushes this single cell upward and then divides it into two cells. The flow pattern and number of vortices at this Rayleigh number is similar to those at $Ra = 10^5$, expect the value of $|\Psi_{max}|$ is greater at $Ra = 10^6$. It should be noted that all of the computations in the case of $Ra = 10^6$ are obtained using an unsteady solution and averaging for a certain period of time.

Variation of the local Nusselt number Nu_{loc} over the cold wall and between $\varsigma =$ 90° and 270° is depicted in Fig. 2.11. In general, effects of an increasing A on Nu_{loc} are less pronounced at higher Rayleigh numbers. In addition, the variation of Nu_{loc} is greater for higher values of the undulation amplitude, whereas this variation decreases when the gap between the hot inner and cold outer cylinders increases. The number of extermum the local Nusselt number profiles corresponds to the

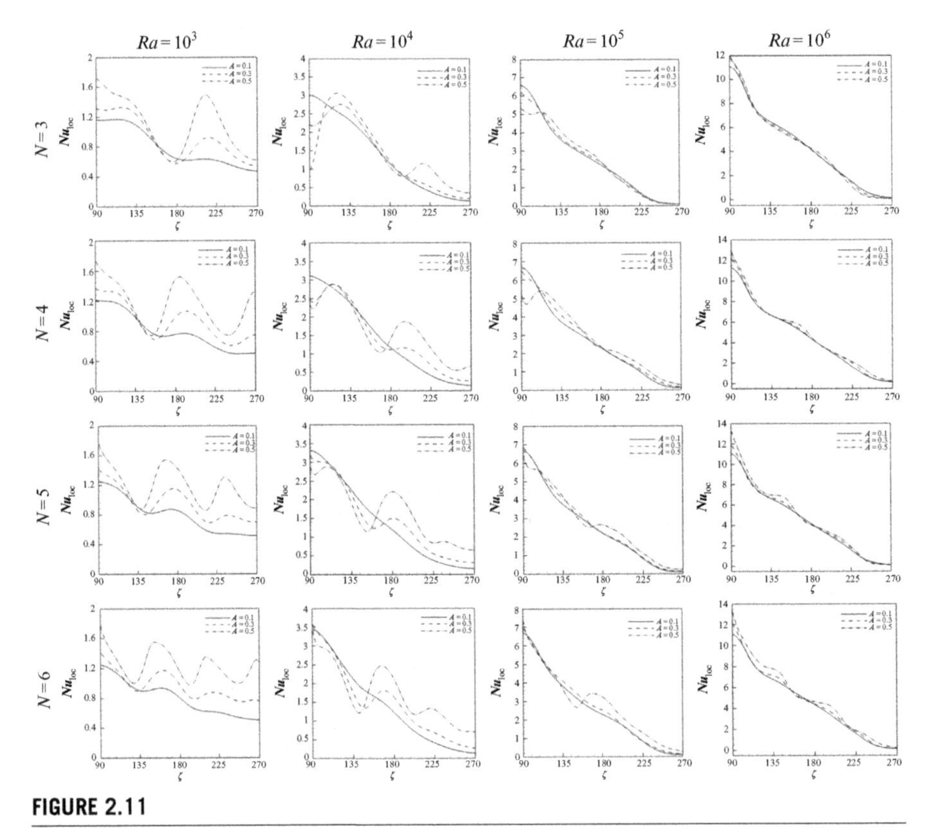

FIGURE 2.11

Effects of the amplitude A, number of undulations N, and Rayleigh number Ra on the local Nusselt number Nu_{loc}.

number of undulations of the inner cylinder. For $N = 3$, 5 and 6 there exist two, three and four local maximum.

Figure 2.11 also shows that Nu_{loc} decreases as the angle ς increases from $\varsigma = 90°$ to 270°; at the top of the inner cylinder ($\varsigma = 90°$), the flow of the cold fluid impinges upon the upper cold surface, which results in a thicker thermal boundary layer on the enclosure's wall. In the case of $\varsigma = 270°$, however, the hot surface is located at the top of the cold wall and the conduction mechanism is more pronounced. Thus Nu_{loc} decreases noticeably. At $Ra = 10^4$, 10^5, and 10^6, the Nu_{loc} profiles uniformly reaches from its maximum value at $\varsigma = 90°$ and is reduced to its minimum value near $\varsigma = 270°$ when $A = 0.1$. For $A = 0.3$ and 0.5, the Nu_{loc} profiles have local extremes, which are related to the thermal plumes and crests over the inner cylinder.

Figure 2.12 shows the effects of A, N, and Ra on the average Nusselt number Nu_{ave}. As Ra increases, the value of Nu_{ave} also increases because of the convective heat transfer mechanism at higher values of Ra. At $Ra = 10^3$, the average Nusselt

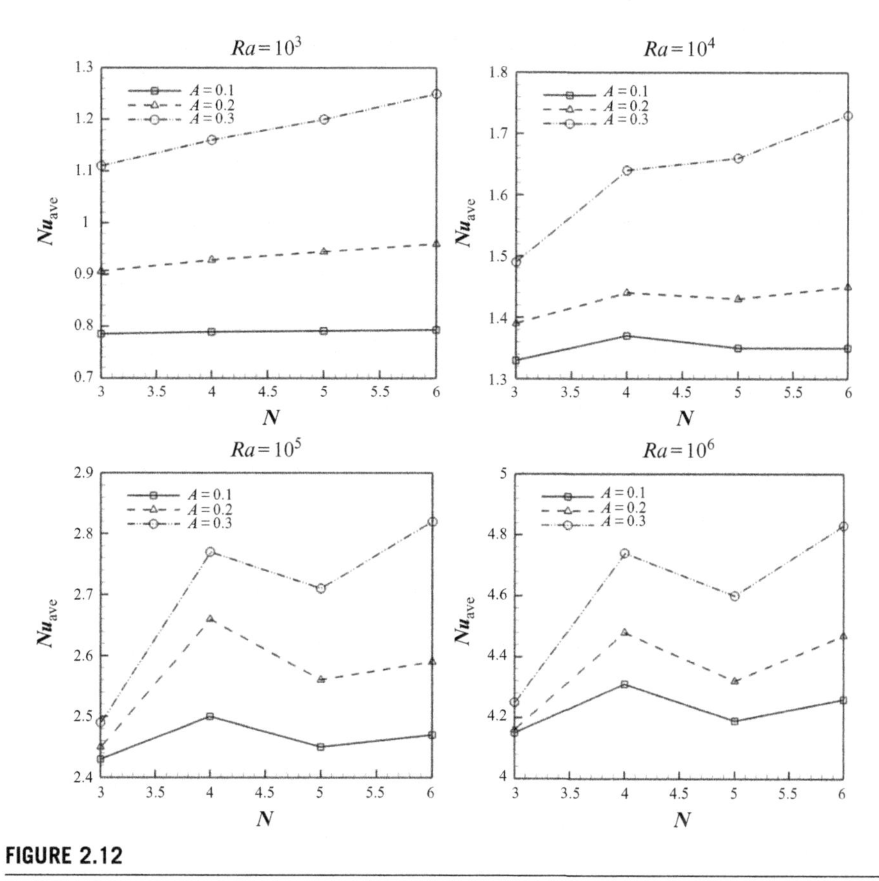

FIGURE 2.12

Effects of the amplitude A, number of undulations N, and Rayleigh number Ra on the average Nusselt number Nu_{ave}.

number is approximately constant at $A = 0.1$ for various numbers of undulations, whereas it increases uniformly at $A = 0.3$ and 0.5 because of the decreasing the space between the hot and cold surfaces. At $Ra = 10^4$, the maximum value of Nu_{ave} is obtained for $N = 4$ and $A = 0.1$, due to more available space. The Nu_{ave} profiles at $Ra = 10^5$ indicate that the minimum values of Nu_{ave} are obtained at $N = 3$, whereas the maximum values of Nu_{ave} occur at $N = 4$ for $A = 0.1$, 0.3 and at $N = 6$ for $A = 0.5$. It is worth pointing out that for higher Rayleigh numbers the profiles of the average Nusselt number have a local minimum at $N = 5$ for all values of A.

REFERENCES

[1] V.R. Voller, Basic Control Volume Finite Element Methods for Fluids and Solids, World Scientific Publishing Co. Pte. Ltd. 5, Tohccxxvc, 2009.

[2] M. Sheikholeslami, M. Gorji-Bandpy, I. Pop, S. Soleimani, Numerical study of natural convection between a circular enclosure and a sinusoidal cylinder using control volume based finite element method, Int. J. Therm. Sci. 72 (2013) 147–158.

[3] B.S. Kim, D.S. Lee, M.Y. Ha, H.S. Yoon, A numerical study of natural convection in a square enclosure with a circular cylinder at different vertical locations, Int. J. Heat Mass Transf. 51 (2008) 1888–1906.

CHAPTER

Nanofluid flow and heat transfer in an enclosure

3

3.1 INTRODUCTION

Nanofluids are produced by dispersing nanometer-scale solid particles into base liquids with low thermal conductivity, such as water, ethylene glycol, and oils. In many energy systems controlling heat transfer is crucial because of the increase in energy prices. In recent years nanofluids technology has been proposed and experimentally or numerically studied by some researchers to control heat transfer in a process. A nanofluid can be applied to engineering problems, such as heat exchangers, cooling electronic equipment, and chemical processes. Almost all of the researchers treated nanofluids as the common pure fluid and used conventional equations of mass, momentum, and energy; the only effects of a nanofluid is its thermal conductivity and viscosity, which are obtained from theoretical models or experimental data. These researchers assumed that nanoparticles are in thermal equilibrium and there are no slip velocities between the nanoparticles and fluid molecules; thus they have a uniform mixture of nanoparticles. Khanafer et al. [1] conducted a numerical investigation of heat transfer enhancement due to adding nanoparticles in a differentially heated enclosure. Abu-Nada et al. [2] investigated natural convection heat transfer enhancement by nanofluid in a horizontal concentric annuli field. At low Rayleigh numbers, they found nanoparticles with higher thermal conductivity enhance heat transfer more. Jou and Tzeng [3] numerically studied the natural convection heat transfer enhancement of nanofluids within a two-dimensional enclosure. Their results showed that increasing the buoyancy parameter and volume fraction of nanofluids causes an increase in the average heat transfer coefficient. Sheikholeslami et al. [4] performed a numerical study to investigate natural convection in a square cavity with curve boundaries filled with a copper-water nanofluid. Natural convection in a nanofluid-filled concentric annulus between an outer square cylinder and an inner elliptic cylinder was studied by Sheikholeslami et al. [5]. They concluded that the minimum value of heat transfer enhancement occurs at eccentricity (ϵ) equal to 0.95 for $Ra = 10^5$, but for other Rayleigh numbers it occurred at an eccentricity (ϵ) equals to 0.65. Sheikholeslami et al. [6] used the lattice Boltzmann method to examine free convection of nanofluids. Copper as the nanoparticle leads to the highest enhancement for this problem. Asymmetric laminar flow and heat transfer of a nanofluid between contracting rotating disks was investigated by Hatami et al. [7]. Their results indicated that the temperature profile becomes more flat near the middle of two disks with an increase of injection, but the opposite trend is observed with an increase in the expansion ratio. Sheikholeslami et al. [8] analyzed magnetohydrodynamic (MHD) nanofluid flow and heat transfer between two horizontal plates in a rotating system. Their results indicated that, for both suction and injection, the Nusselt number has a direct relationship with the nanoparticle volume fraction. Sheikholeslami et al. [9] investigated the boundary-layer flow of nanofluid over a permeable stretching wall. Their results indicated that an increase in the nanoparticle volume fraction decreases the momentum boundary-layer thickness and entropy generation rate, whereas the thermal boundary-layer thickness increases. The problem of laminar nanofluid flow in a semiporous channel in the presence of a transverse magnetic field was investigated analytically by Sheikholeslami et al. [10]. Their results showed that velocity boundary-layer thickness decreases with an increase in Reynolds number, and it increases as the Hartmann number increases. Squeezing, unsteady nanofluid flow and heat transfer was studied by Sheikholeslami et al. [11]. They showed that for a case in which two plates are moving together, the Nusselt number increases with an increase in the nanoparticle volume fraction and Eckert number, whereas it decreases with an increase in the squeeze number. Heat transfer of a nanofluid flow that is squeezed between parallel plates was investigated analytically using the homotopy perturbation method by Sheikholeslami and Ganji [12].They reported that the Nusselt number has a direct relationship with the nanoparticle volume fraction, the squeeze number, and the Eckert number when two plates are separated, but it has the reverse relationship with the squeeze number when two plates are squeezed. Ashorynejad et al. [13] studied the flow and heat transfer of a nanofluid over a stretching cylinder in the presence of a magnetic field. Natural convection of a non-Newtonian copper-water nanofluid between two infinite, parallel, vertical, flat plates was investigated by Domairry et al. [14]. They concluded that thermal boundary layer thickness decreases with increase of nanoparticle volume fraction. Hatami et al. [15] investigated the MHD Jeffery-Hamel nanofluid flow in nonparallel walls. They found that the skin friction coefficient is an increasing function of the Reynolds number, the opening angle, and the nanoparticle volume friction but a decrease function of the Hartmann number. Sheikholeslami and Ganji [16] studied the MHD flow in a permeable channel filled with nanofluid. They showed that velocity boundary-layer thickness decreases with an increase in the Reynolds number and nanoparticle volume fraction, and it increases as the Hartmann number increases. Hatami et al. [17] simulated the flow and heat transfer of nanofluid between two parallel plates. They showed that to reach a maximum Nusselt number, copper should be used as the nanoparticle. Sheikholeslami et al. [18] applied Group Method of Data Handling (GMDH) to investigate heat transfer of a copper-water nanofluid over a stretching cylinder in the presence of a magnetic field. Their results indicate that GMDH type NN in comparison with fourth-order Runge-Kutta integration scheme. Rashidi et al. [19] studied the effects of the magnetic interaction number, slip factor, and relative temperature difference on velocity and temperature profiles, as well as entropy generation, in MHD flow of a fluid with variable properties over a rotating disk. Rashidi et al. [20] considered the analysis of the second law of thermodynamics applied to an electrically conducting incompressible nanofluid flowing over a porous rotating disk. They concluded that using magnetic rotating disk drives has important applications in enhancing heat transfer in renewable energy systems. Ellahi [21] studied the MHD flow of a non-Newtonian nanofluid in a pipe. Sheikholeslami Kandelousi [22] used KKL correlation for simulation of nanofluid flow and heat transfer in a permeable channel. He found that heat transfer enhancement has direct relationship with Reynolds number when power law index is equals to zero but opposite trend is observed for other values of power law index. Mohammed and Narrein [23] studied the effects of using different geometrical parameters with a combination of nanofluids on heat transfer and fluid flow characteristics in a helically coiled tube heat exchanger. They found that a counterflow configuration produced better results compared with a parallel-flow configuration.

All the above-mentioned studies assume that there are no slip velocities between nanoparticles and fluid molecules. It is believed that in natural convection of nanofluids, the nanoparticles cannot accompany fluid molecules because of some slip mechanisms such as Brownian motion and thermophoresis, so the volume fraction of nanofluids may no longer be uniform and there may be a variable concentration of nanoparticles in a mixture. Nanofluid flow and heat transfer characteristics between two horizontal parallel plates in a rotating system were investigated by Sheikholeslami et al. [24]. They proved that Nusselt number increases with increase of nanoparticle volume fraction and Reynolds number but it decreases with increase of Eckert number, magnetic and rotation parameters. Khan and Pop [25] studied boundary-layer flow of a nanofluid past a stretching sheet. They indicated that the reduced Nusselt number is a decreasing function of each dimensionless number. Cheng [26] studied the natural convection boundary-layer flow over a truncated cone embedded in a porous medium saturated by a nanofluid with a constant wall temperature. He showed that an increase in the thermophoresis parameter or the Brownian parameter tends to decrease the local Nusselt number. Mixed convection of a nanofluid consisting of water and silicon dioxide in an inclined enclosure cavity was studied numerically by Alinia et al. [27] using a two-phase mixture model. They found that effect of the inclination angle is more pronounced at higher Richardson numbers. Three-dimensional heat and mass transfer in a rotating system using nanofluid was investigated by Sheikholeslami and Ganji [28]. They also investigated two-phase modeling of a nanofluid in a rotating system with a permeable sheet [29]. They found that the Nusselt number has a direct relationship with the Reynolds number and injection parameter, whereas it has the reverse relationship with the rotation parameter, Schmidt number, thermophoretic parameter, and Brownian parameter.

Free convective heat transfer has attracted much attention in recent years because of its wide applications in areas such as aircraft cabin insulation, cooling electronic equipment, and heating and ventilation control in building design. In engineering applications, however, enclosures are often more complicated than simple enclosures. Natural convective heat transfer in horizontal annuli between two concentric circular cylinders has been well studied in previous literatures. Cesini et al. [30] numerically and experimentally analyzed natural convection from a horizontal cylinder enclosed in a rectangular cavity. Their results showed that the average heat transfer coefficients increase with an increase in the Rayleigh number. Sheikholeslami et al. [31] used Lattice Boltzmann method in order to simulate nanofluid flow and heat transfer in a horizontal cylindrical enclosure with an inner triangular cylinder. They found that effect of adding nanoparticle become more obvious with increase of Lorentz forces. Ghaddar [32] reported the numerical results of natural convection from a uniformly heated horizontal cylinder placed in a large, rectangular, air-filled enclosure. He observed that flow and thermal behavior depended on heat fluxes imposed on the inner cylinder within the isothermal enclosure. Moukalled and Acharya [33] studied numerically the natural convective heat transfer from a heated horizontal cylinder placed concentrically inside a square enclosure. The governing equations in their work are solved in a body-filled coordinate system using a control volume-based numerical procedure.

3.2 NANOFLUID

3.2.1 DEFINITION OF NANOFLUID

Low thermal conductivity of conventional heat transfer fluids such as water, oil, and ethylene glycol is a serious limitation in improving the performance and compactness of many types of engineering equipment, such as heat exchangers and electronic devices. To overcome this disadvantage, there is strong motivation to develop advanced heat transfer fluids with substantially higher conductivity. An innovative way of improving the thermal conductivities of fluids is to suspend small solid particles in the fluid. Various types of powders, such as metallic, nonmetallic, and polymeric particles, can be added to fluids to form slurries. The thermal conductivities of fluids with suspended particles are expected to be higher than that of common fluids. Nanofluids are a new kind of heat transfer fluid containing a small quantity of nano-sized particles (usually <100 nm) that are uniformly and stably suspended in a liquid. The dispersion of a small amount of solid nanoparticles in conventional fluids changes their thermal conductivity remarkably. Compared with existing techniques for enhancing heat transfer, nanofluids show a superior potential for increasing heat transfer rates in a variety of cases [1–3].

3.2.2 MODEL DESCRIPTION

In the literature convective heat transfer with nanofluids can be modeled using mainly the two-phase or single-phase approach. In the two-phase approach the velocity between the fluid and particles might not be zero [34] because of several factors such as gravity, friction between the fluid and the solid particles, Brownian forces, Brownian diffusion, sedimentation, and dispersion. In the single-phase approach, the nanoparticles can be easily fluidized, and therefore one may assume that the motion slip between the phases, if any, would be negligible [35]. The latter approach is simpler and more computationally efficient.

3.2.3 CONSERVATION EQUATIONS

3.2.3.1 Single-phase model

Although nanofluids are solid-liquid mixtures, the approach conventionally used in most studies of natural convection considers the nanofluid as a single-phase (homogenous) fluid. In fact, because of the extreme size and low concentration of suspended nanoparticles, they are assumed to move at the same velocity as the fluid. Also, by considering the local thermal equilibrium, the solid particle-liquid mixture may then be approximately considered to behave as a conventional single-phase fluid with properties that are to be evaluated as functions of those of the constituents. The governing equations for a homogenous analysis of natural convection are continuity, momentum, and energy equations with their density, specific heat, thermal conductivity, and viscosity modified for nanofluid application. The specific governing equations for various studied enclosures are not shown here; they can be found in different publications [36–39]. It should be mentioned that sometimes this assumption is not correct. For example, according to Ding and Wen [40], this assumption may not always remain true for a nanofluid. They investigated particle migration in a nanofluid used in a pipe flow and stated that at Peclet numbers exceeding 10 the particle distribution is significantly nonuniform. Figure 3.1 shows this relationship between concentration and Peclet number for a pipe with a 0.05 particle volume fraction, illustrating that the homogenous flow assumption is conditional.

Nevertheless, many studies have performed numerical simulation using the single-phase assumption and reported acceptable results for the heat transfer and hydrodynamic properties of the flow.

3.2.3.2 Two-phase model

Several authors have tried to establish convective transport models for nanofluids [41–43]. Nanofluid is a two-phase mixture in which the solid phase consists of nanosized particles. In view of the nanoscale size of the particles, whether the theory of a conventional two-phase flow can be applied in describing the flow characteristics of

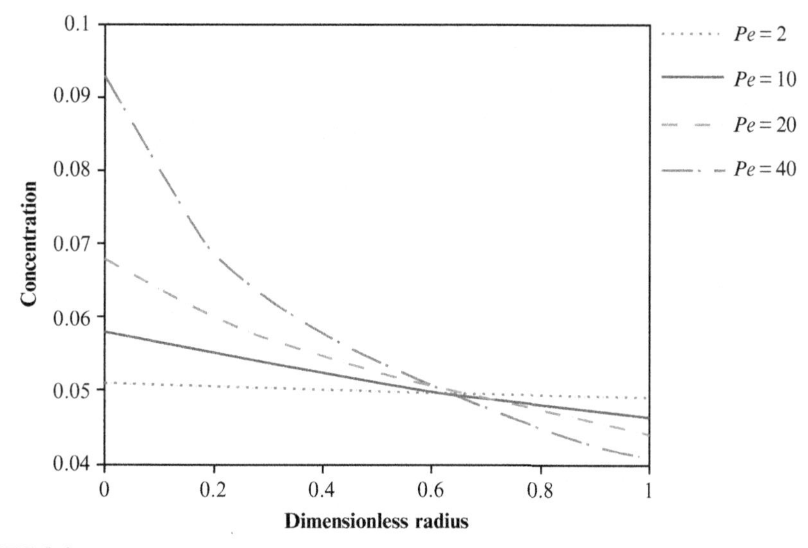

FIGURE 3.1

Relationship between particle concentration distribution and Peclet number (*Pe*) [40]

nanofluid may be questionable. On the other hand, several factors such as gravity, friction between the fluid and solid particles and Brownian forces, the phenomena of Brownian diffusion, sedimentation, and dispersion may affect a nanofluid flow. Consequently, the slip velocity between the fluid and particles cannot be neglected when simulating nanofluid flows. Since the two-phase approach considers the movement between the solid and fluid molecule, it may have better prediction in nanofluid study. To fully describe and predict the flow and behavior of complex flows, different multiphase theories have been proposed and used. A large number of published articles concerning multiphase flows typically used the mixture theory to predict the behavior of nanofluids [41–43]. Buongiorno [44] performed comprehensive survey of convective transport in nanofluids using a model accounting for Brownian motion and thermophoresis. Buongiorno developed a two-component, four-equation, nonhomogeneous equilibrium model for mass, momentum, and heat transfer in nanofluids. The nanofluid is treated as a two-component mixture (base fluid and nanoparticles) with the following assumptions: no chemical reactions, negligible external forces, dilute mixture ($\phi = 1$), negligible viscous dissipation, negligible radiative heat transfer, and the nanoparticle and base fluid locally in thermal equilibrium. Invoking the above assumptions, the equations below represent the mathematical formulation of the nonhomogenous single-phase model for the governing equations as formulated by Buongiorno.

3.2.3.2.1 Continuity equation

$$\nabla \cdot v = 0 \tag{3.1}$$

where v is the velocity.

3.2.3.2.2 Nanoparticle continuity equation

$$\frac{\partial \phi}{\partial t} + v \cdot \nabla \phi = \nabla \cdot \left(D_B \nabla \phi + D_T \frac{\nabla T}{T} \right) \tag{3.2}$$

Here, ϕ is the nanoparticle volume fraction and D_B is the Brownian diffusion coefficient given by the Einstein-Stokes equation:

$$D_B = \frac{k_B T}{3\pi \mu d_p}, \tag{3.3}$$

where μ is the viscosity of the fluid, d_p is the nanoparticle diameter, $k_B = 1.385 \times 10^{-23}$ is the Boltzmann constant, and D_T is the thermophoretic diffusion coefficient, which is defined as

$$D_T = \left(\frac{\mu}{\rho}\right) \left(0.26 \frac{k}{k + k_p}\right). \tag{3.4}$$

In Eq. (3.4), k and k_p are the thermal conductivity of the fluid and of the particle materials, respectively.

3.2.3.2.3 Momentum equation

$$v \cdot \nabla v = -\frac{1}{\rho_{\mathrm{nf}}} \nabla p + \nabla \cdot \tau + g \tag{3.5}$$

where

$$\tau = -\mu_{\mathrm{nf}} \left(\nabla v + (\nabla v)^{\mathrm{t}} \right), \tag{3.6}$$

where the t indicates the transpose of ∇v. Also p is pressure.

3.2.3.2.4 Energy equation

$$v \cdot \nabla T = \nabla (\alpha_{\mathrm{nf}} \nabla T) + \frac{\rho_{\mathrm{p}} c_{\mathrm{p}}}{\rho_{\mathrm{nf}} c_{\mathrm{nf}}} \left(D_{\mathrm{B}} \nabla \phi \cdot \nabla T + D_{\mathrm{T}} \frac{\nabla T \cdot \nabla T}{T} \right) \tag{3.7}$$

where ϕ and T are nanoparticle concentration and the temperature of the nanofluid, respectively.

This nanofluid model can be characterized as a "two-fluid" (nanoparticles and the base fluid), four-equation (continuity, mass, momentum, energy), nonhomogeneous (nanoparticle/fluid slip velocity is allowed) equilibrium (nanoparticle/fluid temperature differences are not allowed) model. Note that the conservation equations are strongly coupled. That is, v depends on ϕ via viscosity; ϕ depends on T mostly because of thermophoresis; T depends on ϕ via thermal conductivity, as well as via the Brownian and thermophoretic terms in the energy equation; ϕ and T obviously depend on v because of the convection terms in the nanoparticle continuity and energy equations, respectively.

A numerical study by Behzadmehr et al. [45] implemented for the first time a two-phase mixture model to investigate the behavior of a copper-water nanofluid in a tube, and the results were compared with previous works using a single-phase approach. The authors claimed that the simulation done by assuming that base fluid and particles behave separately provided results that are more precise compared with those of the previous computational modeling. They implemented the mixture theory in their work, suggesting that the continuity, momentum, and energy equations be written for a mixture of a fluid and a solid phase. Some assumptions also were stated for the model, such as a strong coupling between two phases and the fluid being closely followed by the particles, with each phase owning a different velocity, leading to a term called *slip velocity* of nanoparticles, as in Eq. (3.8):

$$V_{\mathrm{pf}} = V_{\mathrm{P}} - V_{\mathrm{f}} = \frac{\rho_{\mathrm{p}} d_{\mathrm{p}}^2}{18 \mu_{\mathrm{f}} f_{\mathrm{drag}}} \frac{\left(\rho_{\mathrm{p}} - \rho_{\mathrm{m}} \right)}{\rho_{\mathrm{p}}} a, \quad a = g - (V_{\mathrm{m}} \cdot \nabla) V_{\mathrm{m}}, \quad f_{\mathrm{drag}} = \begin{cases} 1 + 0.15 Re_{\mathrm{p}}^{0.687} \\ 0.0183 Re_{\mathrm{p}} \end{cases} \tag{3.8}$$

The conservation equations (continuity, momentum, and energy, respectively) are written for the mixture as follows:

$$\nabla \cdot (\rho_{\mathrm{m}} V_{\mathrm{m}}) = 0 \tag{3.9}$$

$$\nabla\cdot(\rho_{\rm m}V_{\rm m}V_{\rm m}) = -\nabla P_{\rm m} + \nabla\cdot[\tau-\tau_{\rm t}] + \rho_{\rm m}g + \nabla\cdot\left(\sum_{k=1}^{n}\phi_k\rho_k V_{{\rm dr},k}V_{{\rm dr},k}\right) \tag{3.10}$$

$$\nabla\cdot\left(\phi_{\rm p}V_k(\rho_k h_k + p)\right) = \nabla\cdot(k_{\rm eff}\nabla T - C_{\rm p}\rho_{\rm m}vt) \tag{3.11}$$

where $V_{\rm dr,p}$ is the particle draft velocity that is related to the slip velocity and is defined as:

$$V_{\rm dr,p} = V_{\rm P} - V_{\rm f} = V_{\rm pf} - \frac{\sum_{k=1}^{n}\phi_k\rho_k}{\rho_{\rm m}}V_{{\rm f}k} \tag{3.12}$$

3.2.4 PHYSICAL PROPERTIES OF NANOFLUIDS IN A SINGLE-PHASE MODEL

Properties of a base nanofluid have been published in the literature over the past 20 years. However, only recently have some data on temperature-dependent properties been provided, even though they are for only nanofluid effective thermal conductivity and effective absolute viscosity.

3.2.4.1 Density

In the absence of experimental data for nanofluid densities, values that are constant and temperature-independent, based on nanoparticle volume fraction, are used:

$$\rho_{\rm nf} = \rho_{\rm f}(1-\phi) + \rho_{\rm p}\phi \tag{3.13}$$

3.2.4.2 Specific heat capacity

It has been suggested that the effective specific heat can be calculated using the following equation, as reported by Pak and Cho [46]:

$$(C_{\rm p})_{\rm nf} = (C_{\rm p})_{\rm f}(1-\phi) + (C_{\rm p})_{\rm p}\phi \tag{3.14}$$

Other authors suggest an alternative approach based on the heat capacity concept [47]:

$$(\rho C_{\rm p})_{\rm nf} = (\rho C_{\rm p})_{\rm f}(1-\phi) + (\rho C_{\rm p})_{\rm p}\phi \tag{3.15}$$

These two formulations may, of course, lead to different results for specific heat. Because of the lack of experimental data, both formulations are considered equivalent in estimating nanofluid-specific heat capacity [48].

3.2.4.3 Thermal expansion coefficient

The thermal expansion coefficient of a nanofluid can be obtained as follows [1]:

$$(\rho\beta)_{\rm nf} = (\rho\beta)_{\rm f}(1-\phi) + (\rho\beta)_{\rm p}\phi \tag{3.16}$$

3.2.4.4 Electrical conductivity

The effective electrical conductivity of a nanofluid was presented by Maxwell [49]:

$$\sigma_{nf}/\sigma_f = 1 + 3(\sigma_p/\sigma_f - 1)\phi/((\sigma_p/\sigma_f + 2) - (\sigma_p/\sigma_f - 1)\phi) \tag{3.17}$$

3.2.4.5 Dynamic viscosity

Various suggestions for models of the viscosity of a nanofluid mixture take into account the percentage of nanoparticles suspended in the base fluid. The classic Brinkman model [50] seems to be a proper one that has been extensively used in studies of numerical simulation concerning nanofluids. Equation (3.18) shows the relationship between the nanofluid viscosity, base fluid viscosity, and the nanoparticle concentration in this model.

$$\mu_{nf} = \mu_f/(1-\phi)^{2.5} \tag{3.18}$$

In some recent computational studies, however, other models have been used in the numerical process, such as that described by Abu-Nada and Chamkha [51], who investigated the convection of a copper(II) oxide-ethylene glycol-water nanofluid in an enclosure where the Namburu correlation for viscosity [52] was applied:

$$\log(\mu_{nf}) = Ae^{-BT} \tag{3.19}$$

where

$$\begin{aligned} A &= 1.837\phi^2 - 29.643\phi + 165.65 \\ B &= 4\times 10^{-6}\phi^2 - 0.001\phi + 0.0186 \end{aligned} \tag{3.20}$$

In their study, the results where compared with those of viscosity modeled by Brinkman. These results outlined that as far as a value for a normalized average Nusselt number for the fluid is concerned, for various Rayleigh numbers, the Brinkman model predicts a higher value compared with that predicted by the Namburu model, showing the notable role of the viscosity model used in the calculations. Abu-Nada and Chamkha [51] also stated that a combination of different models may be implemented, showing different dependence on volume concentration, as well as the geometry aspect ratio, yet with the limitation that the models include only the ones mentioned in their study. Other studies also have shown that different models might lead to different results, like that by Arefmanesh and Mahmoodi [53], which investigated a numerical simulation using models described by Maïga et al. [54], as well as Brinkman models, and showed a difference in the results for the heat transfer of aluminum oxide-water nanofluid in a square cavity. The difference was dependent on the Richardson number and the volume concentration of particles. A number of suggested relationships of viscosity models used in numerical studies are presented in Table 3.1.

3.2.4.6 Thermal conductivity

Different nanofluid models based on a combination of the different formulas for thermal conductivity adopted in studies of natural convection are summarized in Table 3.2. Table 3.3 demonstrates values of thermophysical properties for different materials used as suspended particles in nanofluids.

Table 3.1 Different Models Calculating the Viscosity of Nanofluids Used in Simulations

Model	Equation
Brinkman model [50]	$\mu_{nf} = \frac{\mu_f}{(1-\phi)^{2.5}}$
Einstein model [55]	$\mu_{nf} = (2.5\phi + 1)\mu_f,\ \phi < 0.05$
Pak and Cho's correlation [56]	$\mu_{nf} = \mu_f(1 + 39.11\phi + 533.9\phi^2)$
Nguyen et al. model [57] (fitted by Abu-Nada [58])	$\mu_{CuO} = -0.6967 + 15.937/T + 1.238\phi + 1356.14/T^2 - 0.259\phi^2 - 30.88\phi/T - 19652.74/T^3 + 0.01593\phi^3 + 4.38206\phi^2/T + 147.573\phi/T^2$ $\mu_{Al_2O_3} = \exp(3.003 - 0.04203T - 0.5445\phi + 0.0002553T^2 + 0.0524\phi^2 - 1.622\phi^{-1})$
Jang et al. model [59]	$\mu_{nf} = (2.5\phi + 1)\mu_f\left[1 + \eta(d_p/H)^{-2\epsilon}\phi^{2/3}(\epsilon + 1)\right]$
Koo and Kleinstreuer [60]	$\mu_{nf} = 5\times10^4\beta\phi\rho_f\sqrt{\frac{k_s T}{d_p\rho_p}}f(T,\phi),\ \begin{cases}\beta = 0.0137(100\phi)^{-0.5229} & \text{for } \phi < 1\% \\ \beta = 0.0011(100\phi)^{-0.7272} & \text{for } \phi > 1\%\end{cases}$
Maïga model [61]	$\mu_{nf} = \mu_f(1 + 7.3\phi + 123\phi^2)$
Brownian model [62]	$\mu_{nf} = \mu_f(1 + 2.5\phi + 6.17\phi^2)$
Nguyen model [63]	$\mu_{nf} = \mu_f(1 + 0.025\phi + 0.015\phi^2)$
Masoumi et al. [64]	$\mu_{nf} = \mu_f + \rho_p V_B d_p^2/(72C\delta)$
Gherasim et al. [65]	$\mu_{nf} = \mu_f 0.904 e^{14.8\phi}$

3.3 SIMULATION OF NANOFLUID IN VORTICITY STREAM FUNCTION FORM

3.3.1 MATHEMATICAL MODELING OF A SINGLE-PHASE MODEL

3.3.1.1 Natural convection

Flow is considered to be steady, two-dimensional, and laminar. Neglecting displacement currents, dissipation, and using the Boussinesq approximation, the governing equations of nanofluid heat transfer and fluid flow can be obtained in dimensional form:

$$\frac{\partial u}{\partial x} + \frac{\partial v}{\partial y} = 0 \tag{3.21}$$

$$u\frac{\partial u}{\partial x} + v\frac{\partial u}{\partial y} = -\frac{1}{\rho_{nf}}\frac{\partial P}{\partial x} + \upsilon_{nf}\left(\frac{\partial^2 u}{\partial x^2} + \frac{\partial^2 u}{\partial y^2}\right) \tag{3.22}$$

$$u\frac{\partial v}{\partial x} + v\frac{\partial v}{\partial y} = -\frac{1}{\rho_{nf}}\frac{\partial P}{\partial y} + \upsilon_{nf}\left(\frac{\partial^2 v}{\partial x^2} + \frac{\partial^2 v}{\partial y^2}\right) + \beta_{nf}g(T - T_c) \tag{3.23}$$

Table 3.2 Different Models of Thermal Conductivity of Nanofluids Used in Simulations

Model	Equation
Wasp model [36]	$\frac{k_{nf}}{k_f}=\frac{k_p+2k_f-2\phi(k_f-k_p)}{k_p+2k_f+\phi(k_f-k_p)}$
Jang and Choi model [66]	$\frac{k_{nf}}{k_f}=(1-\phi)+Bk_p\phi+18\times10^6\frac{3d_f}{d_p}k_f Re_{dp}^2 Pr\,\phi$
Bruggeman model [67]	$k_{nf}=0.25k_f(3\phi-1)k_p/k_f+[3(1-\phi)-1]+\sqrt{\Delta_B}$ $\Delta_B=[(3\phi-1)k_p/k_f+(3(1-\phi)-1)]^2+8k_p/k_f$
Chon et al. model [68]	$k_{nf}=k_f\left[1+64.7\phi^{0.7640}(d_f/d_p)^{0.369}(k_f/k_p)^{0.7476}Pr_T^{0.9955}Re^{1.2321}\right]$
Koo and Kleinstreuer [60]	$k_{nf}=k_f\left[\frac{k_p+2k_f-2\phi(k_f-k_p)}{k_p+2k_f+\phi(k_f-k_p)}\right]+5\times10^4\beta\phi\rho_f(C_p)_f\sqrt{\frac{k_BT}{d_p\rho_p}}f(T,\phi)$
Charuyakorn et al. [69]	$\frac{k_{nf}}{k_f}=\left[\frac{k_p+2k_f-2\phi(k_f-k_p)}{k_p+2k_f+\phi(k_f-k_p)}\right]\left(1+b\phi Pe_p^m\right)$
Stationary model [70]	$k_{nf}=k_f[1+k_p\phi d_f/(k_f(1-\phi)d_p)]$
Yu and Choi [71]	$k_{nf}=k_f\left[k_p+2k_f-2\phi(k_f-k_p)(1+\eta)^3\right]/\left[k_p+2k_f+\phi(k_f-k_p)(1+\eta)^3\right]$
Patel et al. [72]	$\frac{k_{nf}}{k_f}=1+\frac{k_p d_f\phi}{k_f d_p(1-\phi)}\left[1+c\frac{2k_BTd_p}{\pi\alpha_f\mu_f d_p^2}\right]$
Mintsa et al. [73]	$k_{nf}=k_f(1.72\phi+1.0)$

Table 3.3 Thermophysical Properties of Nanofluids

	ρ (kg/m^3)	C_p (J/kg K)	k (W/m K)	β (K^{-1})	σ (Ω m)$^{-1}$
Pure water	997.1	4179	0.613	21×10^{-5}	0.05
Copper (Cu)	8933	385	401	1.67×10^{-5}	5.95×10^7
Silver (Ag)	10,500	235	429	1.89×10^{-5}	3.60×10^7
Alumina (Al_2O_3)	3970	765	40	0.85×10^{-5}	1×10^{-10}
Titanium oxide (TiO_2)	4250	686.2	8.9538	0.9×10^{-5}	1×10^{-12}

$$u\frac{\partial T}{\partial x}+v\frac{\partial T}{\partial y}=\alpha_{nf}\left(\frac{\partial^2 T}{\partial x^2}+\frac{\partial^2 T}{\partial y^2}\right) \tag{3.24}$$

where ϕ is the solid volume fraction of nanoparticles and nf is nanofluid properties.

The stream function and vorticity are defined as follows:

$$u=\frac{\partial\psi}{\partial y},\quad v=-\frac{\partial\psi}{\partial x},\quad \text{and}\quad \omega=\frac{\partial v}{\partial x}-\frac{\partial u}{\partial y} \tag{3.25}$$

The stream function satisfies the continuity Eq. (3.21). The vorticity equation is obtained by eliminating the pressure between the two momentum equations, that is, by taking the y-derivative of Eq. (3.22) and subtracting from it the x-derivative of Eq. (3.23). This gives:

$$\frac{\partial\psi}{\partial y}\frac{\partial\omega}{\partial x}-\frac{\partial\psi}{\partial x}\frac{\partial\omega}{\partial y}=\upsilon_{\mathrm{nf}}\left(\frac{\partial^2\omega}{\partial x^2}+\frac{\partial^2\omega}{\partial y^2}\right)+\beta_{\mathrm{nf}}g\left(\frac{\partial T}{\partial x}\right) \tag{3.26}$$

$$\frac{\partial\psi}{\partial y}\frac{\partial T}{\partial x}-\frac{\partial\psi}{\partial x}\frac{\partial T}{\partial y}=\alpha_{\mathrm{nf}}\left(\frac{\partial^2 T}{\partial x^2}+\frac{\partial^2 T}{\partial y^2}\right) \tag{3.27}$$

$$\frac{\partial^2\psi}{\partial x^2}+\frac{\partial^2\psi}{\partial y^2}=-\omega \tag{3.28}$$

By introducing the following nondimensional variables:

$$X=\frac{x}{L},\quad Y=\frac{y}{L},\quad \Omega=\frac{\omega L^2}{\alpha_{\mathrm{f}}},\quad \Psi=\frac{\psi}{\alpha_{\mathrm{f}}},\quad U=\frac{uL}{\alpha_{\mathrm{f}}},\quad V=\frac{vL}{\alpha_{\mathrm{f}}},\quad \Theta=\frac{T-T_{\mathrm{c}}}{\Delta T}\quad \begin{cases}\Delta T=T_{\mathrm{h}}-T_{\mathrm{c}}\\ \Delta T=(q''L/k_{\mathrm{f}})\end{cases} \tag{3.29}$$

where $\Delta T=T_{\mathrm{h}}-T_{\mathrm{c}}$ and $\Delta T=(q''L/k_{\mathrm{f}})$ are used for constant temperature and constant heat flux boundary conditions, respectively. Using the dimensionless parameters, the equations now become:

$$\frac{\partial\Psi}{\partial Y}\frac{\partial\Omega}{\partial X}-\frac{\partial\Psi}{\partial X}\frac{\partial\Omega}{\partial Y}=Pr\frac{(\mu_{\mathrm{nf}}/\mu_{\mathrm{nf}})}{(\rho_{\mathrm{nf}}/\rho_{\mathrm{nf}})}\left(\frac{\partial^2\Omega}{\partial X^2}+\frac{\partial^2\Omega}{\partial Y^2}\right)+RaPr\left[\frac{\beta_{\mathrm{nf}}}{\beta_{\mathrm{f}}}\right]\left(\frac{\partial\Theta}{\partial X}\right) \tag{3.30}$$

$$\frac{\partial\Psi}{\partial Y}\frac{\partial\Theta}{\partial X}-\frac{\partial\Psi}{\partial X}\frac{\partial\Theta}{\partial Y}=\left[\frac{\frac{k_{\mathrm{nf}}}{k_{\mathrm{f}}}}{\frac{(\rho C_{\mathrm{p}})_{\mathrm{nf}}}{(\rho C_{\mathrm{p}})_{\mathrm{f}}}}\right]\left(\frac{\partial^2\Theta}{\partial X^2}+\frac{\partial^2\Theta}{\partial Y^2}\right) \tag{3.31}$$

$$\frac{\partial^2\Psi}{\partial X^2}+\frac{\partial^2\Psi}{\partial Y^2}=-\Omega \tag{3.32}$$

where the Rayleigh number and the Prandtl number are defined as follows:

$$Ra=g\beta_{\mathrm{f}}L^3\Delta T/(\alpha_{\mathrm{f}}\upsilon_{\mathrm{f}}),\quad Pr=\upsilon_{\mathrm{f}}/\alpha_{\mathrm{f}} \tag{3.33}$$

3.3.1.2 Force convection

Similar to the previous section, the equations governing nanofluid heat transfer and fluid flow can be obtained in dimensional form:

$$\frac{\partial u}{\partial x}+\frac{\partial v}{\partial y}=0 \tag{3.34}$$

$$u\frac{\partial u}{\partial x}+v\frac{\partial u}{\partial y}=-\frac{1}{\rho_{\mathrm{nf}}}\frac{\partial P}{\partial x}+\upsilon_{\mathrm{nf}}\left(\frac{\partial^2 u}{\partial x^2}+\frac{\partial^2 u}{\partial y^2}\right) \tag{3.35}$$

$$u\frac{\partial v}{\partial x}+v\frac{\partial v}{\partial y}=-\frac{1}{\rho_{\rm nf}}\frac{\partial P}{\partial y}+\upsilon_{\rm nf}\left(\frac{\partial^2 v}{\partial x^2}+\frac{\partial^2 v}{\partial y^2}\right) \tag{3.36}$$

$$u\frac{\partial T}{\partial x}+v\frac{\partial T}{\partial y}=\alpha_{\rm nf}\left(\frac{\partial^2 T}{\partial x^2}+\frac{\partial^2 T}{\partial y^2}\right) \tag{3.37}$$

by introducing the following nondimensional variables:

$$X=\frac{x}{L},\quad Y=\frac{y}{L},\quad \Omega=\frac{\omega L}{u_{\rm r}},\quad \Psi=\frac{\psi}{u_{\rm r}L},\quad U=\frac{u}{u_{\rm r}},\quad V=\frac{v}{u_{\rm r}},\quad \Theta=\frac{T-T_{\rm c}}{\Delta T}\quad \begin{cases}\Delta T=T_{\rm h}-T\\ \Delta T=(q''L/k_{\rm f})\end{cases} \tag{3.38}$$

where in Eq. (3.38) $L=r_{\rm out}-r_{\rm in}=r_{\rm in}$. Also, $\Delta T=T_{\rm h}-T_{\rm c}$ and $\Delta T=(q''L/k_{\rm f})$ are used for constant temperature and constant heat flux boundary conditions, respectively. The vorticity equation is obtained by eliminating the pressure between the two momentum equations, that is, by taking the y-derivative of Eq. (3.35) and subtracting from it the x-derivative of Eq. (3.36). Using the dimensionless parameters, the equations now become:

$$\frac{\partial\Psi}{\partial Y}\frac{\partial\Omega}{\partial X}-\frac{\partial\Psi}{\partial X}\frac{\partial\Omega}{\partial Y}=\frac{1}{Re}\frac{(\mu_{\rm nf}/\mu_{\rm nf})}{(\rho_{\rm nf}/\rho_{\rm nf})}\left(\frac{\partial^2\Omega}{\partial X^2}+\frac{\partial^2\Omega}{\partial Y^2}\right)+\frac{Gr}{Re^2}\frac{\beta_{\rm nf}}{\beta_{\rm f}}\frac{\partial\Theta}{\partial X} \tag{3.39}$$

$$\frac{\partial\Psi}{\partial Y}\frac{\partial\Theta}{\partial X}-\frac{\partial\Psi}{\partial X}\frac{\partial\Theta}{\partial Y}=\frac{1}{RePr}\left(\frac{k_{\rm nf}}{k_{\rm f}}\right)\Big/\left(\frac{(\rho C_{\rm p})_{\rm nf}}{(\rho C_{\rm p})_{\rm f}}\right)\left(\frac{\partial^2\Theta}{\partial X^2}+\frac{\partial^2\Theta}{\partial Y^2}\right) \tag{3.40}$$

$$\frac{\partial^2\Psi}{\partial X^2}+\frac{\partial^2\Psi}{\partial Y^2}=-\Omega \tag{3.41}$$

where $Re=\frac{\rho_{\rm f}u_{\rm r}L}{\mu_{\rm f}}$ is the Reynolds number.

3.3.1.3 Mixed convection

To reach mixed convection equations buoyancy forces should be added to Eq. (3.39). So, the governing equations in vorticity stream form are:

$$\frac{\partial\Psi}{\partial Y}\frac{\partial\Omega}{\partial X}-\frac{\partial\Psi}{\partial X}\frac{\partial\Omega}{\partial Y}=\frac{1}{Re}\frac{(\mu_{\rm nf}/\mu_{\rm nf})}{(\rho_{\rm nf}/\rho_{\rm nf})}\left(\frac{\partial^2\Omega}{\partial X^2}+\frac{\partial^2\Omega}{\partial Y^2}\right)+\frac{Gr}{Re^2}\frac{\beta_{\rm nf}}{\beta_{\rm f}}\frac{\partial\Theta}{\partial X} \tag{3.42}$$

$$\frac{\partial\Psi}{\partial Y}\frac{\partial\Theta}{\partial X}-\frac{\partial\Psi}{\partial X}\frac{\partial\Theta}{\partial Y}=\frac{1}{RePr}\left(\frac{k_{\rm nf}}{k_{\rm f}}\right)\Big/\left(\frac{(\rho C_{\rm p})_{\rm nf}}{(\rho C_{\rm p})_{\rm f}}\right)\left(\frac{\partial^2\Theta}{\partial X^2}+\frac{\partial^2\Theta}{\partial Y^2}\right) \tag{3.43}$$

$$\frac{\partial^2\Psi}{\partial X^2}+\frac{\partial^2\Psi}{\partial Y^2}=-\Omega \tag{3.44}$$

where $Gr=g\beta\Delta TL^3/\upsilon^2$ is the Grashof number and the Richardson number is defined as: $Ri=\frac{Gr}{Re^2}$.

In all of the equations above the properties of a nanofluid should be determined via one of the models that are presented in previous section.

3.3.2 CVFEM FOR NANOFLUID FLOW AND HEAT TRANSFER (SINGLE-PHASE MODEL)

3.3.2.1 Natural convection heat transfer in a nanofluid-filled, inclined, L-shaped enclosure

3.3.2.1.1 Problem definition

The physical model, along with important geometrical parameters and the mesh of the enclosure used in the present control volume finite element method (CVFEM) program, are shown in Fig. 3.2 [74]. The width and height of the enclosure are H. The right and top walls of the enclosure are maintained at constant cold temperatures T_c, whereas the inner circular wall is maintained at constant hot temperature T_h and the two bottom and left walls, with a length of $H/2$, are thermally insulated. In all cases the $T_h > T_c$ condition is maintained. To assess the shape of the inner circular and outer rectangular boundaries, which consist of the right and top walls, a super-elliptic function can be used:

$$\left(\frac{X}{a}\right)^{2n} + \left(\frac{Y}{b}\right)^{2n} = 1 \tag{3.45}$$

When $a = b$ and $n = 1$, the geometry becomes a circle. As n increases from 1, the geometry approaches a rectangle for $a \neq b$ and a square for $a = b$.

The local Nusselt number of the nanofluid along the hot wall can be expressed as:

$$Nu_{\text{local}} = \left(\frac{k_{\text{nf}}}{k_{\text{f}}}\right)\frac{\partial \Theta}{\partial r}, \tag{3.46}$$

where r is the radial direction. The average Nusselt number on the hot circular wall is evaluated as:

$$Nu_{\text{ave}} = \frac{1}{\gamma}\int_0^{\gamma} Nu_{\text{loc}}(\zeta)\text{d}\zeta \tag{3.47}$$

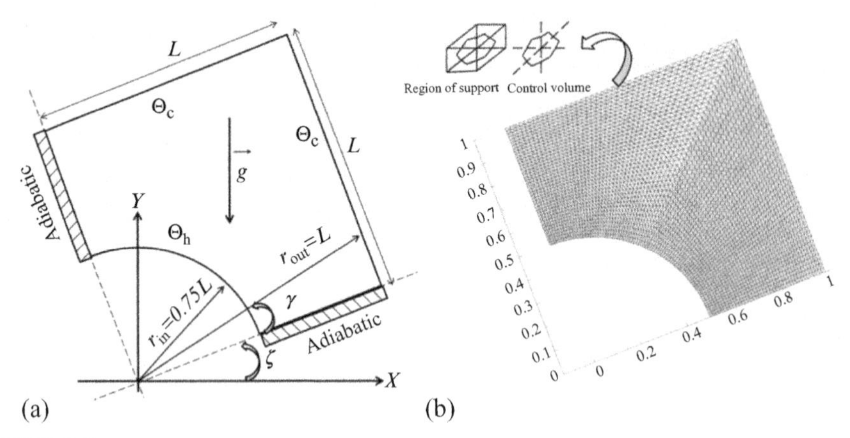

FIGURE 3.2

Geometry and the boundary conditions (a) with the mesh of enclosure (b) considered in this work.

To estimate the enhancement of heat transfer between the case of $\phi = 0.06$ and a pure fluid (base fluid), the enhancement is defined as:

$$E = \frac{Nu(\phi = 0.06) - Nu(\text{basefluid})}{Nu(\text{basefluid})} \times 100 \tag{3.48}$$

The present FORTRAN code is validated by comparing the results obtained for $Pr = 0.7$ with other works reported in the literature [1,75]. As shown in Table 3.4, The obtained results are in good agreements with previous publications. In addition, Fig. 3.3 illustrates excellent agreement between the present calculations and the

Table 3.4 Comparison of the Present Results with Previous Works for Different Rayleigh Numbers When $Pr = 0.7$

Rayleigh Number (*Ra*)	Present Work	Khanafer et al. [1]	De Vahl Davis [75]
10^3	1.1432	1.118	1.118
10^4	2.2749	2.245	2.243
10^5	4.5199	4.522	4.519

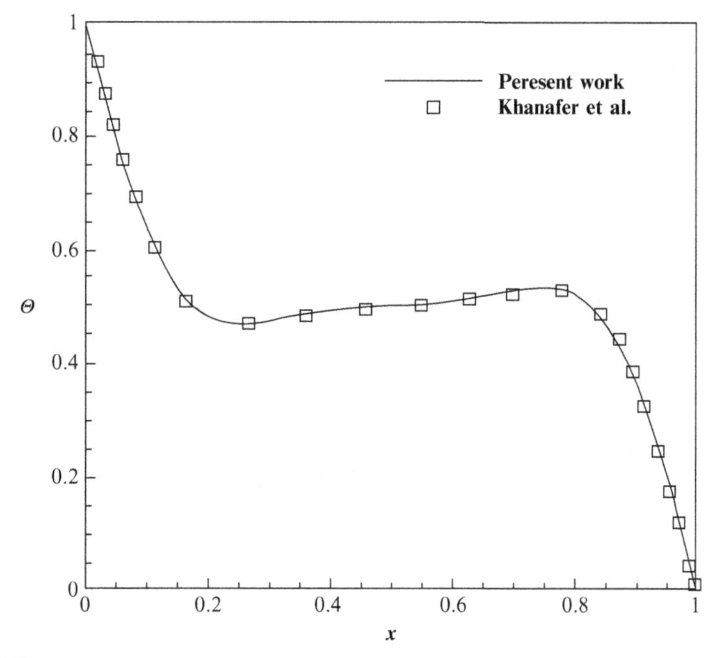

FIGURE 3.3

Comparison of the temperature on the axial midline between the present results and the numerical results obtained by Khanafer et al. [1] for $Gr = 10^4$, $\phi = 0.1$, and $Pr = 6.2$ (copper-water).

results of Khanafer et al. [1] for natural convection in an enclosure filled with a copper-water nanofluid.

3.3.2.1.2 Effect of active parameters

In this study natural convection heat transfer in an L-shaped, inclined enclosure filled with a nanofluid is investigated numerically using the CVFEM. Calculations are made for various values of the volume fraction of nanoparticles ($\phi = 0\%$, 2%, 4%, and 6%), the Rayleigh number ($Ra = 10^3$, 10^4, and 10^5), different inclination angles of turn ($\gamma = -45°$, $0°$, and $45°$), and a constant Prandtl number ($Pr = 6.2$).

A comparison of the isotherm and streamline contours for different Rayleigh numbers and inclined angles at $\phi = 0.06$ is shown in Fig. 3.4. When $\zeta = 0°$, for a low Rayleigh number ($Ra = 10^3$), the isotherms are parallel to each other and take the shape of the enclosure, which is the main characteristic of the conduction heat transfer mechanism. As the Rayleigh number increases the isotherms become more distorted and the stream function values enhance, which is caused by the domination of the convective heat transfer mechanism at higher Rayleigh numbers. At Rayleigh numbers $Ra = 10^3$ and 10^4, the maximum value of the stream function occurs at $\zeta = -45°$, while this maximum value corresponds to $\zeta = 0°$ at $Ra = 10^5$. At $\zeta = 45°$, the temperature counters and streamlines are symmetric with respect to the vertical lines that pass through the corner of the enclosure. This is due to the existence of a symmetrical boundary condition and the geometry of the cavity with respect to this line. At $\gamma = -45°$ and $0°$, an increase in the Rayleigh number causes the thermal boundary layer on the hot circular wall to decrease near the bottom wall of the enclosure; hence it can be predicted that the local Nusselt number obtains its minimum value at this area. The isotherms show that at $Ra = 10^4$ a thermal plume appears over the hot surface at $\gamma = 45°$. At this inclination angle, when Ra increases up to 10^5, the thermal plume grows and forces the hot fluid into contact with the cold walls.

Figure 3.5 shows the distribution of local Nusselt numbers along the surface of the inner circular wall for different inclined angles, Rayleigh numbers, and nanoparticle volume fractions. For all values of ζ, increasing the nanoparticle volume fraction and Rayleigh number leads to an increase in the local Nusselt number. At $Ra = 10^3$, because of the domination of the conduction heat transfer mechanism, the distribution of the local Nusselt numbers along the surface of the inner circular wall shows a symmetrical shape. In addition, as the Rayleigh number increases the local Nusselt number increases due to an increasing convection effect. In addition, as the Rayleigh number increases, the location of the minimum local Nusselt number approaches $\zeta = 45°$.

Figure 3.6 shows the effects of the Rayleigh number and the inclined angle forcopper-water nanofluids on the average Nusselt number. When $Ra = 10^3$, the variation of the average Nusselt number with respect to ζ is small. At $Ra = 10^4$, the variation of the average Nusselt number is more sensible than at $Ra = 10^3$. For this Rayleigh number the maximum and minimum average Nusselt numbers correspond to $\zeta = -45°$ and $45°$, respectively, whereas the opposite trend is observed at $Ra = 10^5$. This observation occurs because of the domination of the conduction heat transfer mechanism in these cases.

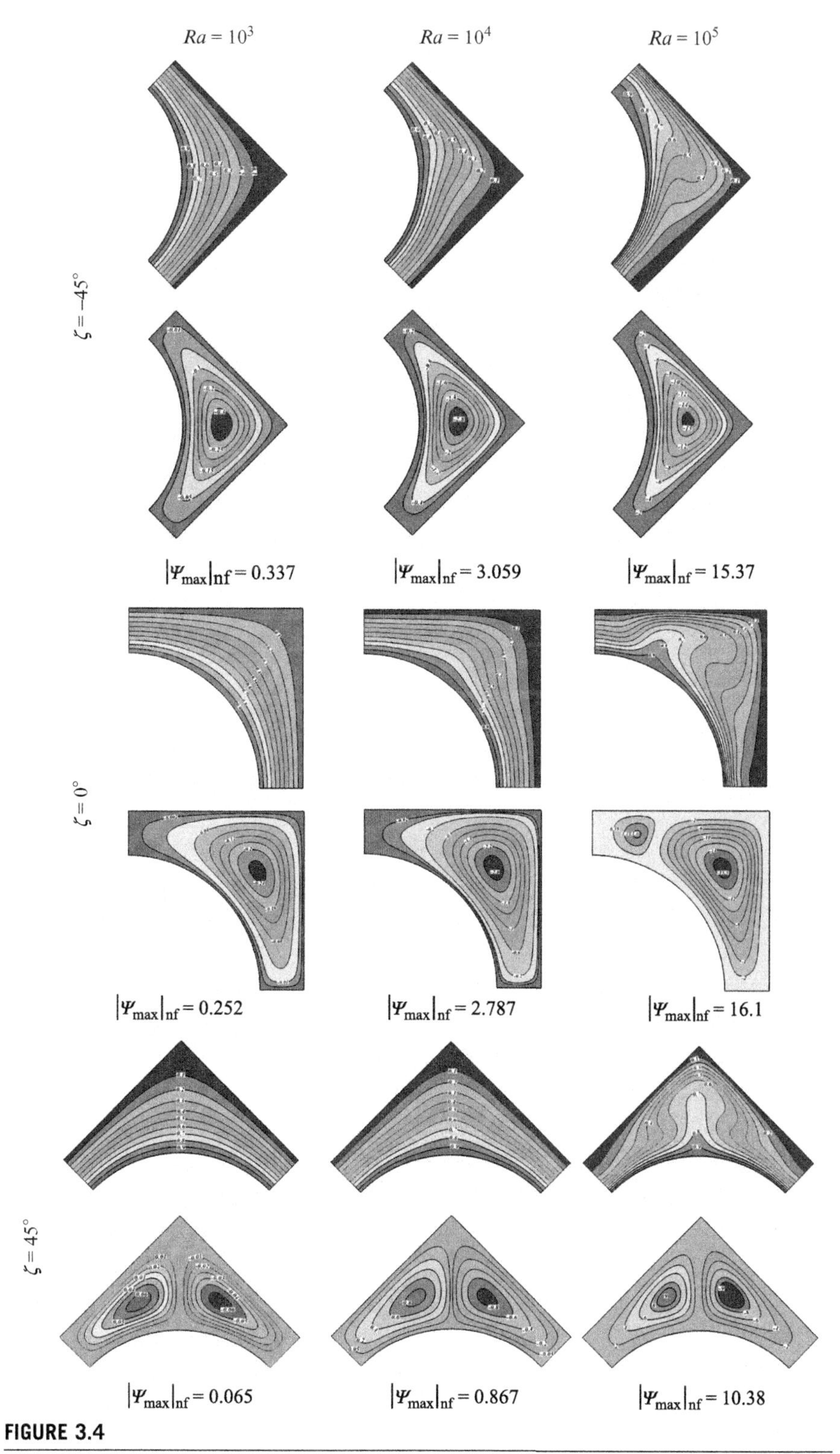

FIGURE 3.4

Comparison of the isotherm (up) and streamline (down) contours for different Rayleigh numbers and inclined angles at $\phi = 0.06$.

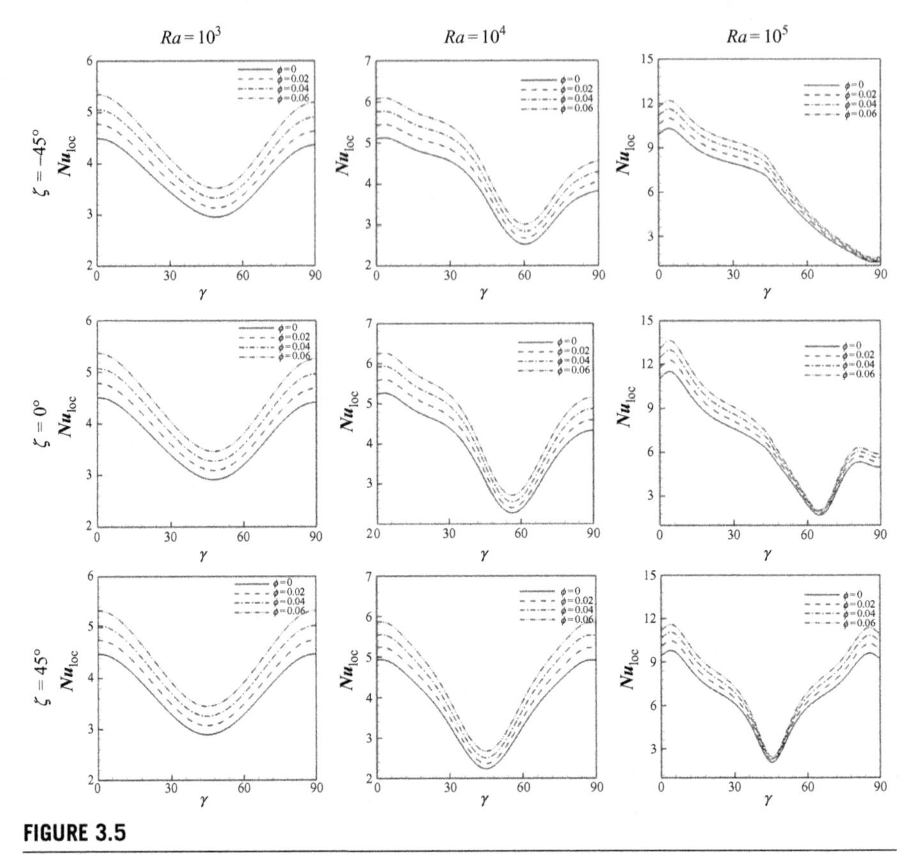

FIGURE 3.5

Effects of the nanoparticle volume fraction, Rayleigh number, and inclined angle of copper-water nanofluids on the local Nusselt number.

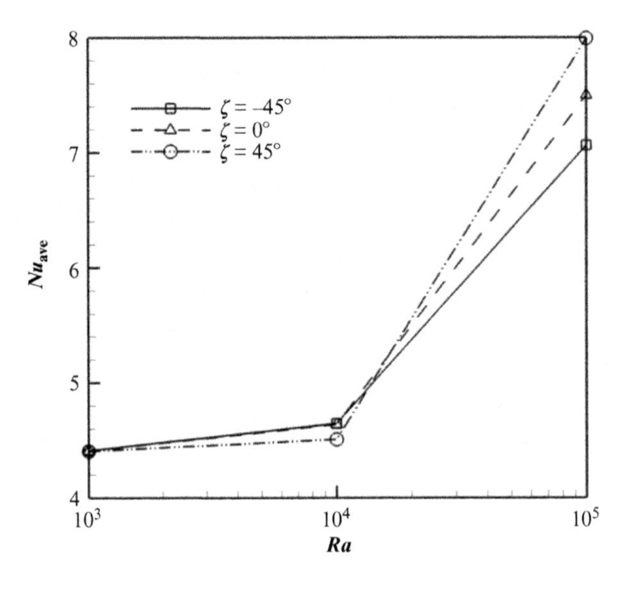

FIGURE 3.6

Effects of the Rayleigh number and the inclined angle of copper-water nanofluids on the average Nusselt number at $\phi = 0.06$.

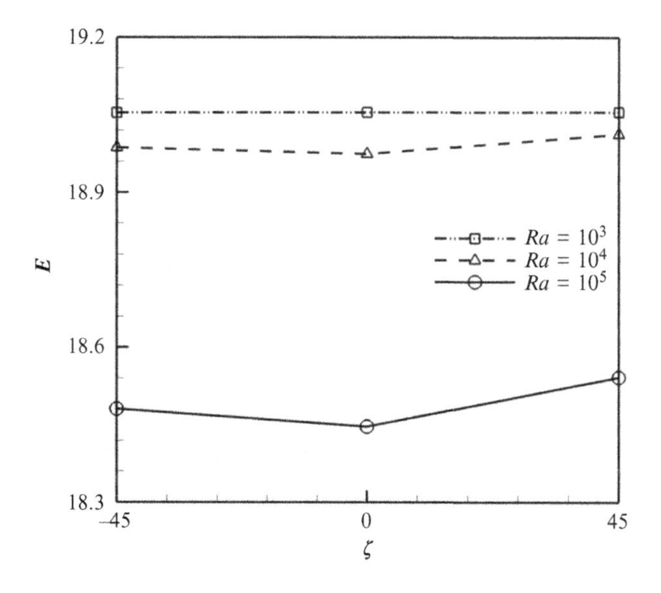

FIGURE 3.7

Effects of the Rayleigh number and the inclined angle of copper-water nanofluids on the ratio of heat transfer enhancement attributable to the addition of nanoparticles.

The heat transfer enhancement ratio affected by the addition of nanoparticles for different values of ζ and Ra is shown in Fig. 3.7. The effect of nanoparticles is more pronounced at a low Rayleigh number than at a high Rayleigh number because of the greater enhancement rate. This observation can be explained by noting that at a low Rayleigh number the heat transfer is dominant by conduction. Therefore, the addition of nanoparticles with high thermal conductivity increases the conduction and makes the enhancement more effective. Interestingly, the enhanced heat transfer at $Ra=10^3$ is the same at all inclination angles. Also, for $Ra=10^4$ and 10^5, the minimum value of enhancement is obtained at $\zeta=-0°$.

3.3.2.2 Natural convection heat transfer in a nanofluid-filled, semiannulus enclosure

3.3.2.2.1 Problem definition

The schematic diagram and the mesh of the semiannulus enclosure used in the present CVFEM program are shown in Fig. 3.8 [76]. The inner and outer walls are maintained at constant temperatures T_h and T_c, respectively, while the two other walls are thermally insulated.

The local Nusselt number of the nanofluid along the hot wall can be expressed as:

$$Nu_{\text{local}} = \left(\frac{k_{\text{nf}}}{k_{\text{f}}}\right)\frac{\partial \Theta}{\partial r} \tag{3.49}$$

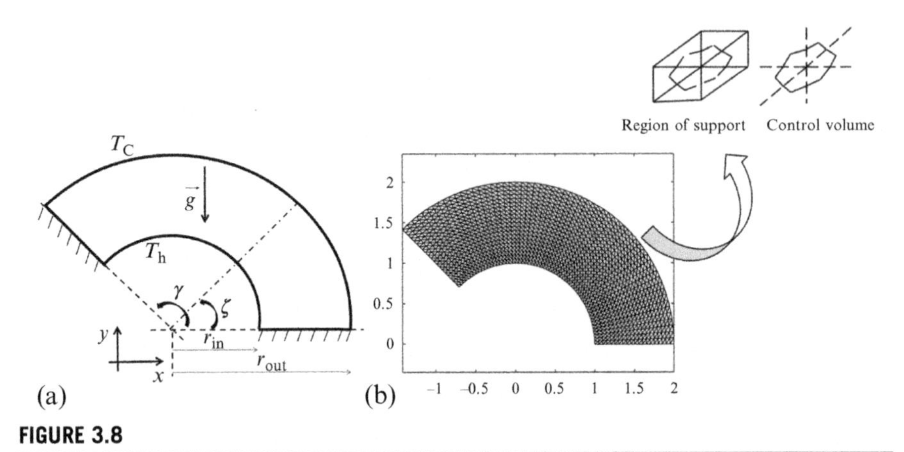

FIGURE 3.8

Geometry and the boundary conditions (a) with the mesh of semiannulus enclosure (b) considered in this work.

where r is the radial direction. The average Nusselt number on the hot circular wall is evaluated as:

$$Nu_{ave} = \frac{1}{\gamma}\int_0^{\gamma} Nu_{local}(\zeta)\mathrm{d}\zeta \tag{3.50}$$

3.3.2.2.2 Effect of active parameters

In this study natural convection heat transfer in a semiannulus enclosure filled with a nanofluid is investigated numerically using the CVFEM. The fluid in the enclosure is a copper-water nanofluid. Calculations are made for various volume fractions of the nanoparticles ($\phi = 0\%$, 2%, 4%, and 6%), Rayleigh number ($Ra = 10^3$, 10^4, and 10^5), different angles of turn ($\gamma = 45°$, $90°$, $135°$, and $180°$), and a constant Prandtl number ($Pr = 6.2$).

Effects of nanoparticles on streamlines and isotherms are shown in Fig. 3.9. As can be seen, the velocity components increase with an increase in the nanoparticle volume fraction, which enhances the energy transport within the fluid. Thus the absolute values of stream functions indicate that the strength of the flow increases with an increasing volume fraction of the nanofluid. The sensitivity of the thermal boundary-layer thickness to the nanoparticle volume fraction is related to the increased thermal conductivity of the nanofluid. In fact, higher values of thermal conductivity are accompanied by higher values of thermal diffusivity. The high value of thermal diffusivity causes a drop in the temperature gradients and, accordingly, increases the boundary thickness. This increase in thermal boundary-layer thickness reduces the Nusselt number; however, according to Eq. (3.50), the Nusselt number is calculated by multiplying the temperature gradient and the thermal conductivity ratio (conductivity of the nanofluid to the conductivity of the base fluid). Since the

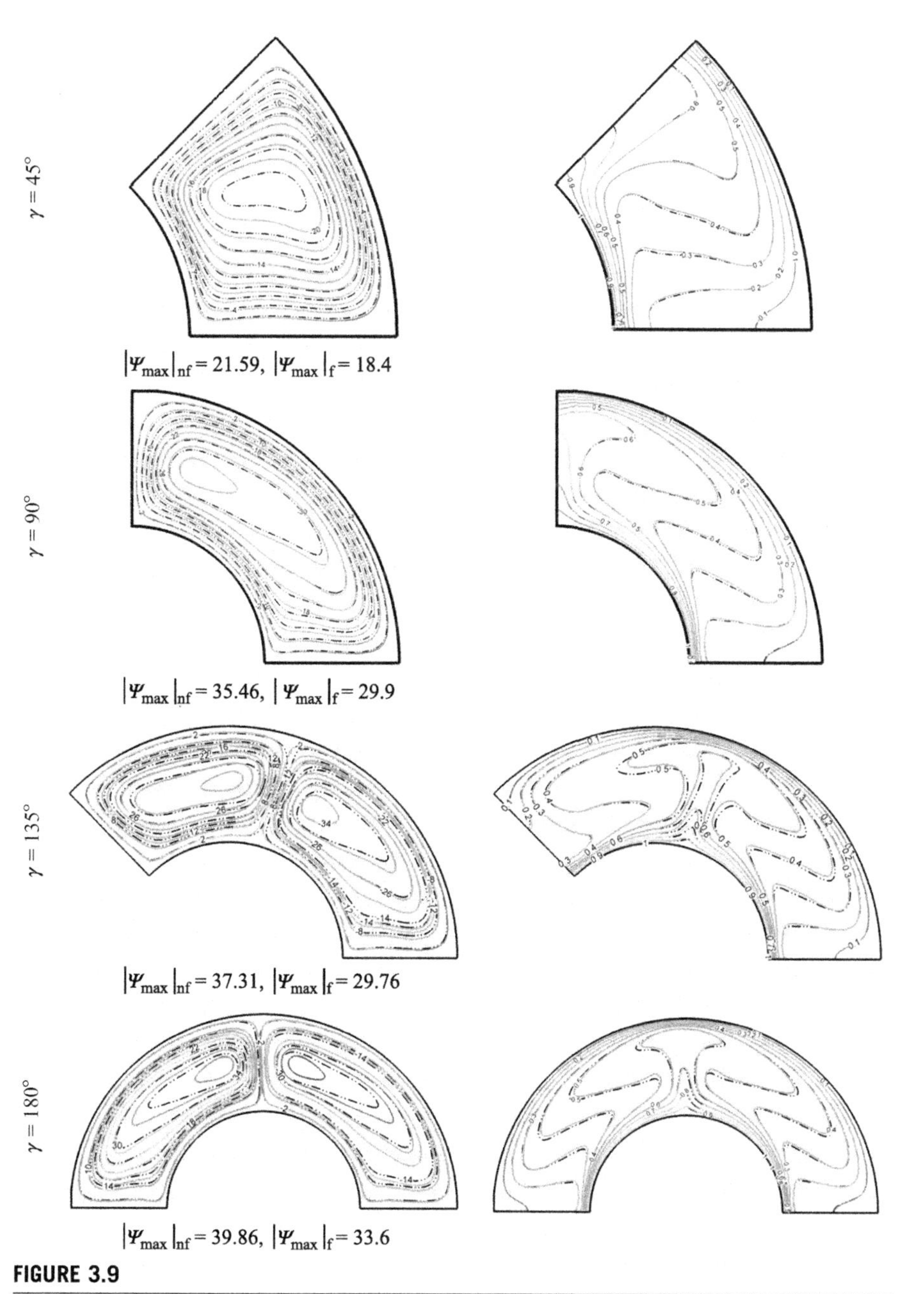

FIGURE 3.9

Comparison of the streamline (left) and isotherm (right) contours between a nanofluid ($\phi = 0.06$) (—) and pure fluid ($\phi = 0$) (- - -) for different values of γ at $Ra = 10^5$ and $Pr = 6$.

reduction in temperature gradient caused by the presence of nanoparticles is much smaller than the thermal conductivity ratio, the Nusselt number is enhanced by increasing the volume fraction of nanoparticles.

Figure 3.10 shows the temperature and stream function distribution of a nanofluid for $\gamma = 45°$ and $90°$, $\phi = 0.06$, and $Ra = 10^3$, 10^4, and 10^5. For $Ra = 10^3$, conduction,

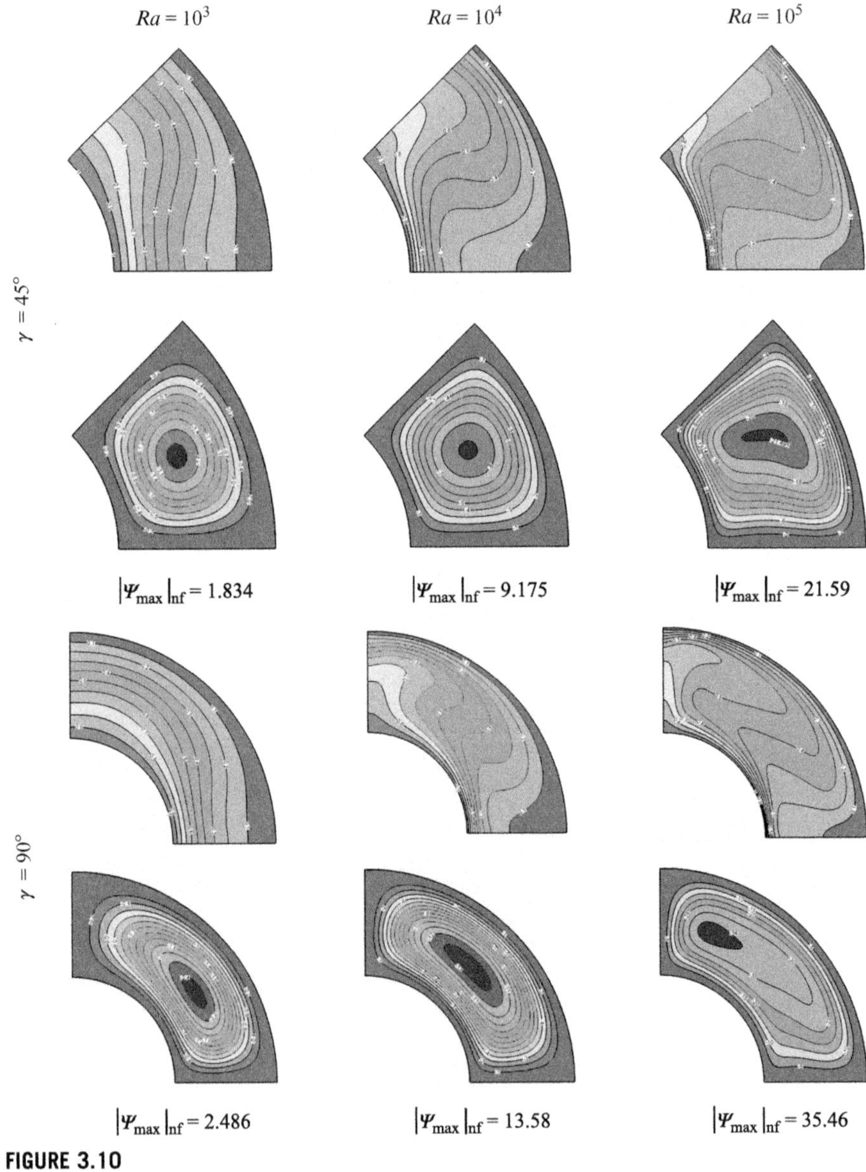

FIGURE 3.10

Comparison of the isotherm (up) and streamline (down) contours for different Rayleigh numbers, γ values (45° and 90°), and $\phi = 0.06$.

compared with convection, is the dominating mechanism of heat transfer. By increasing the Rayleigh number, however, the buoyancy forces increase and overcome the viscous forces, and the heat transfer is dominated by convection at high Rayleigh numbers. Moreover, the isotherms are more distorted at higher Rayleigh numbers because of the stronger convection effects. At $\gamma = 45°$ the central vortex of the main eddy is almost circular at low Rayleigh numbers, whereas it stretches horizontally at $Ra = 10^5$. At $\gamma = 90°$, as the Rayleigh number increases, the center of the inner vortex moves upward and approaches the left wall of the enclosure because of the convection effect. As a result, the isotherms are more distorted on the upper left part of the enclosure and the gradient of isotherms decreases in this location. The circulation of the flow shows two overall rotating eddies with two inner vortices for $\gamma = 135°$ and $\gamma = 180°$, as shown in Fig. 3.11. As the Rayleigh number increases up to 10^4, the role of convection in heat transfer becomes more significant

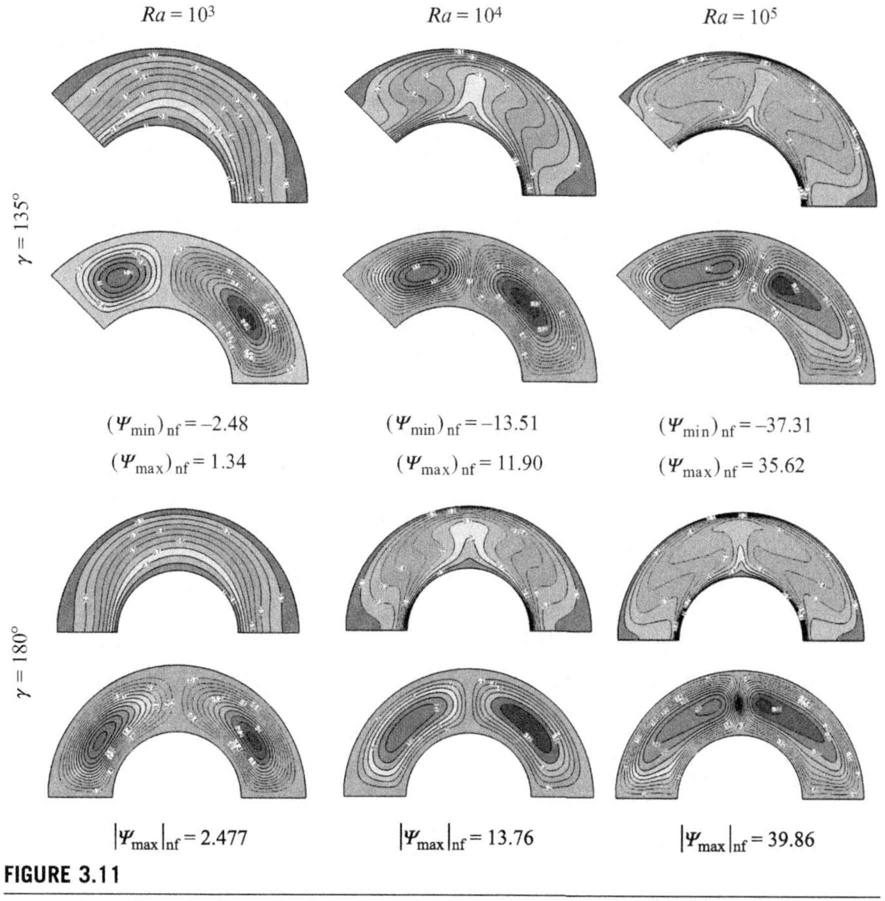

FIGURE 3.11

Comparison of the isotherm (up) and streamline (down) contours for different Rayleigh numbers, γ values (135° and 180°), and at $\phi = 0.06$.

and, consequently, the thermal boundary layer on the surface of the inner circular wall becomes thinner. Also, a plume starts to appear on the top of the inner circular wall. At $Ra = 10^5$, a strong plume drives the flow to strongly impact the outer wall of the enclosure. For $\gamma = 135°$, the size of the clockwise eddy located at the right side is larger than that at the left side. As the Rayleigh number increases, the cores of two vortices approach each other. Based on the maximum and minimum values of the stream function, it is clear that the right side vortex is stronger that the left one. This may be because there is more available space in the right side of the enclosure. It is worthwhile to mention that with an increase in the Rayleigh number, the absolute value of the maximum-to-minimum stream function ratio $((\Psi_{max})_{nf}/(\Psi_{min})_{nf})$ decreases from 1.87 to 1.04, which indicates that the effect of increasing convection velocity becomes more pronounced compared with the effect of available space. As the Rayleigh number increases, at $\gamma = 135°$ the plume appears at the top of the inner curved boundary slanting to the right, whereas the figures show that at $\gamma = 180°$, the streamlines and isotherms are symmetrical about the vertical centerline of the enclosure because of the existence of symmetrical geometry and boundary conditions. The streamlines also depict that for this angle of turn, the eddies' cores move upward with an increasing Rayleigh number.

Figure 3.12 shows the distribution of local Nusselt numbers along the surface of the inner circular wall for different angles of turn, Rayleigh numbers, and nanoparticle volume fractions. For all values of γ, increasing the nanoparticle volume fraction leads to an increase in the local Nusselt number. In addition, as the Rayleigh number increases the local Nusselt number increases because of an increasing convection effect. Also, as the Rayleigh number increases the location of the maximum local Nusselt number approaches the bottom adiabatic wall of the enclosure at $\gamma = 45°$ and $90°$. As stated before, at higher Rayleigh numbers the flow becomes stronger and the thermal boundary layer becomes thinner, causes the flow to detach from the cold wall and move to the proximity of the bottom adiabatic wall, and reaches the bottom left corner of the enclosure at the lower position. At $\gamma = 45°$, the local Nusselt number decreases with an increasing ζ. The patterns of local Nusselt number variation are similar to those at $\gamma = 45°$, but more variation occurs at $\gamma = 90°$, as shown in Fig. 3.12.

At $\gamma = 135°$, the minimum Nu_{local} is located at $\zeta = 80°$ because of the existence of thermal plumes on the surface of the inner circular wall. At $\gamma = 135°$, the profile of the local Nusselt number is almost symmetrical to the vertical centerline, and the minimum value of Nu_{local} is obtained at $\gamma = 90°$, where the isotherms are coarsest because of the occurrence of the plume; the maximum Nu_{local} is located near the two adiabatic walls, where the isotherms become denser because of the returning flow accompanying the cold fluid.

Figure 3.13a depicts the average Nusselt number as a function of angles of turn and nanoparticle volume fractions at different Rayleigh numbers. For all values of γ, the average Nusselt number increases with an increase in the nanoparticle volume fraction and Rayleigh number. When $Ra = 10^3$, the variation of the average Nusselt number with respect to γ is small. At this Rayleigh number, the minimum average

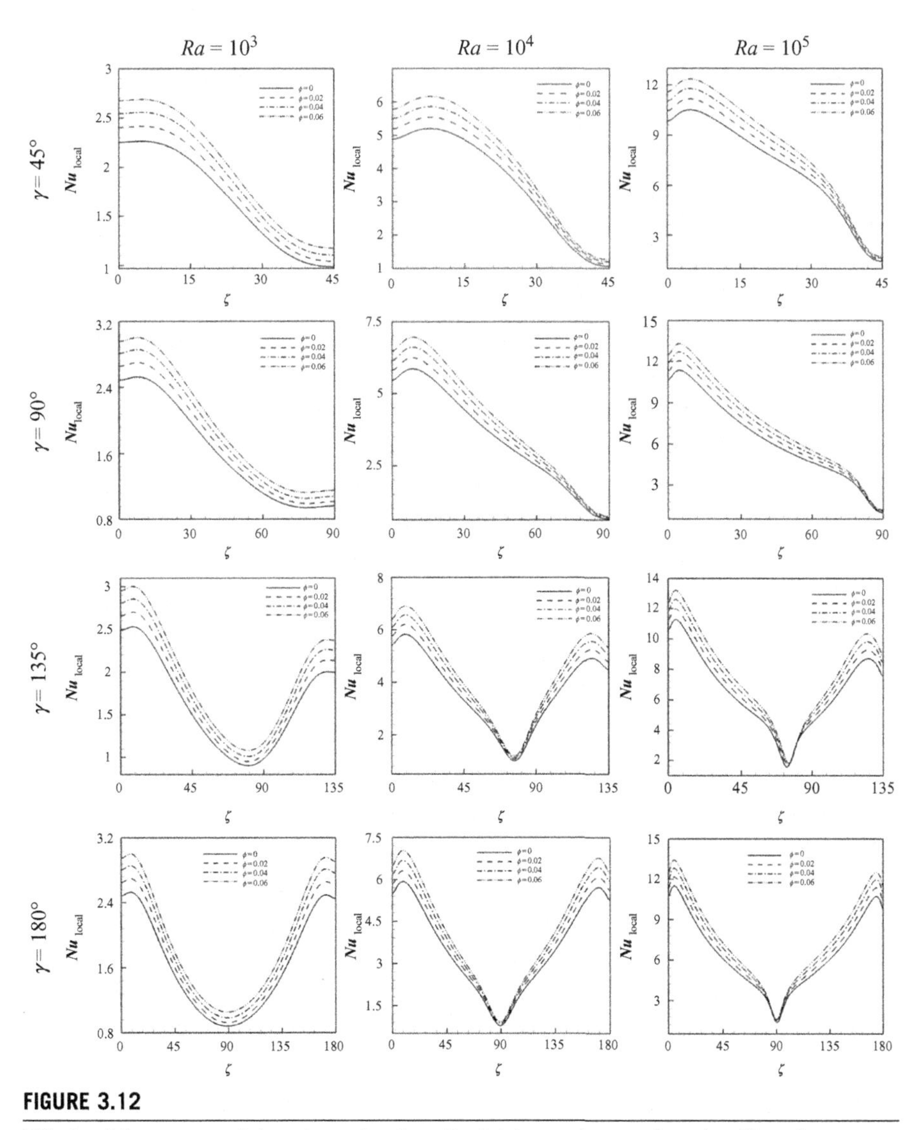

FIGURE 3.12

Effects of the nanoparticle volume fraction, Rayleigh number, and angle of turn for copper-water nanofluids on the local Nusselt number.

Nusselt number is obtained at $\gamma = 90°$, but the maximum value is related to $\gamma = 135°$. At $Ra = 10^4$ the variation of the average Nusselt number is more sensible than at $Ra = 10^3$. For this Rayleigh number the maximum and minimum average Nusselt numbers correspond to $\gamma = 90°$ and $135°$, respectively. As the Rayleigh number increases up to $Ra = 10^5$, the minimum and maximum average Nusselt numbers are obtained for $\gamma = 90°$ and $45°$, respectively.

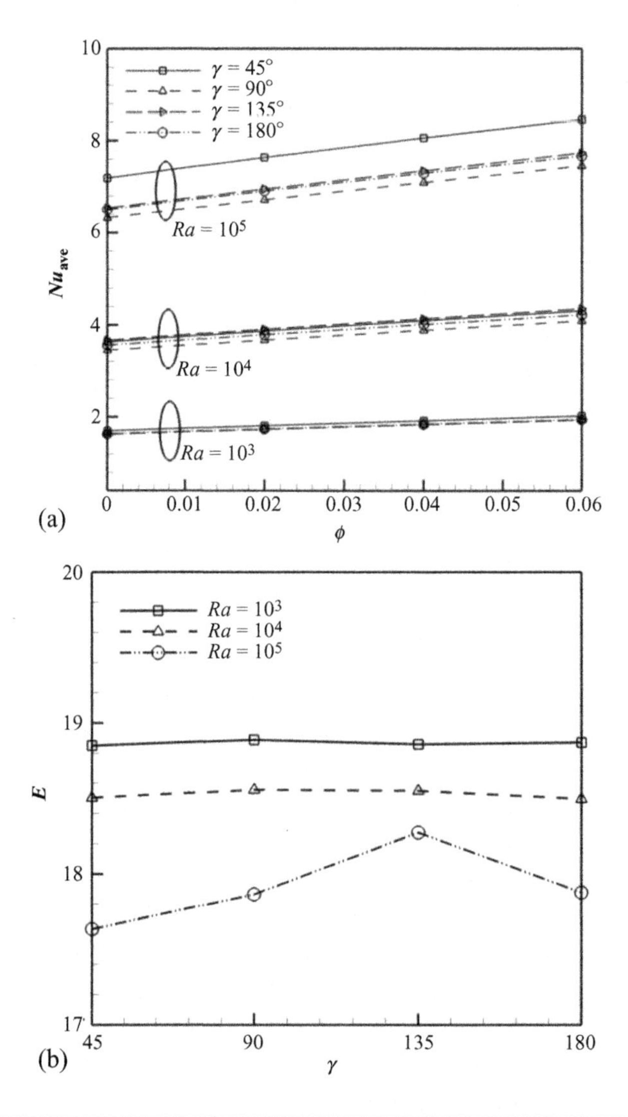

FIGURE 3.13

(a) Effects of the nanoparticle volume fraction, Rayleigh number, and angle of turn on the average Nusselt number. (b) Effects of γ and Ra on the ratio of heat transfer enhancement due to the addition of nanoparticles when $Pr=6.2$ (in the case of a copper-water nanofluid).

The heat transfer enhancement ratio after the addition of nanoparticles for different values of γ and Ra is shown in Fig. 3.13b. The figure shows that the heat transfer enhancement ratio behaves differently at various Rayleigh numbers. That the enhancements in heat transfer at $Ra=10^3$ and 10^4 are almost constant for all angles of turn is an interesting observation. When $Ra=10^5$ by increasing γ from 45° to 180°, at first the percentage of heat transfer enhancement increases and then decreases. At this Rayleigh number the maximum heat transfer enhancement ratio is obtained at $\gamma = 135°$. In addition, the effect of nanoparticles is more pronounced at low Rayleigh numbers than at high Rayleigh numbers because of the greater enhancement rate. This observation can be explained by noting that at low Rayleigh numbers heat transfer is dominated by conduction. Therefore the addition of nanoparticles with high thermal conductivity increases the conduction and makes the enhancement more effective.

3.3.3 TWO-PHASE MODEL

3.3.3.1 Natural convection

A nanofluid's density ρ is

$$\begin{aligned}\rho &= \phi\rho_p + (1-\phi)\rho_f \\ &\cong \phi\rho_p + (1-\phi)\{\rho_{f_0}(1-\beta(T-T_c)\}\end{aligned} \tag{3.51}$$

where ρ_f is the base fluid's density, T_c is a reference temperature, ρ_{f_0} is the base fluid's density at the reference temperature, and β is the volumetric coefficient of expansion. Taking the density of the base fluid as that of the nanofluid, as adopted by Yadav et al. [77], the density ρ in Eq. (3.51) thus becomes

$$\rho \cong \phi\rho_p + (1-\phi)\{\rho_0(1-\beta(T-T_c)\} \tag{3.52}$$

where ρ_0 is the density of the nanofluid at the reference temperature.

The continuity, momentum under Boussinesq approximation, and energy equations for laminar and steady-state natural convection in a two-dimensional enclosure can be written in dimensional form as follows:

$$\frac{\partial u}{\partial x} + \frac{\partial v}{\partial y} = 0 \tag{3.53}$$

$$\rho_f\left\{u\frac{\partial u}{\partial x} + v\frac{\partial u}{\partial y}\right\} = -\frac{\partial P}{\partial x} + \mu\left(\frac{\partial^2 u}{\partial x^2} + \frac{\partial^2 u}{\partial y^2}\right) + \sigma B^2\left(v\sin\lambda\cos\lambda - u\sin^2\lambda\right) \tag{3.54}$$

$$\rho_f\left\{u\frac{\partial v}{\partial x} + v\frac{\partial v}{\partial y}\right\} = -\frac{\partial P}{\partial y} + \mu\left(\frac{\partial^2 v}{\partial x^2} + \frac{\partial^2 v}{\partial y^2}\right) - (\phi-\phi_c)\left(\rho_p - \rho_{f_0}\right)g + (1-\phi_c)\rho_{f_0}(T-T_c)g \tag{3.55}$$

$$u\frac{\partial T}{\partial x} + v\frac{\partial T}{\partial y} = \alpha\left(\frac{\partial^2 T}{\partial x^2} + \frac{\partial^2 T}{\partial y^2}\right) + \frac{(\rho c)_p}{(\rho c)_f}\left[D_B\left\{\frac{\partial \phi}{\partial x}\cdot\frac{\partial T}{\partial x} + \frac{\partial \phi}{\partial y}\cdot\frac{\partial T}{\partial y}\right\} + (D_T/T_c)\left\{\left(\frac{\partial T}{\partial x}\right)^2 + \left(\frac{\partial T}{\partial y}\right)^2\right\}\right] \tag{3.56}$$

$$u\frac{\partial \phi}{\partial x} + v\frac{\partial \phi}{\partial y} = D_B\left\{\frac{\partial^2 \phi}{\partial x^2} + \frac{\partial^2 \phi}{\partial y^2}\right\} + \left(\frac{D_T}{T_c}\right)\left\{\frac{\partial^2 T}{\partial x^2} + \frac{\partial^2 T}{\partial y^2}\right\} \tag{3.57}$$

The stream function and vorticity are defined as follows:

$$u=\frac{\partial\psi}{\partial y},\quad v=-\frac{\partial\psi}{\partial x},\quad \omega=\frac{\partial v}{\partial x}-\frac{\partial u}{\partial y} \tag{3.58}$$

The stream function satisfies the continuity Eq. (3.53). The vorticity equation is obtained by eliminating the pressure between the two momentum equations, that is, by taking the y-derivative of Eq. (3.54) and subtracting from it the x-derivative of Eq. (3.55). In addition, the following nondimensional variables should be introduced:

$$X=\frac{x}{L},\quad Y=\frac{y}{L},\quad \Omega=\frac{\omega L^2}{\alpha},\quad \Psi=\frac{\psi}{\alpha},\quad \Theta=\frac{T-T_c}{T_h-T_c},\quad \Phi=\frac{\phi-\phi_c}{\phi_h-\phi_c},\quad U=\frac{uL}{\alpha},\quad V=\frac{vL}{\alpha} \tag{3.59}$$

Using these dimensionless parameters the equations become:

$$\left[\frac{\partial\Psi}{\partial Y}\frac{\partial\Omega}{\partial X}-\frac{\partial\Psi}{\partial X}\frac{\partial\Omega}{\partial Y}\right]=Pr\left(\frac{\partial^2\Omega}{\partial X^2}+\frac{\partial^2\Omega}{\partial Y^2}\right)+PrRa\left(\frac{\partial\Theta}{\partial X}-Nr\frac{\partial\Theta}{\partial X}\right) \tag{3.60}$$

$$\frac{\partial\Psi}{\partial Y}\frac{\partial\Theta}{\partial X}-\frac{\partial\Psi}{\partial X}\frac{\partial\Theta}{\partial Y}=\left(\frac{\partial^2\Theta}{\partial X^2}+\frac{\partial^2\Theta}{\partial Y^2}\right)+Nb\left(\frac{\partial\Phi}{\partial X}\frac{\partial\Theta}{\partial X}+\frac{\partial\Phi}{\partial Y}\frac{\partial\Theta}{\partial Y}\right)+Nt\left(\left(\frac{\partial\Theta}{\partial X}\right)^2+\left(\frac{\partial\Theta}{\partial Y}\right)^2\right) \tag{3.61}$$

$$\frac{\partial\Psi}{\partial Y}\frac{\partial\Phi}{\partial X}-\frac{\partial\Psi}{\partial X}\frac{\partial\Phi}{\partial Y}=\frac{1}{Le}\left(\frac{\partial^2\Phi}{\partial X^2}+\frac{\partial^2\Phi}{\partial Y^2}\right)+\frac{Nt}{NbLe}\left(\frac{\partial^2\Theta}{\partial X^2}+\frac{\partial^2\Theta}{\partial Y^2}\right) \tag{3.62}$$

$$\frac{\partial^2\Psi}{\partial X^2}+\frac{\partial^2\Psi}{\partial Y^2}=-\Omega \tag{3.63}$$

where the thermal Rayleigh number, the buoyancy ratio, the Prandtl number, the Brownian motion parameter, the thermophoretic parameter, and the Lewis number are defined as follows:

$$Ra=(1-\phi_c)\rho_{f_0}g\beta L^3(T_h-T_c)/(\mu\alpha),\; Nr=\left(\rho_p-\rho_0\right)(\phi_h-\phi_c)/[(1-\phi_c)\rho_{f_0}\beta L(T_h-T_c)],$$
$$Pr=\mu/\rho_f\alpha,\; Nb=(\rho c)_p D_B(\phi_h-\phi_c)/((\rho c)_f\alpha), Nt=(\rho c)_p D_T(T_h-T_c)/[(\rho c)_f\alpha T_c] \text{ and}$$
$$Le=\alpha/D_B \tag{3.64}$$

3.3.3.2 Force convection

To obtain the governing equations for force convection, the nondimensional variables should be introduced as:

$$X=\frac{x}{L},\quad Y=\frac{y}{L},\quad \Omega=\frac{\omega L}{u_r},\quad \Psi=\frac{\psi}{u_r L},\quad U=\frac{u}{u_r},\quad V=\frac{v}{u_r},\quad \Theta=\frac{T-T_c}{T_h-T_c},\quad \Phi=\frac{\phi-\phi_c}{\phi_h-\phi_c} \tag{3.65}$$

So, the governing equations are as follows:

$$\left[\frac{\partial\Psi}{\partial Y}\frac{\partial\Omega}{\partial X}-\frac{\partial\Psi}{\partial X}\frac{\partial\Omega}{\partial Y}\right]=\frac{1}{Re}\left(\frac{\partial^2\Omega}{\partial X^2}+\frac{\partial^2\Omega}{\partial Y^2}\right) \tag{3.66}$$

$$\begin{aligned}\frac{\partial\Psi}{\partial Y}\frac{\partial\Theta}{\partial X}-\frac{\partial\Psi}{\partial X}\frac{\partial\Theta}{\partial Y}&=\frac{1}{RePr}\left(\frac{\partial^2\Theta}{\partial X^2}+\frac{\partial^2\Theta}{\partial Y^2}\right)\\&+\frac{Nb}{Re}\left(\frac{\partial\Phi}{\partial X}\frac{\partial\Theta}{\partial X}+\frac{\partial\Phi}{\partial Y}\frac{\partial\Theta}{\partial Y}\right)+\frac{Nt}{Re}\left(\left(\frac{\partial\Theta}{\partial X}\right)^2+\left(\frac{\partial\Theta}{\partial Y}\right)^2\right)\end{aligned} \tag{3.67}$$

$$\frac{\partial\Psi}{\partial Y}\frac{\partial\Phi}{\partial X}-\frac{\partial\Psi}{\partial X}\frac{\partial\Phi}{\partial Y}=\frac{1}{LeRe}\left(\frac{\partial^2\Phi}{\partial X^2}+\frac{\partial^2\Phi}{\partial Y^2}\right)+\frac{Nt}{NbLeRe}\left(\frac{\partial^2\Theta}{\partial X^2}+\frac{\partial^2\Theta}{\partial Y^2}\right) \tag{3.68}$$

$$\frac{\partial^2\Psi}{\partial X^2}+\frac{\partial^2\Psi}{\partial Y^2}=-\Omega \tag{3.69}$$

where the thermal Reynolds number, the buoyancy ratio number, the Prandtl number, the Brownian motion parameter, the thermophoretic parameter, and the Lewis number are defined as

$$\begin{aligned}&Re=\frac{\rho_f L u_r}{\mu},\quad Nr=\left(\rho_p-\rho_0\right)(\phi_h-\phi_c)/\left[(1-\phi_c)\rho_{f_0}\beta L(T_h-T_c)\right],\\&Pr=\mu/\rho_f\alpha,\quad Nb=(\rho c)_p D_B(\phi_h-\phi_c)/\left((\rho c)_f\upsilon\right),\ \text{and}\ Le=\upsilon/D_B\end{aligned} \tag{3.70}$$

3.3.3.3 Mixed convection

To obtain mixed convection equations, buoyancy forces should be added to Eq. (3.60). So, the governing equations in vorticity stream form are

$$\left[\frac{\partial\Psi}{\partial Y}\frac{\partial\Omega}{\partial X}-\frac{\partial\Psi}{\partial X}\frac{\partial\Omega}{\partial Y}\right]=\frac{1}{Re}\left(\frac{\partial^2\Omega}{\partial X^2}+\frac{\partial^2\Omega}{\partial Y^2}\right)+\frac{Gr}{Re^2}\left(\frac{\partial\Theta}{\partial X}-Nr\frac{\partial\Theta}{\partial X}\right) \tag{3.71}$$

$$\begin{aligned}\frac{\partial\Psi}{\partial Y}\frac{\partial\Theta}{\partial X}-\frac{\partial\Psi}{\partial X}\frac{\partial\Theta}{\partial Y}&=\frac{1}{Re\,\mathrm{Pr}}\left(\frac{\partial^2\Theta}{\partial X^2}+\frac{\partial^2\Theta}{\partial Y^2}\right)+\frac{Nb}{Re}\left(\frac{\partial\Phi}{\partial X}\frac{\partial\Theta}{\partial X}+\frac{\partial\Phi}{\partial Y}\frac{\partial\Theta}{\partial Y}\right)\\&+\frac{Nt}{Re}\left(\left(\frac{\partial\Theta}{\partial X}\right)^2+\left(\frac{\partial\Theta}{\partial Y}\right)^2\right)\end{aligned} \tag{3.72}$$

$$\frac{\partial\Psi}{\partial Y}\frac{\partial\Phi}{\partial X}-\frac{\partial\Psi}{\partial X}\frac{\partial\Phi}{\partial Y}=\frac{1}{LeRe}\left(\frac{\partial^2\Phi}{\partial X^2}+\frac{\partial^2\Phi}{\partial Y^2}\right)+\frac{Nt}{NbLeRe}\left(\frac{\partial^2\Theta}{\partial X^2}+\frac{\partial^2\Theta}{\partial Y^2}\right) \tag{3.73}$$

$$\frac{\partial^2\Psi}{\partial X^2}+\frac{\partial^2\Psi}{\partial Y^2}=-\Omega \tag{3.74}$$

where $Gr=(1-\phi_c)\rho_{f_0}g\beta L^3\rho_f(T_h-T_c)/(\mu^2)$ is the Grashof number and the Richardson number is defined as $Ri=Gr/Re^2$.

3.3.4 CVFEM FOR NANOFLUID FLOW AND HEAT TRANSFER (TWO-PHASE MODEL)

3.3.4.1 Two-phase simulation of nanofluid flow and heat transfer using heatline analysis

3.3.4.1.1 Problem definition

The physical model and the corresponding triangular elements used in the present CVFEM program are shown in Fig. 3.14 [78]. The inner and outer walls are maintained at constant temperatures T_h and T_c, respectively. The shape of the inner cylinder profile is assumed to mimic the following pattern:

$$r = r_{in} + A\cos(N(\zeta)) \tag{3.75}$$

in which r_{in} is the radius of the base circle, r_{out} is the radius of the outer cylinder, and A and N are the amplitude and number of undulations, respectively. ζ is the rotation angle. In this study A and N equal 0.2 and 8, respectively.

The local Nusselt number on the cold circular wall can be expressed as

$$r = r_{in} + A\cos(N(\zeta)) \tag{3.76}$$

where n is the normal direction of the outer cylinder surface. The average Nusselt number on the cold circular wall is evaluated as:

$$Nu_{ave} = \frac{1}{0.5\pi}\int_0^{0.5\pi} Nu_{loc}(\zeta)\,d\zeta \tag{3.77}$$

Heatlines are adequate tools for visualizing and analyzing two-dimensional convection heat transfer; in this case the heat flux line concept is extended to include advection terms. Heat function (H) is defined in terms of the energy equation as

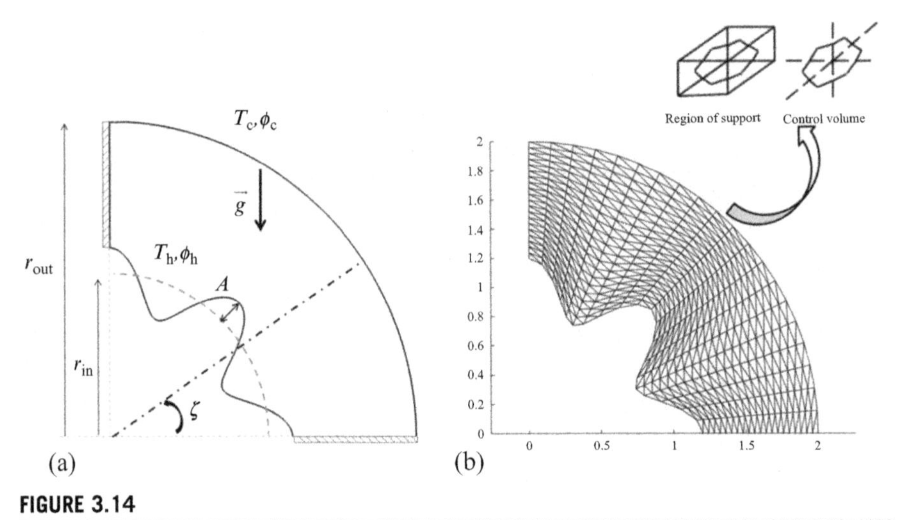

FIGURE 3.14

Geometry and the boundary conditions (a) and the mesh of enclosure (b) considered in this work.

$$\frac{\partial H}{\partial Y} = U\Theta - \frac{\partial \Theta}{\partial X}, \quad -\frac{\partial H}{\partial X} = V\Theta - \frac{\partial \Theta}{\partial Y} \tag{3.78}$$

3.3.4.1.2 Effect of active parameters

In this study CVFEM is applied to investigate natural convection heat transfer in an enclosure filled with a nanofluid. Effects of the thermal Rayleigh number ($Ra = 10^3$, 10^4, and 10^5), buoyancy ratio number ($Nr = 0.1$-4), and Lewis number ($Le = 1, 2, 4, 6,$ and 8) on flow and heat transfer characteristics are examined. The Brownian motion parameter of nanofluids ($Nb = 0.5$), the thermophoretic parameter of nanofluids ($Nt = 0.5$), and the Prandtl number ($Pr = 10$) are fixed.

The isotherm, streamline, isoconcentration, and heatline contours for different thermal Rayleigh numbers are compared in Fig. 3.15. At $Ra = 10^3$, the conduction heat transfer mechanism is more pronounced. For this reason, the isotherms are parallel to each other. As Ra increases, the distribution of the isotherm contours increases. The streamline has two eddies within the enclosure; the lower one is stronger. With increase in Ra, the role of convection in heat transfer becomes more significant. Also, as Ra increases the distribution of the isoconcentration contours increases. The heat flow within the enclosure is displayed using the heat function obtained from conductive heat fluxes ($\partial\Theta/\partial X$, $\partial\Theta/\partial Y$), as well as convective heat fluxes ($V\Theta$, $U\Theta$). Heatlines emanate from hot regimes and end on cold regimes, illustrating the path of heat flow. The domination of conduction heat transfer at low Rayleigh numbers can be observed from the heatline patterns; no passive area exists. The increase in Ra causes the clustering of heatlines from the hot to the cold wall and generates an area of passive heat transfer in which heat is rotated without having a significant effect on heat transfer between walls.

The isotherm, streamline, isoconcentration, and heatline contours at different buoyancy ratios are compared in Fig. 3.16. It should be mentioned that negative Nr values (opposing buoyancy forces) showed more complex and interesting flow patterns, such as multicells, which is worth presenting and discussing. Therefore, in this study results for a positive Nr (aiding buoyancy forces), where the temperature- and species-induced buoyancy forces aide each other, are not considered. For $Nr = 0$, the species-induced buoyancy force has no effect on flow; the flow is driven solely by the thermal buoyancy force. However, the effect of species-induced buoyancy increases as the Nr value increases and reaches a certain value, where the effect of thermally induced buoyancy becomes negligible in comparison with the solutal one. For small Nr values, the flow is driven mainly by the thermal buoyancy force. When $Nr = 0.1$, two upwelling plumes appear on the crests of the inner cylinder. A third plume with the opposite direction appears above the outer circular cylinder; this directional change results from the counterclockwise eddy existing between two other clockwise eddies in this region. For $Nr > 1$, the flow is driven mainly by species concentration-induced buoyancy force. So, as Nr increases the thermal plumes disappears. The number of passive areas increases with an increase in solute forces.

The isotherm, streamline, isoconcentration, and heatline contours at different Lewis numbers are compared in Fig. 3.17. The Lewis number (Le) is a dimensionless number defined as the ratio of thermal diffusivity to mass diffusivity. It is used to characterize fluid flows in which there is simultaneous heat and mass transfer by convection. When

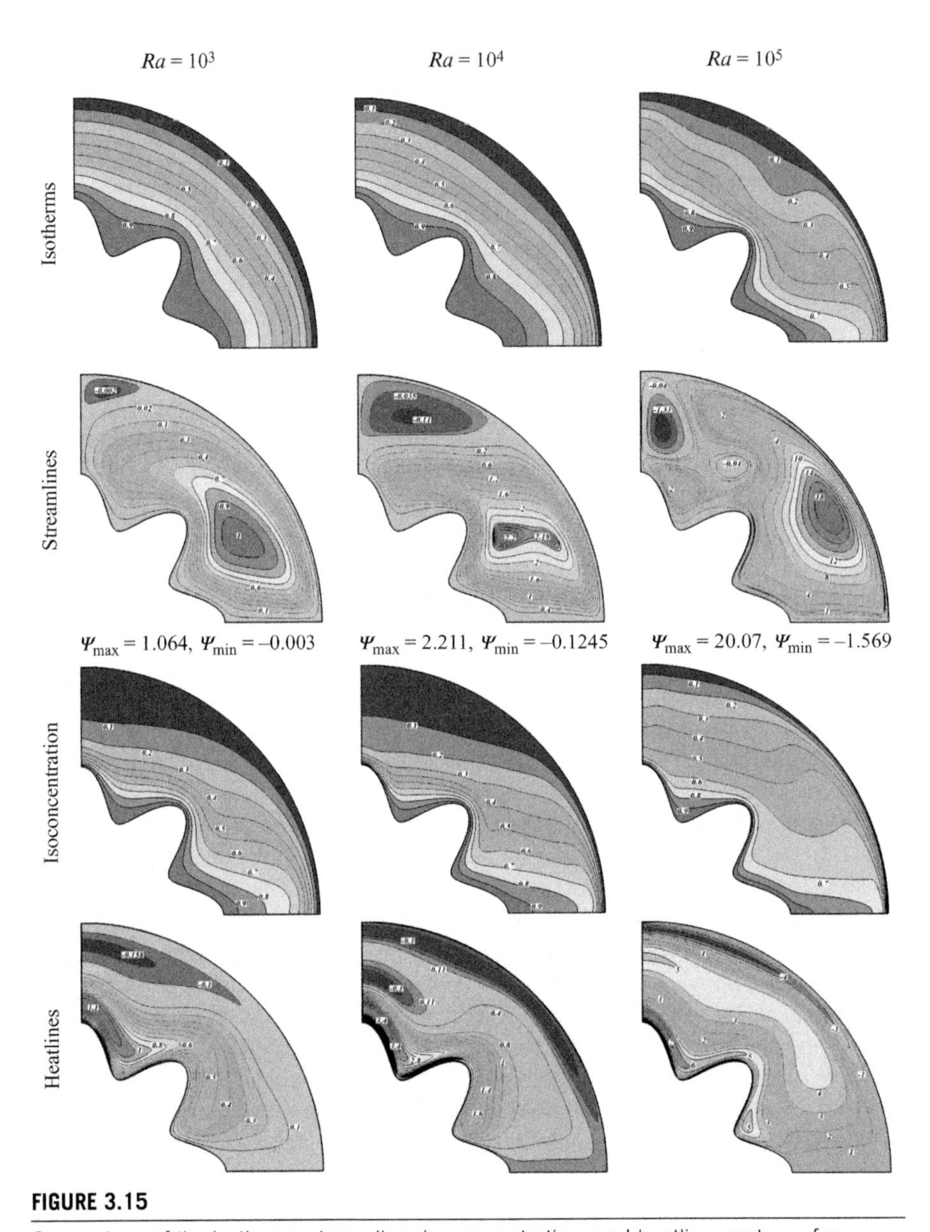

FIGURE 3.15

Comparison of the isotherm, streamline, isoconcentration, and heatline contours for different thermal Rayleigh numbers when $Nr=4$, $Le=8$, $Nt=Nb=0.5$, and $Pr=10$.

$Le=2$, two equal eddies with the same direction are observed. As Le increases to 8, these eddies are converted into four eddies. The mass flow is given by $\psi_{max} \approx \delta_s v$, where the solutal boundary-layer thickness is given by $\delta_s \approx (RaLeNr)^{-1/4}$ and $v \approx (RaLeNr)^{1/2}$, so $\psi_{max} \approx (RaLeNr)^{1/4}$. The solutal boundary layer δ_s becomes thinner by increasing either Ra, Le, or both. Heatlines are more distorted as Le increases.

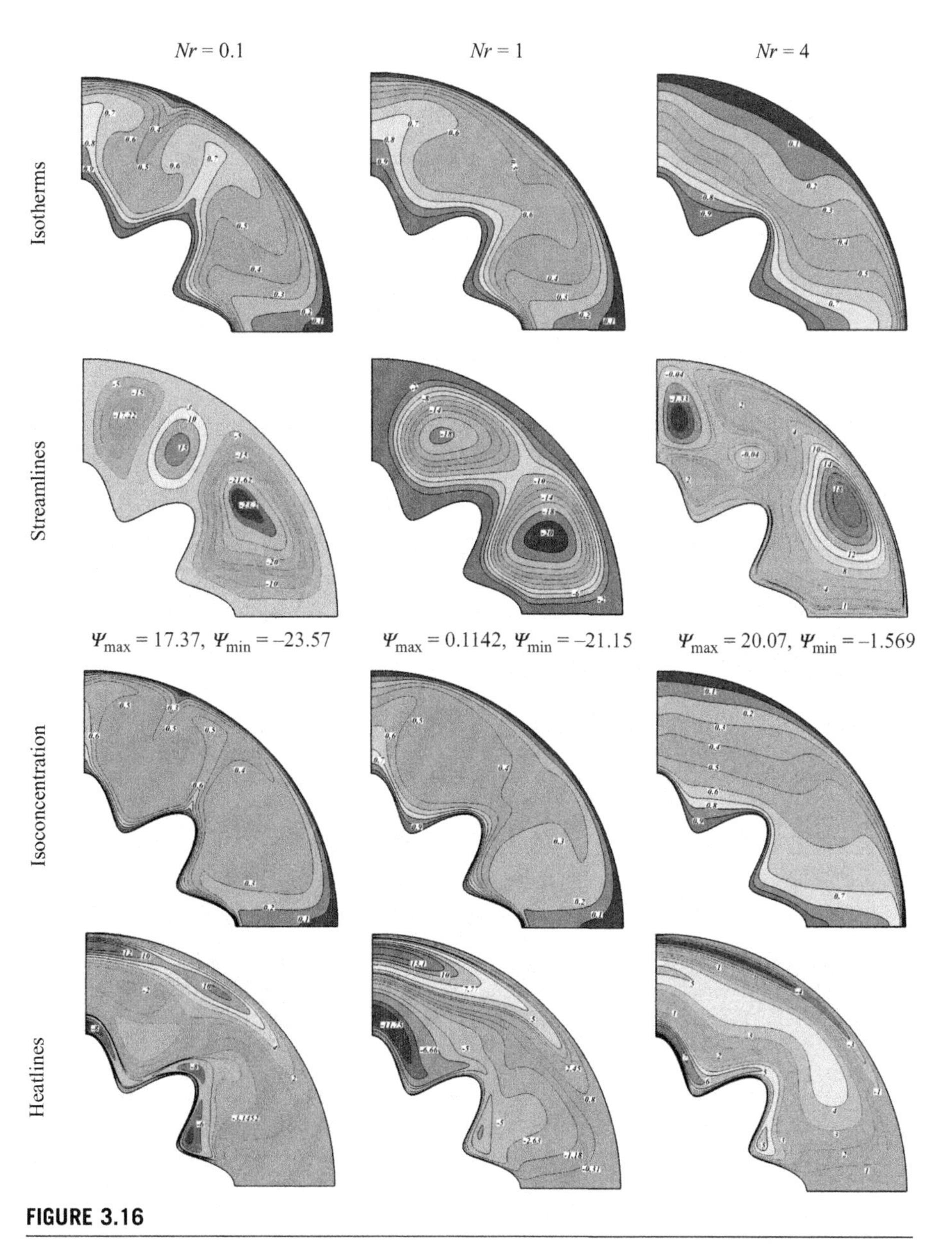

FIGURE 3.16

Comparison of the isotherm, streamline, isoconcentration, and heatline contours for different buoyancy ratios when $Ra = 10^5$, $Le = 8$, $Nt = Nb = 0.5$, and $Pr = 10$.

Figure 3.18 depicts the effects of the thermal Rayleigh number, buoyancy ratio, and Lewis number on the local Nusselt number. A change in a flow structure has a significant effect on the local Nusselt number. The number of extermum in the local Nusselt number profile corresponds to the number of thermal plumes on the cold

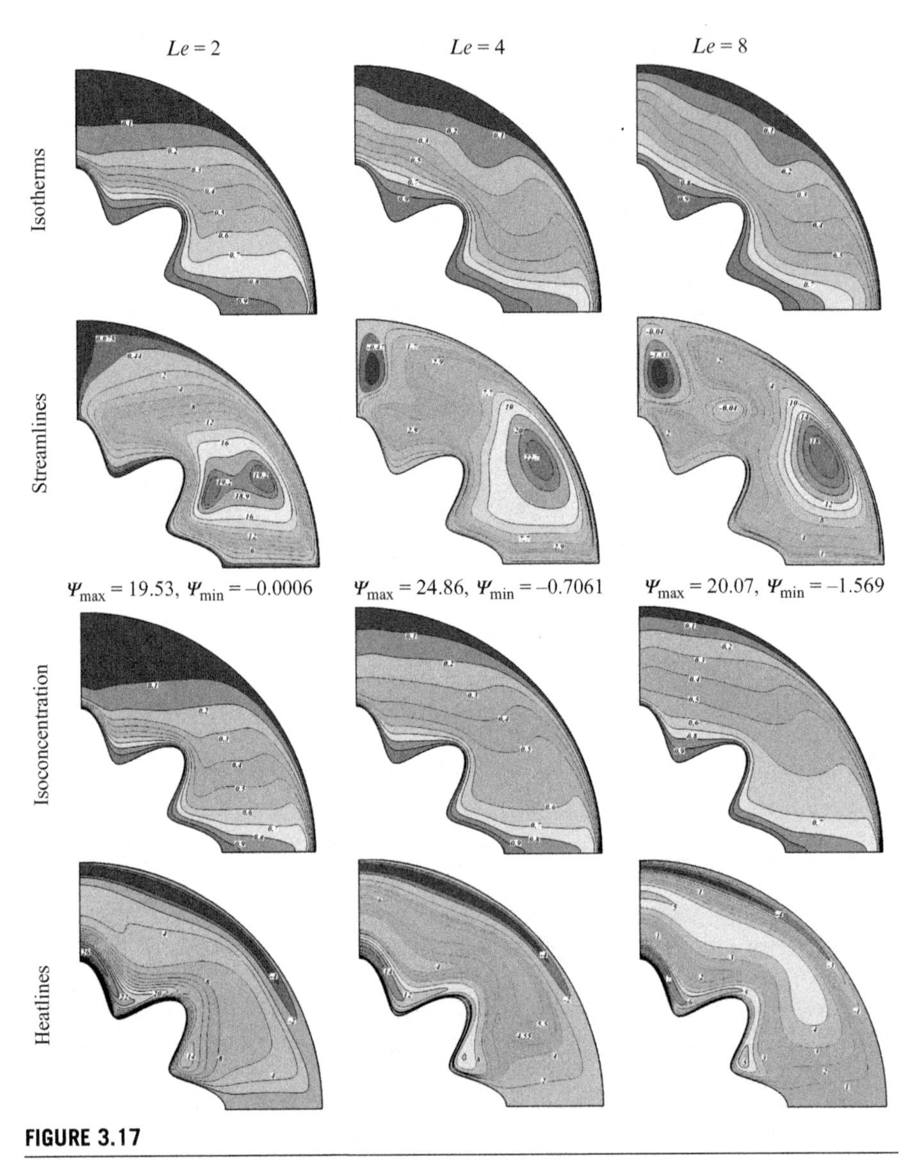

FIGURE 3.17

Comparison of the isotherm, streamline, isoconcentration, and heatline contours for different Lewis numbers when $Ra = 10^5$, $Nr = 4$, $Nt = Nb = 0.5$, and $Pr = 10$.

wall. Maxima for Nu_{loc} occur because of dense heatlines, based on the conductive heat transport occurring at those respective portions. For low values of Nr (<1), the maximum rate of heat transfer takes place at $\zeta = 90°$, where the thermal boundary layer is thin. For $Nr > 1$, however, the reverse is true, that is, the rate of heat transfer is high at $\zeta = 0°$. In fact, the trend reflects the physics of flow reversal as a function of Nr.

Figure 3.19 illustrates the effects of the thermal Rayleigh number, buoyancy ratio, and Lewis number on the average Nusselt number. As a general trend, the average

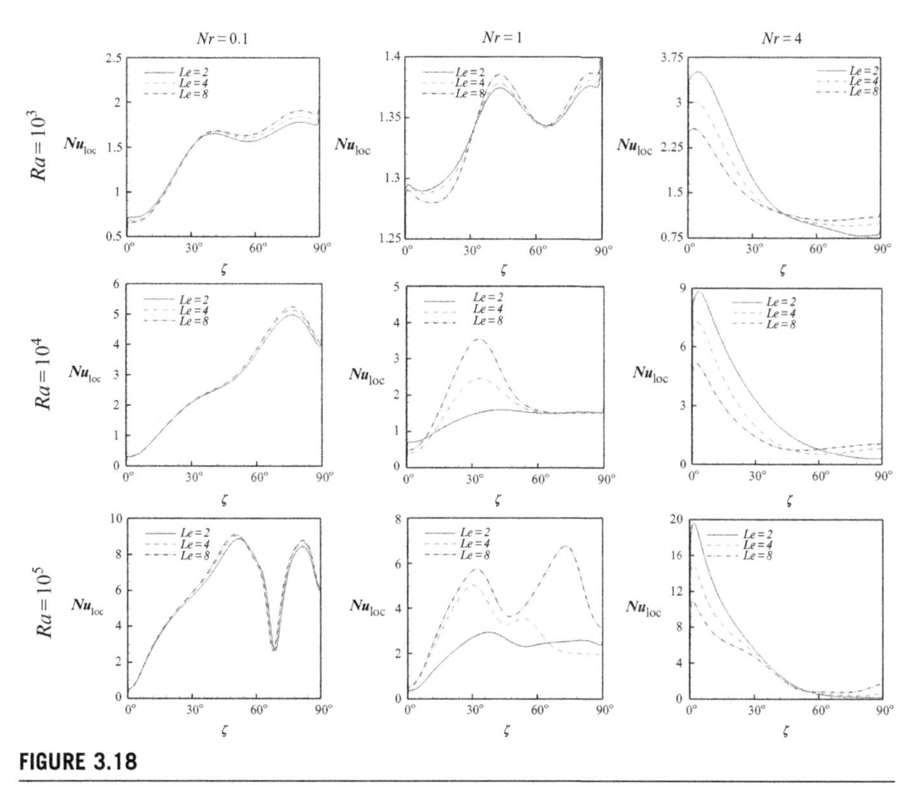

FIGURE 3.18

Effects of the thermal Rayleigh number, buoyancy ratio, and Lewis number on the local Nusselt number at $Nt = Nb = 0.5$ and $Pr = 10$.

Nusselt number decreases as Nr increases until it reaches a minimum value, and then it starts increasing. As Nr increases the effect of opposing buoyancy (species) increases; hence the force driving the net fluid flow decreases. The Nu_{ave} reaches a minimum value when the thermal buoyancy equals the species-induced buoyancy. The thermal boundary-layer thickness differs from the species boundary-layer thickness because Le is not at unity. The minimum Nu_{ave} is obtained for buoyancy ratios (Nr_{cr}) shifting slightly to a higher Nr as Le increases. As the Le increases the thickness of the species concentration boundary layer decreases compared with the thermal boundary-layer thickness. Hence the resulting species-induced force decreases as the Le increases and higher Nr is needed to compensate for the decrease in the species concentration boundary layer. As expected, the flow is dominated by species-induced buoyancy for $Nr > Nr_{cr}$, and the rate of transport increases as Nr increases. The heat transfer increase with Le for $Nr > Nr_{cr}$ and decreases for $Nr < Nr_{cr}$ with a fixed Nr. Such a trend is a consequence of a separate scale on which the thermal and solutal effect act, and such a separate scale increases as the Le value increase. It is interesting to note that the rate of decrease in Nu_{ave} is more prominent for Nr values between 0 and Nr_{cr} compared with Nr values greater than Nr_{cr}. This can be explained by the fact that the results are obtained for $Le \geq 2$; hence the thermal boundary layer is thicker than the species concentration boundary layer.

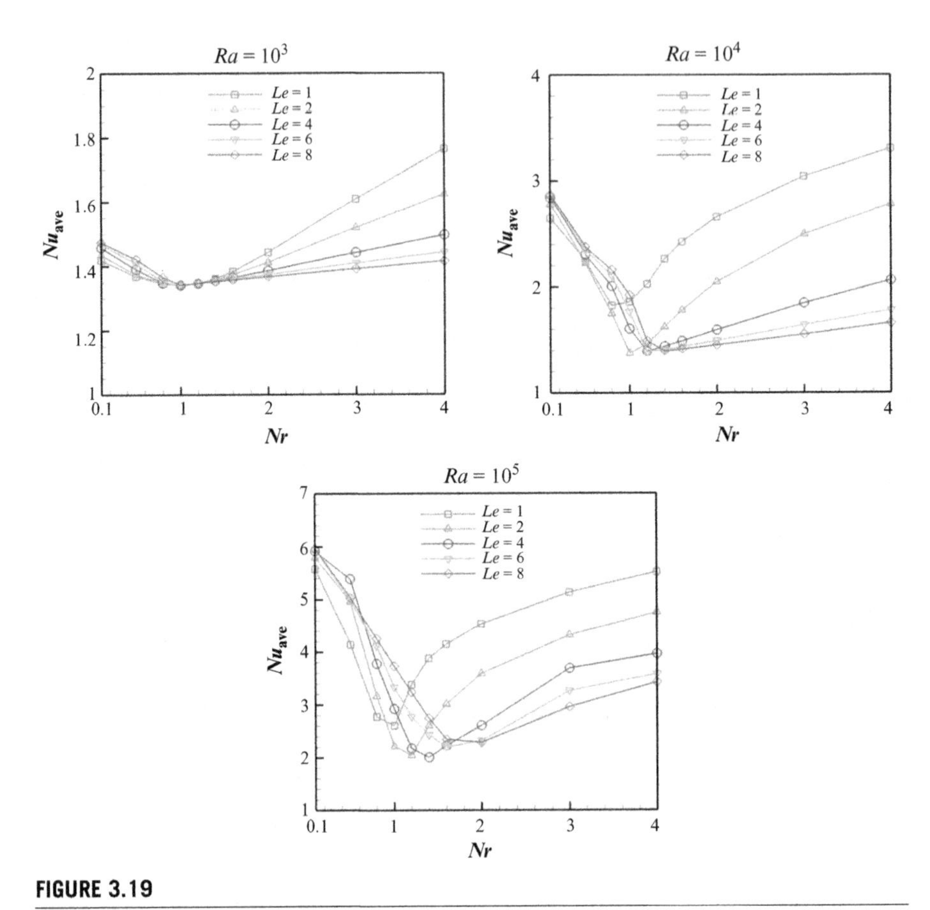

FIGURE 3.19

Effects of the thermal Rayleigh number, buoyancy ratio, and Lewis number on the average Nusselt number at $Nt = Nb = 0.5$ and $Pr = 10$.

3.3.4.2 Thermal management for free convection of a nanofluid using a two-phase model

3.3.4.2.1 Problem definition

The schematic diagram and the mesh of the semiannulus enclosure used in the present CVFEM program are shown in Fig. 3.20 [79]. The inner and outer walls are maintained at constant temperatures T_h and T_c, respectively, while the two other walls are thermally insulated. The boundary conditions of concentration are similar to those of temperature.

3.3.4.2.2 Effect of active parameters

Natural convection heat transfer in an enclosure filled with a nanofluid is investigated numerically using the CVFEM. Effects of buoyancy ratio ($Nr = 0.1$-4), Lewis number ($Le = 2$, 4, 6, and 8), and angle of turn for the enclosure ($\gamma = 30°$, $60°$, to $90°$) on flow and heat transfer characteristics are examined. The Brownian motion parameter of

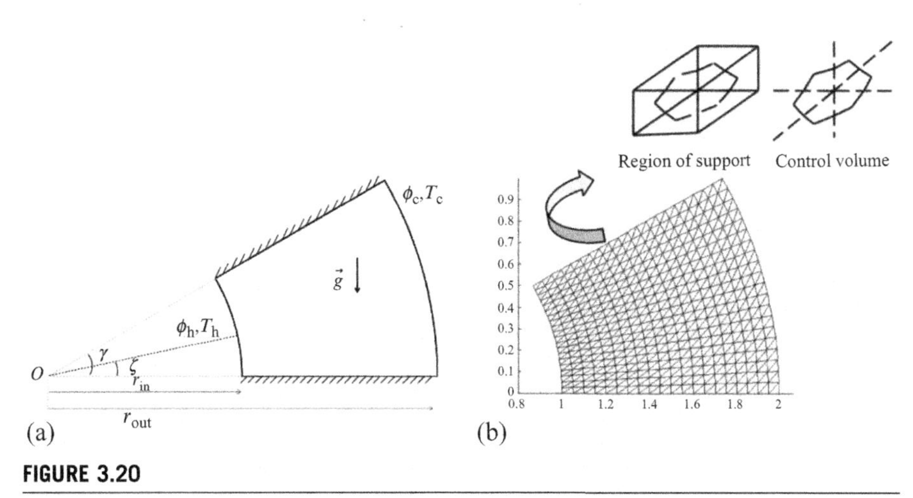

FIGURE 3.20

Geometry and the boundary conditions (a) and the mesh of the enclosure (b) considered in this work.

nanofluid ($Nb=0.5$), thermophoretic parameter of nanofluid ($Nt=0.5$), thermal Rayleigh number ($Ra=10^5$), and Prandtl number ($Pr=10$) are fixed.

The effects of angle of turn, Lewis number, and buoyancy ratio on isotherms, streamlines, and isoconcentrations are shown in Figs. 3.21 and 3.22. It should be mentioned that positive Nr values (opposing buoyancy forces) showed more complex and interesting flow patterns, such as multicells, which is worth presenting and discussing. So, in this study results for a negative Nr (aiding buoyancy forces), where the temperature- and species-induced buoyancy forces aide each other, are not considered. For $Nr=0$, the species-induced buoyancy force has no effect on flow; the flow is driven solely by the thermal buoyancy force. However, the effect of species-induced buoyancy increases as the Nr value increases and reaches a certain value, where the effect of thermally induced buoyancy becomes negligible in comparison with that induced by the solutal buoyancy. For a small Nr value, the flow is driven mainly by the thermal buoyancy force. When Nr increases a reverse thermal plume appears at $\zeta=90°$ when $\gamma=90°$. This phenomena occurs because of the existence of one counterclockwise eddy in this region. The isoconcentrations are more distorted as the solutal forces increase.

The mass flow is given by $\psi_{max} \approx \delta_s v$, where the solutal boundary-layer thickness is given by $\delta_s \approx (RaLeNr)^{-1/4}$ and $v \approx (RaLeNr)^{1/2}$, so $\psi_{max} \approx (RaLeNr)^{1/4}$. The solutal boundary layer δ_s becomes thinner as Le increases. At a low Lewis number, the isoconcentration lines are similar to those of isotherms. By increasing Le, the difference between these counters becomes more visible. The core of the eddy moves upward with an increase in the angle of turn. Thermal boundary-layer thickness increases by augmenting the angle of turn. That one counterclockwise eddy is generated at $\gamma=90°$ is an interesting observation.

Effects of angle of turn, buoyancy ratio, and Lewis number on the local Nusselt number are shown in Fig. 3.23. The local Nusselt number increases as the buoyancy

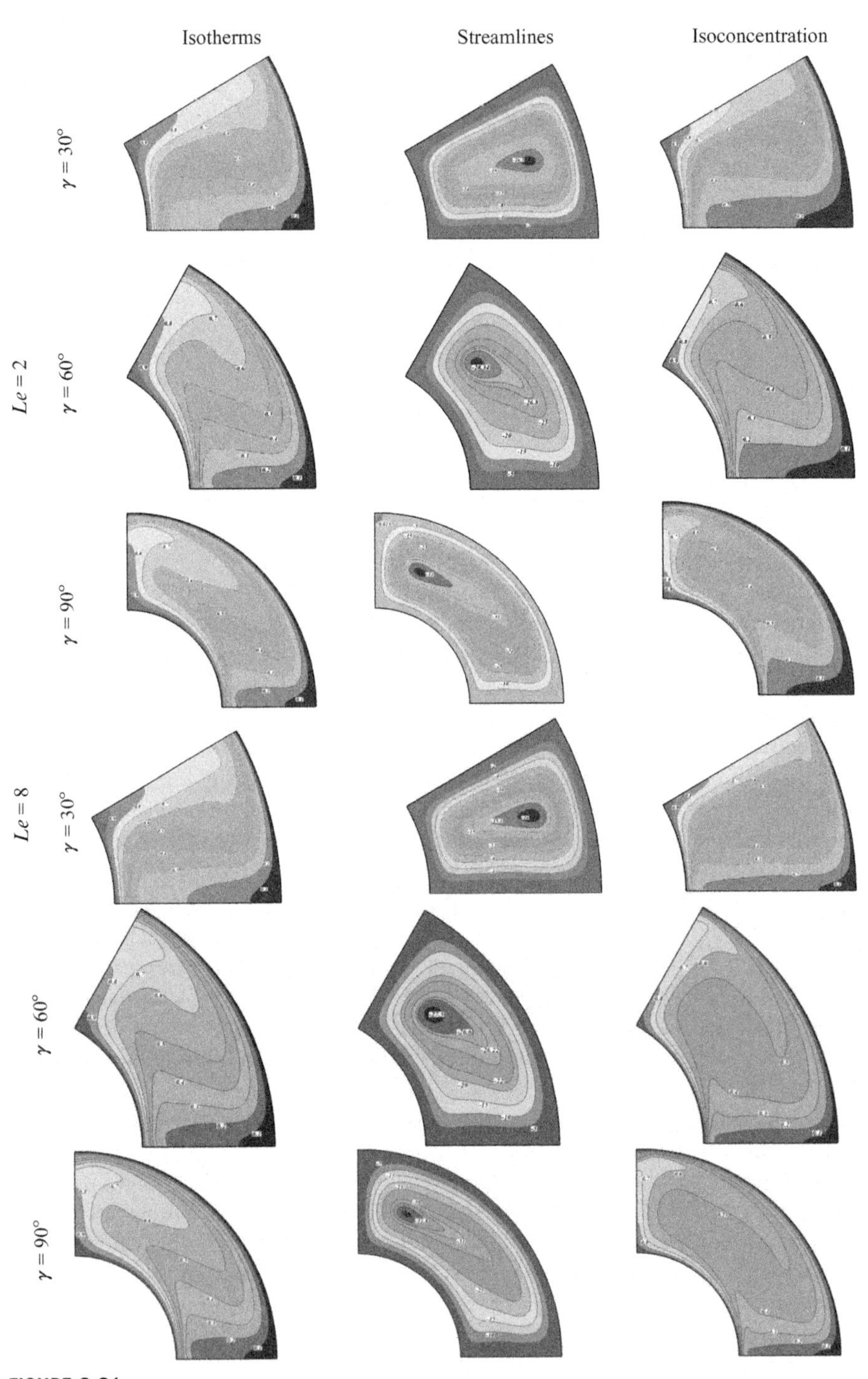

FIGURE 3.21

Comparison of the isotherms, streamlines, and isoconcentrations for different angles of turn and Lewis numbers when $Nr = 0.1$, $Nt = Nb = 0.5$, $Ra = 10^5$, and $Pr = 10$.

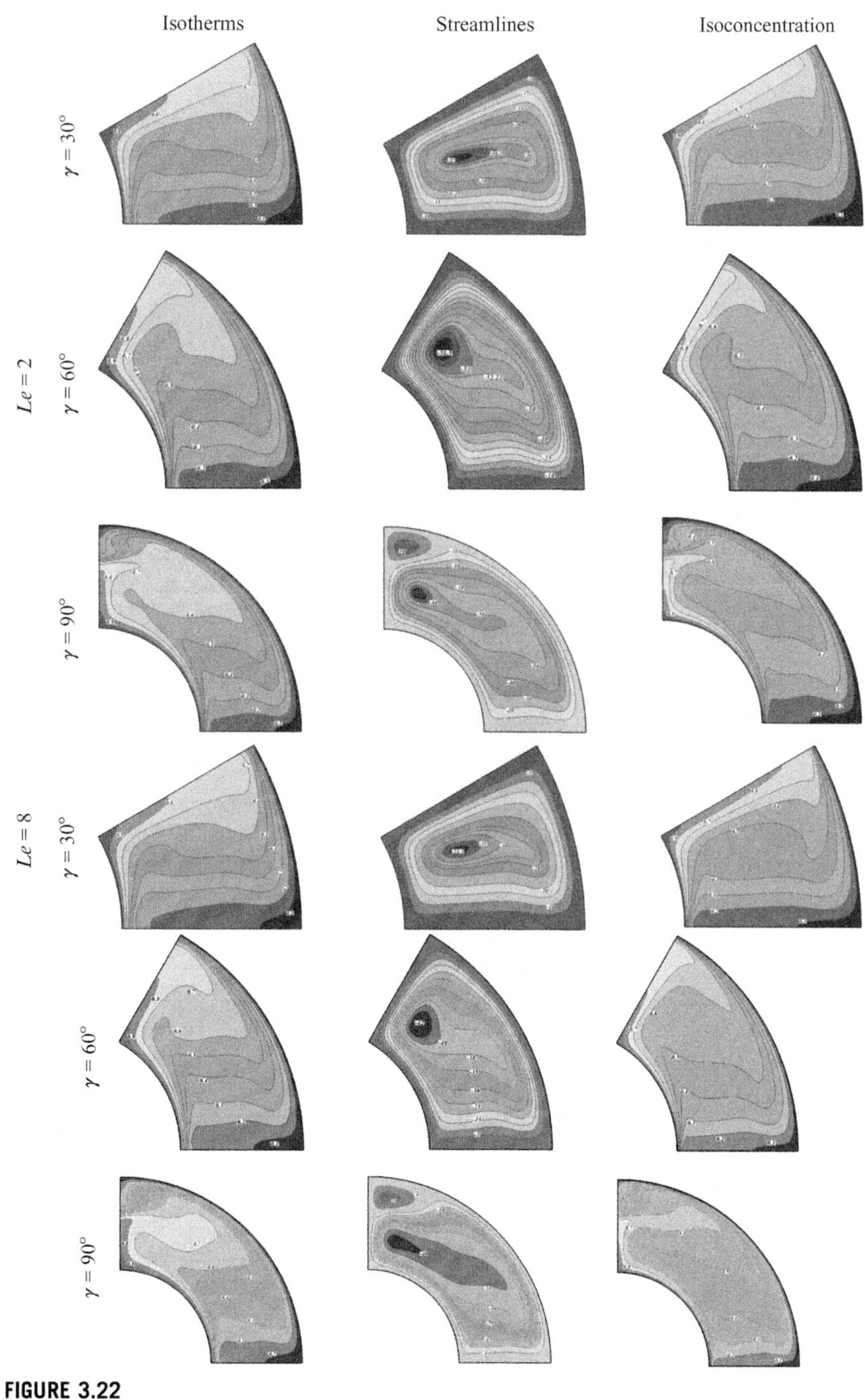

FIGURE 3.22

Comparison of the isotherms, streamlines, and isoconcentrations for different angles of turn and Lewis numbers when $Nr=4$, $Nt=Nb=0.5$, $Ra=10^5$, and $Pr=10$.

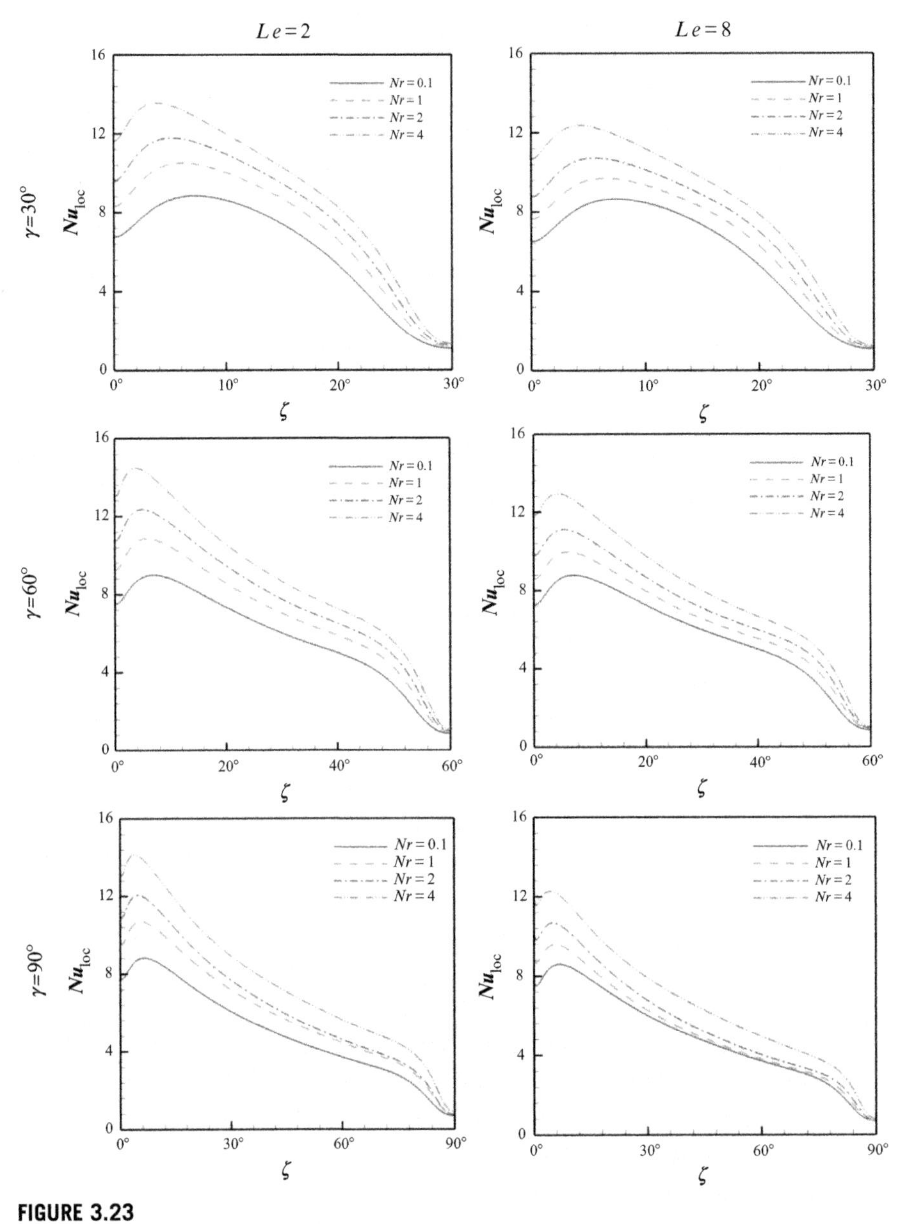

FIGURE 3.23

Effects of the enclosure's angle of turn, the Hartmann number, buoyancy ratio, and Lewis number on the local Nusselt number at $Nt = Nb = 0.5$, $Ra = 10^5$, and $Pr = 10$.

ratio increases, but it decreases as the Lewis number increases. In general, as ζ increases the local Nusselt number decreases because of the increase in the thermal boundary-layer thickness. The local Nusselt number profile has a minimum point at $\zeta = 90°$ because of the thermal plume in this region. The local Nusselt number also increases as the angle of turn is augmented.

Effects of buoyancy ratio, angle of turn, and Lewis number on the average Nusselt number are shown in Fig. 3.24. As the Lewis number increases the thermal boundary-layer thickness increases and, in turn, the Nusselt number decreases. In addition, it can be concluded that the Nusselt number increases as the buoyancy ratio increases, but it decreases with and increase in the angle of turn. The corresponding

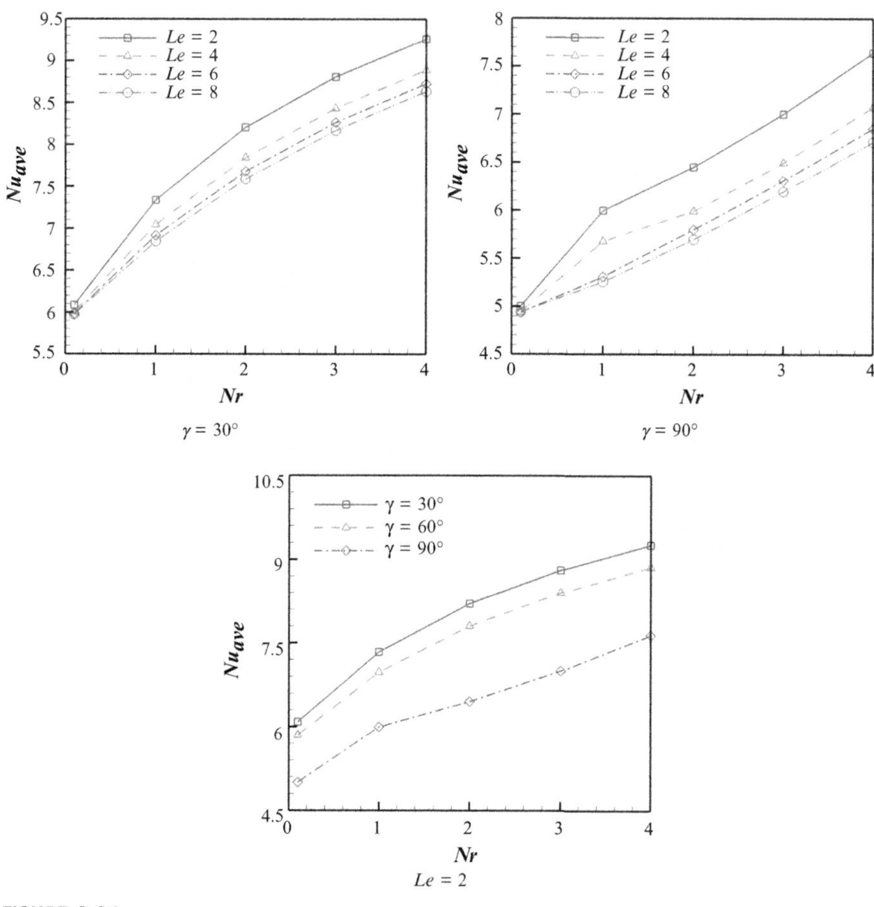

FIGURE 3.24

Effects of the enclosure's angle of turn, buoyancy ratio, and Lewis number on the average Nusselt number at $Nt = Nb = 0.5$, $Ra = 10^5$, and $Pr = 10$.

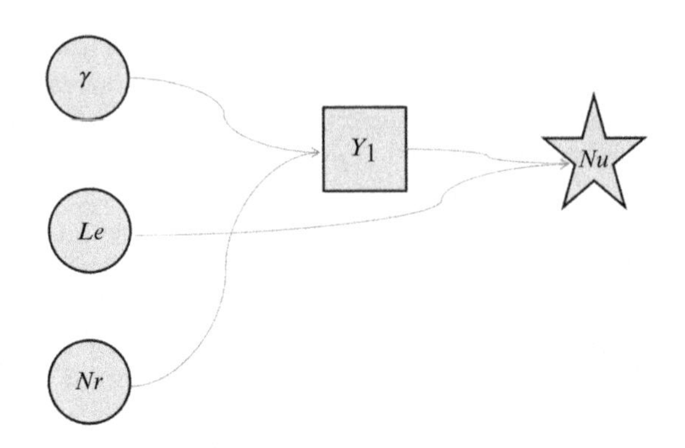

FIGURE 3.25

Evolved structure of correlation for the Nusselt number.

Table 3.5 Constant Coefficients (a_{ij}) for Using Eq. (3.79)

	$i=1$	$i=2$	$i=3$	$i=4$	$i=5$	$i=6$
$j=1$	5.334872	2.065756	1.111228	−1.5454	−0.06732	−0.18785
$j=2$	0.191916	1.077506	−0.12466	0.001021	0.015991	−0.01835

polynomial representation of such a model of the Nusselt number is shown in Eq. (3.79) (Fig. 3.25):

$$Nu = a_{12} + a_{22}Y_1 + a_{32}Le + a_{42}Y_1{}^{*2} + a_{52}Le^2 + a_{62}Y_1Le$$
$$Y_1 = a_{11} + a_{21}\gamma + a_{31}Nr + a_{41}\gamma^2 + a_{51}Nr^2 + a_{61}\gamma Nr \quad (3.79)$$

Also, a_{ij} can be found in Table 3.5. For example, a_{21} equals 2.065756.

REFERENCES

[1] K. Khanafer, K. Vafai, M. Lightstone, Buoyancy-driven heat transfer enhancement in a two-dimensional enclosure utilizing nanofluids, Int. J. Heat Mass Transfer 46 (2003) 3639–3653.

[2] E. Abu-Nada, Z. Masoud, A. Hijazi, Natural convection heat transfer enhancement in horizontal concentric annuli using nanofluids, Int. Commun. Heat Mass Transfer 35 (2008) 657–665.

[3] R.Y. Jou, S.C. Tzeng, Numerical research of nature convective heat transfer enhancement filled with nanofluids in rectangular enclosures, Int. Commun. Heat Mass Transfer 33 (2006) 727–736.

[4] M. Sheikholeslami, M. Gorji-Bandpy, S.M. Seyyedi, D.D. Ganji, H.B. Rokni, S. Soleimani, Application of LBM in simulation of natural convection in a nanofluid filled square cavity with curve boundaries, Powder Technol. 247 (2013) 87–94.

[5] M. Sheikholeslami, M. Gorji-Bandpy, D.D. Ganji, Natural convection in a nanofluid filled concentric annulus between an outer square cylinder and an inner elliptic cylinder, Sci. Iran. Trans. B: Mech. Eng. 20 (4) (2013) 1241–1253.
[6] M. Sheikholeslami, M. Gorji-Bandpy, G. Domairry, Free convection of nanofluid filled enclosure using lattice Boltzmann method (LBM), Appl. Math. Mech. Engl. Ed. 34 (7) (2013) 1–15.
[7] M. Hatami, M. Sheikholeslami, D.D. Ganji, Laminar flow and heat transfer of nanofluid between contracting and rotating disks by least square method, Powder Technol. 253 (2014) 769–779.
[8] M. Sheikholeslami, M. Hatami, D.D. Ganji, Nanofluid flow and heat transfer in a rotating system in the presence of a magnetic field, J. Mol. Liq. 190 (2014) 112–120.
[9] M. Sheikholeslami, R. Ellahi, H.R. Ashorynejad, G. Domairry, T. Hayat, Effects of heat transfer in flow of nanofluids over a permeable stretching wall in a porous medium, J. Comput. Theor. Nanosci. 11 (2014) 1–11.
[10] M. Sheikholeslami, M. Hatami, D.D. Ganji, Analytical investigation of MHD nanofluid flow in a semi-porous channel, Powder Technol. 246 (2013) 327–336.
[11] M. Sheikholeslami, D.D. Ganji, H.R. Ashorynejad, Investigation of squeezing unsteady nanofluid flow using ADM, Powder Technol. 239 (2013) 259–265.
[12] M. Sheikholeslami, D.D. Ganji, Heat transfer of Cu-water nanofluid flow between parallel plates, Powder Technol. 235 (2013) 873–879.
[13] H.R. Ashorynejad, M. Sheikholeslami, I. Pop, D.D. Ganji, Nanofluid flow and heat transfer due to a stretching cylinder in the presence of magnetic field, Heat Mass Transfer 49 (2013) 427–436.
[14] D. Domairry, M. Sheikholeslami, H.R. Ashorynejad, R.S.R. Gorla, M. Khani, Natural convection flow of a non-Newtonian nanofluid between two vertical flat plates, Proc. IMechE N: J. Nanoeng. Nanosyst. 225 (3) (2012) 115–122, http://dx.doi.org/10.1177/1740349911433468.
[15] M. Hatami, M. Sheikholeslami, M. Hosseini, D.D. Ganji, Analytical investigation of MHD nanofluidflow in non-parallel walls, J. Mol. Liq. 194 (2014) 251–259.
[16] M. Sheikholeslami, D.D. Ganji, Magnetohydrodynamic flow in a permeable channel filled with nanofluid, Sci. Iran. B 21 (1) (2014) 203–212.
[17] M. Hatami, M. Sheikholeslami, D.D. Ganji, Nanofluid flow and heat transfer in an asymmetric porous channel with expanding or contracting wall, J. Mol. Liq. 195 (2014) 230–239.
[18] M. Sheikholeslami, F.B. Sheykholeslami, S. Khoshhal, H. Mola-Abasi, D.D. Ganji, H.B. Rokni, Effect of magnetic field on Cu–water nanofluid heat transfer using GMDH-type neural network, Neural Comput. Appl. 25 (2014) 171–178.
[19] M.M. Rashidi, N. Kavyani, S. Abelman, Investigation of entropy generation in MHD and slip flow over a rotating porous disk with variable properties, Int. J. Heat Mass Transfer 70 (2014) 892–917.
[20] M.M. Rashidi, S. Abelman, N. Freidooni Mehr, Entropy generation in steady MHD flow due to a rotating porous disk in a nanofluid, Int. J. Heat Mass Transfer 62 (2013) 515–525.
[21] R. Ellahi, The effects of MHD and temperature dependent viscosity on the flow of non-Newtonian nanofluid in a pipe: analytical solutions, Appl. Math. Modell. 37 (3) (2013) 1451–1467.
[22] M. Sheikholeslami Kandelousi, KKL correlation for simulation of nanofluid flow and heat transfer in a permeable channel, Phys. Lett. A 378 (45) (2014) 3331–3339.

[23] H.A. Mohammed, K. Narrein, Thermal and hydraulic characteristics of nanofluid flow in a helically coiled tube heat exchanger, Int. Commun. Heat Mass Transfer 39 (2012) 1375–1383.
[24] M. Sheikholeslami, S. Abelman, D.D. Ganji, Numerical simulation of MHD nanofluid flow and heat transfer considering viscous dissipation, Int. J. Heat Mass Tran. 79 (2014) 212–222.
[25] W.A. Khan, I. Pop, Boundary-layer flow of a nanofluid past a stretching sheet, Int. J. Heat Mass Transfer 53 (2010) 2477–2483.
[26] C.-Y. Cheng, Natural convection boundary layer flow over a truncated cone in a porous medium saturated by a nanofluid, Int. Commun. Heat Mass Transfer 39 (2012) 231–235.
[27] M. Alinia, D.D. Ganji, M. Gorji-Bandpy, Numerical study of mixed convection in an inclined two sided lid driven cavity filled with nanofluid using two-phase mixture model, Int. Commun. Heat Mass Transfer 38 (2011) 1428–1435.
[28] M. Sheikholeslami, D.D. Ganji, Three dimensional heat and mass transfer in a rotating system using nanofluid, Powder Technol. 253 (2014) 789–796.
[29] M. Sheikholeslami, D.D. Ganji, Numerical investigation for two phase modeling of nanofluid in a rotating system with permeable sheet, J. Mol. Liq. 194 (2014) 13–19.
[30] G. Cesini, M. Paroncini, G. Cortellab, M. Manzan, Natural convection from a horizontal cylinder in a rectangular cavity, Int. J. Heat Mass Transfer 42 (1999) 1801–1811.
[31] M. Sheikholeslami, M. Gorji-Bandpy, K. Vajravelu, Lattice Boltzmann simulation of magnetohydrodynamic natural convection heat transfer of Al_2O_3-water nanofluid in a horizontal cylindrical enclosure with an inner triangular cylinder, Int. J. Heat Mass Tran. 80 (2015) 16–25.
[32] N.K. Ghaddar, Natural convection heat transfer between a uniformly heated cylindrical element and its rectangular enclosure, Int. J. Heat Mass Transfer 35 (1992) 2327–2334.
[33] F. Moukalled, S. Acharya, Natural convection in the annulus between concentric horizontal circular and square cylinders, J. Thermophys. Heat Transfer 10 (3) (1996) 524–531.
[34] Y. Xuan, Q. Li, Heat transfer enhancement of nanofluids, Int. J. Heat Fluid Flow 21 (2000) 58–64.
[35] Y. Xuan, W. Roetzel, Conceptions for heat transfer correlations of nanofluids, Int. J. Heat Mass Transfer 43 (2000) 3701–3707.
[36] F.J. Wasp, Solid-Liquid Slurry Pipeline Transportation, Trans Tech, Berlin, 1977.
[37] E. Abu-Nada, H.F. Oztop, Effects of inclination angle on natural convection in enclosures filled with Cu-water nanofluid, Int. J. Heat Fluid Flow 30 (2009) 669–678.
[38] O. Abouali, A. Falahatpisheh, Numerical investigation of natural convection ofAl_2O_3 nanofluids in vertical annuli, J. Heat Mass Transfer 46 (2009) 15–23.
[39] S.M. Aminossadati, B. Ghasemi, Natural convection of water-CuO nanofluid in a cavity with two pairs of heat source-sink, Int. Commun. Heat Mass Transfer 38 (2011) 672–678.
[40] Y. Ding, D. Wen, Particle migration in a flow of nanoparticle suspensions, Powder Technol. 149 (2-3) (2005) 84–92.
[41] V. Bianco, O. Manca, S. Nardini, Numerical investigation on nanofluids turbulent convection heat transfer inside a circular tube, Int. J. Therm. Sci. 29 (2009) 3632–3642.
[42] S. Mirmasoumi, A. Behzadmehr, Numerical study of laminar mixed convection of a nanofluid in a horizontal tube using two-phase mixture model, Appl. Therm. Eng. 28 (2008) 717–727.
[43] A. Akbarinia, R. Laur, Investigation the diameter of solid particles effects on a laminar nanofluid flow in a curved tube using a two phase approach, Int. J. Heat Fluid Flow 30 (2009) 706–714.

[44] J. Buongiorno, Convective transport in nanofluids, ASME J. Heat Transfer 128 (2006) 240–250.

[45] A. Behzadmehr, M. Saffar-Avval, N. Galanis, Prediction of turbulent forced convection of a nanofluid in a tube with uniform heat flux using a two phase approach, Int. J. Heat Fluid Flow 28 (2) (2007) 211–219.

[46] B.C. Pak, Y.I. Cho, Hydrodynamic and heat transfer study of dispersed fluids with submicron metallic oxide particles, Exp. Heat Transfer 11 (1998) 151–170.

[47] J.A. Eastman, S.U.S. Choi, S. Li, G. Soyez, L.J. Thompson, R.J. Di Melfi, Novel thermal properties of nanostructured materials, Mater. Sci. Forum 312-314 (1999) 629–634.

[48] S.J. Palm, G. Roy, C.T. Nguyen, Heat transfer enhancement with the use of nanofluids in radial flow cooling systems considering temperature dependent properties, Appl. Therm. Eng. 26 (2006) 2209–2218.

[49] J.C. Maxwell, A Treatise on Electricity and Magnetism, second ed., Oxford University Press, Cambridge, 1904, pp. 435-441.

[50] H. Brinkman, The viscosity of concentrated suspensions and solutions, J. Chem. Phys. 20 (1952) 571.

[51] E. Abu-Nada, A.J. Chamkha, Effect of nanofluid variable properties on natural convection in enclosures filled with a CuO-EG-water nanofluid, Int. J. Therm. Sci. 49 (12) (2010) 2339–2352.

[52] P.K. Namburu, D.P. Kulkarni, D. Misra, D.K. Das, Viscosity of copper oxide nanoparticles dispersed in ethylene glycol and water mixture, Exp. Therm. Fluid Sci. 32 (2) (2007) 397–402.

[53] A. Arefmanesh, M. Mahmoodi, Effects of uncertainties of viscosity models for Al_2O_3-water nanofluid on mixed convection numerical simulations, Int. J. Therm. Sci. 50 (9) (2011) 1706–1719.

[54] S.E.B. Maïga, C.T. Nguyen, N. Galanis, G. Roy, Heat transfer behaviours of nanofluids in a uniformly heated tube, Superlattices Microstruct. 35 (3-6) (2004) 543–557.

[55] A. Einstein, Investigation on the Theory of Brownian Motion, Dover, New York, 1956.

[56] B.C. Pak, Y. Cho, Hydrodynamic and heat transfer study of dispersed fluids with submicron metallic oxide particle, Exp. Heat Transfer 11 (1998) 151–170.

[57] C.T. Nguyen, F. Desgranges, G. Roy, N. Galanis, T. Mare, S. Boucher, M.H. Angue, Temperature and particle-size dependent viscosity data for water-based nanofluids—hysteresis phenomenon, Int. J. Heat Fluid Flow 28 (2007) 1492–1506.

[58] E. Abu-Nada, Effects of variable viscosity and thermal conductivity of Al_2O_3-water nanofluid on heat transfer enhancement in natural convection, Int. J. Heat Fluid Flow 30 (2009) 679–690.

[59] S.P. Jang, J.H. Lee, K.S. Hwang, S.U.S. Choi, Particle concentration and tube size dependence of viscosities of Al_2O_3-water nanofluids flowing through micro and minitubes, Appl. Phys. Lett. 91 (2007) 243112.

[60] J. Koo, C. Kleinstreuer, A new thermal conductivity model for nanofluids, J. Nanopart. Res. 6 (2004) 577–588.

[61] S.E.B. Maïga, S.M. Palm, C.T. Nguyen, G. Roy, N. Galanis, Heat transfer enhancement by using nanofluids in forced convection flows, Int. J. Heat Fluid Flow 26 (2005) 530–546.

[62] D. Orozco, Hydrodynamic behavior of suspension of polar particles, Encycl. Surf. Colloid Sci. 4 (2005) 2375–2396.

[63] C.T. Nguyen, F. Desgranges, N. Galanis, G. Roy, T. Mare, S. Boucher, H.A. Mintsa, Viscosity data for Al_2O_3-water nanofluid-hysteresis: is heat transfer enhancement using nanofluids reliable? Int. J. Therm. Sci. 47 (2008) 103–111.

[64] N. Masoumi, N. Sohrabi, A. Behzadmehr, A new model for calculating the effective viscosity of nanofluids, J. Phys. D: Appl. Phys. 42 (2009) 055501.

[65] I. Gherasim, G. Roy, C.T. Nguyen, D. Vo-Ngoc, Experimental investigation of nanofluids in confined laminar radial flows, Int. J. Therm. Sci. 48 (2009) 1486–1493.

[66] S.P. Jang, S.U.S. Choi, The role of Brownian motion in the enhanced thermal conductivity of nanofluids, Appl. Phys. Lett. 84 (2004) 4316–4318.

[67] B.X. Wang, L.R. Zhou, X.F. Peng, A fractal model for predicting the effective thermal conductivity of liquid with suspension of nanoparticles, Int. J. Heat Mass Transfer 46 (2003) 2665–2672.

[68] C.H. Chon, K.D. Kihm, S.P. Lee, S.U.S. Choi, Empirical correlation finding the role of temperature and particle size for nanofluid (Al_2O_3) thermal conductivity enhancement, Appl. Phys. Lett. 87 (2005) 153107 (article ID 153107).

[69] P. Charuyakorn, S. Sengupta, S.K. Roy, Forced convection heat transfer in micro encapsulated phase change material slurries, Int. J. Heat Mass Transfer 34 (1991) 819–833.

[70] J.A. Eastman, S.R. Phillpot, U.S. Choi, P. Keblinski, Thermal transport in nanofluids, Annu. Rev. Mater. Res. 34 (2000) 219–246.

[71] W. Yu, S.U.S. Choi, The role of interfacial layers in the enhanced thermal conductivity of nanofluids: a renovated Maxwell model, J. Nanopart. Res. 5 (2003) 167–171.

[72] H.E. Patel, T. Sundarrajan, T. Pradeep, A. Dasgupta, N. Dasgupta, S.K. Das, A microconvection model for thermal conductivity of nanofluid, Pramana J. Physiol. 65 (2005) 863–869.

[73] H. Angue Mintsa, G. Roy, C.T. Nguyen, D. Doucet, New temperature dependent thermal conductivity data for water-based nanofluids, Int. J. Therm. Sci. 48 (2009) 363–371.

[74] M. Sheikholeslami, M. Gorji-Bandpy, D.D. Ganji, S. Soleimani, Natural convection heat transfer in a nanofluid filled inclined L-shaped enclosure, IJST Trans. Mech. Eng. 38 (2014) 1–5.

[75] G. De Vahl Davis, Natural convection of air in a square cavity, a benchmark numerical solution, Int. J. Numer. Methods Fluids 3 (1962) 249–264.

[76] S. Soleimani, M. Sheikholeslami, D.D. Ganji, M. Gorji-Bandpay, Natural convection heat transfer in a nanofluid filled semi-annulus enclosure, Int. Commun. Heat Mass Transfer 39 (2012) 565–574.

[77] D. Yadav, G.S. Agrawal, R. Bhargava, Thermal instability of rotating nanofluid layer, Int. J. Eng. Sci. 49 (2011) 1171–1184.

[78] M. Sheikholeslami, M. Gorji-Bandpy, S. Soleimani, Two phase simulation of nanofluid flow and heat transfer using heatline analysis, Int. Commun. Heat Mass Transfer 47 (2013) 73–81.

[79] M. Sheikholeslami, M. Gorji-Bandpy, D.D. Ganji, S. Soleimani, Thermal management for free convection of nanofluid using two phase model, J. Mol. Liq. 194 (2014) 179–187.

CHAPTER

Flow heat transfer in the presence of a magnetic field 4

4.1 INTRODUCTION

Magnetohydrodynamics (MHD) has been a focus of intense research for long time because of its vast importance in numerous fields ranging from the study of natural phenomena in geophysics and astrophysics to many engineering applications such as plasma confinement, liquid metal, and electromagnetic casting. Wide-ranging investigations were conducted by researchers in MHD natural convection. Ozoe and Okada [1] reported a three-dimensional numerical study of the effect of a magnetic field's path in a cubical enclosure. Rudraiah et al. [2] numerically investigated a magnetic field's effect on natural convection in a rectangular enclosure. They found that the magnetic field decreases the rate of heat transfer. Al-Najem et al. [3] calculated flow and temperature fields under a uniform magnetic field in a tilted square cavity with isothermal vertical and adiabatic horizontal walls. They demonstrated that the suppression effect of the magnetic field on convection currents and heat transfer is more significant at small inclination angles and high Rayleigh numbers. Ece and Buyuk [4] examined steady and laminar natural convection flow in the presence of a magnetic field in an inclined rectangular enclosure heated and cooled on adjacent walls. They found that the magnetic field suppressed the convective flow and the heat transfer rate. A magnetic field's effect on natural convection heat transfer in a curved enclosure was studied by Sheikholeslami et al. [5] using the control volume finite element method (CVFEM). The problem of laminar viscous flow in a semiporous channel in the presence of a transverse magnetic field is studied by Sheikholeslami et al. [6]. They showed that the optimal homotopy asymptotic method was a powerful approach for solving nonlinear differential equations such as this problem.

The above literature review reveals that the existence of a magnetic field has a noticeable effect on the reduction of heat transfer under natural convection, whereas in many engineering applications such as magnetic field sensors, magnetic storage media, and the cooling systems of electronic devices, increasing heat transfer from solid surfaces is a goal. Under this condition, the use of nanofluids with higher thermal conductivity can be considered as a promising solution. Sheikholeslami et al. [7] studied the problem of laminar nanofluid flow in a semiporous channel. They found that that the velocity boundary layer thickness decreases with an increasing Reynolds number and nanoparticle volume fraction, and it increases as the Hartmann number increases. Ashorynejad et al. [8] studied the flow and heat transfer of a nanofluid over a stretching cylinder in the presence of a magnetic field. They found that choosing copper (for its small magnetic parameter) and alumina (for large magnetic parameter values) leads to the highest cooling performance. Sheikholeslami et al. [9] used the CVFEM to simulate the effect of a magnetic field on natural convection in an inclined half-annulus enclosure filled with a copper-water nanofluid. Their results indicated that the Hartmann number and the inclination angle of the enclosure can be considered as control parameters at different Rayleigh numbers. Sheikholeslami et al. [10] studied MHD flow in a nanofluid-filled, inclined enclosure with a sinusoidal wall. They reported that for all values of a Hartmann number, at $Ra=10^4$ and 10^5 the maximum value of E is obtained at $\gamma=60°$ and $0°$, respectively. Sheikholeslami et al. [11] performed a numerical analysis of natural convection heat transfer of a copper-water nanofluid in a cold outer circular enclosure containing a hot inner sinusoidal circular cylinder in the presence of a horizontal magnetic field using the CVFEM. They concluded that in the absence of a magnetic field, the enhancement ratio decreases as the Rayleigh number increases, whereas an opposite trend is observed in the presence of a magnetic field. Sheikholeslami et al. [12] studied natural convection in a concentric annulus between a cold outer square and heated inner circular cylinders in the presence of a static radial magnetic field. They reported that the average Nusselt number is an increasing function of the nanoparticle volume fraction as well as the Rayleigh number, whereas it is a decreasing function of the Hartmann number. Sheikholeslami et al. [13] studied the problem of MHD free convection in an eccentric semiannulus filled with a nanofluid. They showed that the Nusselt number decreases with increase of Rayleigh number. The lattice Boltzmann method was used by Sheikholeslami et al. [14] to investigate MHD flow using a copper-water nanofluid in a concentric annulus. Their results proved that the enhancement ratio increases with a decrease in the Rayleigh number, and it increases when the Hartmann number is augmented. Free convection heat transfer in a concentric annulus between a cold square and heated elliptic cylinders in the presence of a magnetic field was investigated by Sheikholeslami et al. [15]. They found that the enhancement in heat transfer increases as the Hartmann number increases, but it decreases with an increase in the Rayleigh number. The effect of a static radial magnetic field on natural convection heat transfer in a horizontal, cylindrical annulus enclosure filled with a nanofluid was investigated numerically by Ashorynejad et al. [16] using the lattice Boltzmann method. They found that the average Nusselt number increases as the nanoparticle volume fraction and Rayleigh number increase, whereas it decreases as the Hartmann number increases. Sheikholeslami et al. [17] studied the effects of a magnetic field and nanoparticles on Jeffery-Hamel flows. They show that increasing the Hartmann number leads to backflow reduction. The results also show that the momentum boundary layer thickness increases as the nanoparticle volume fraction increases.

Magnetic nanofluid (ferrofluid) is a magnetic colloidal suspension consisting of a base liquid and magnetic nanoparticles with a diameter range of 5-15 nm coated with a surfactant layer. The most often used magnetic material is single-domain particles of magnetite, iron, or cobalt and a base liquid such as water or kerosene. A ferrofluid behaves as a fluid that is affected by an external magnetic field, and externally applied magnetic fields can be used to control and direct the flow of ferrofluids, because of which magnetic field is applicable in various fields such as electronic packing, mechanical engineering, thermal engineering, aerospace, and bioengineering [18–21]. Numerical analysis of the heat transfer enhancement and fluid flow characteristics of a rotating cylinder under the influence of a magnetic dipole in backward-facing step geometry was conducted by Selimefendigil and Oztop [22]. They found that the effect of cylinder rotation on local Nusselt number distribution is more pronounced at low Reynolds numbers. Wrobel et al. [23] numerically and experimentally studied thermomagnetic convective flow of a paramagnetic fluid in an annular enclosure with a round rod core and a cylindrical outer wall. Their results showed that magnetizing force affects the heat transfer rate and a strong magnetic field can control the magnetic convection of a paramagnetic fluid. Tzirtzilakis et al. [24] studied the turbulent and two-dimensional flow of electrical conductive blood. They reported that the influence of the magnetic field decreases in the presence of turbulence. Tzirtzilakis et al. [25] investigated biomagnetic fluid flow in a three-dimensional rectangular duct. Their investigations showed that the flow is appreciably influenced in the presence of a magnetic field. Aminfar et al. [26] investigated the effect of positive and negative magnetic field gradients on the hydrodynamic and thermal behavior of a ferrofluid mixed convection flow in a vertical tube. They showed that negative gradients enhance the Nusselt number, whereas positive gradients decrease it. Free convection of ferrofluid in a cavity heated from below in the presence of an external magnetic field was studied by Sheikholeslami and Gorji-Bandpy [27]. They found that smaller particles have a better ability to dissipate heat, and a larger volume fraction would provide a stronger driving force, which leads to an increase in temperature profile. The hydrothermal behavior of an electrical conductive ferrofluid in the presence of magnetic fields was simulated by Aminfar et al. [28] using a two-phase mixture model. Their results indicated that the negative gradient axial field and uniform transverse field act similarly and enhance both the Nusselt number and the friction factor, whereas a positive gradient axial field decreases them. Lajvardi et al. [29] conducted an experimental work on the convective heat transfer of a ferrofluid flowing through a heated copper tube in the laminar regime in the presence of a magnetic field; they reported a significant increase in the heat transfer of the ferrofluid by applying various orders of magnetic field.

4.2 MHD NANOFLUID FLOW AND HEAT TRANSFER

The existence of a magnetic field has a noticeable effect on heat transfer reduction under natural convection, whereas in many engineering applications increasing heat transfer from solid surfaces is a goal. In these circumstances, the use of nanofluids with higher thermal conductivity can be considered a promising solution. It is also assumed that the uniform magnetic field ($\vec{B}=B_x\vec{e}_x+B_y\vec{e}_y$) of constant magnitude

$B=\sqrt{B_x^2+B_y^2}$ is applied, where $\vec{e}_x$ and $\vec{e}_y$ are unit vectors in the Cartesian coordinate system. The orientation of the magnetic field forms an angle λ with a horizontal axis such that $\lambda=B_x/B_y$. The electric current J and the electromagnetic force F are defined by $J=\sigma\left(\vec{V}\times\vec{B}\right)$ and $F=\sigma\left(\vec{V}\times\vec{B}\right)\times\vec{B}$, respectively. Here we present the governing equation in vorticity stream function form for different cases.

4.2.1 MATHEMATICAL MODELING FOR A SINGLE-PHASE MODEL

4.2.1.1 Natural convection

Flow is two-dimensional, laminar, and incompressible. Radiation, viscous dissipation, induced electric current, and Joule heating are neglected. The magnetic Reynolds number is assumed to be small so that the induced magnetic field can be neglected compared with the applied magnetic field. The flow is considered to be steady, two-dimensional, and laminar. Neglecting displacement currents, induced magnetic field, dissipation, and Joule heating, and using the Boussinesq approximation, the governing equations of nanofluid heat transfer and fluid flow can be obtained in dimensional form:

$$\frac{\partial u}{\partial x}+\frac{\partial v}{\partial y}=0 \tag{4.1}$$

$$\begin{aligned} u\frac{\partial u}{\partial x}+v\frac{\partial u}{\partial y}&=-\frac{1}{\rho_{\mathrm{nf}}}\frac{\partial P}{\partial x}+\upsilon_{\mathrm{nf}}\left(\frac{\partial^2 u}{\partial x^2}+\frac{\partial^2 u}{\partial y^2}\right)\\ &+\frac{\sigma_{\mathrm{nf}}B^2}{\rho_{\mathrm{nf}}}\left(v\sin\lambda\cos\lambda-u\sin^2\lambda\right) \end{aligned} \tag{4.2}$$

$$\begin{aligned} u\frac{\partial v}{\partial x}+v\frac{\partial v}{\partial y}&=-\frac{1}{\rho_{\mathrm{nf}}}\frac{\partial P}{\partial y}+\upsilon_{\mathrm{nf}}\left(\frac{\partial^2 v}{\partial x^2}+\frac{\partial^2 v}{\partial y^2}\right)+\beta_{\mathrm{nf}}g(T-T_{\mathrm{c}})\\ &+\frac{\sigma_{\mathrm{nf}}B^2}{\rho_{\mathrm{nf}}}\left(u\sin\lambda\cos\lambda-v\cos^2\lambda\right) \end{aligned} \tag{4.3}$$

$$u\frac{\partial T}{\partial x}+v\frac{\partial T}{\partial y}=\alpha_{\mathrm{nf}}\left(\frac{\partial^2 T}{\partial x^2}+\frac{\partial^2 T}{\partial y^2}\right) \tag{4.4}$$

where ϕ is the solid volume fraction of the nanoparticles and nf is the nanofluid's properties.

The stream function and vorticity are defined as follows:

$$\begin{aligned} &u=\frac{\partial\psi}{\partial y},\ v=-\frac{\partial\psi}{\partial x}\ \text{represent stream function}\\ &\omega=\frac{\partial v}{\partial x}-\frac{\partial u}{\partial y}\ \text{represent vorticity} \end{aligned} \tag{4.5}$$

The stream function satisfies the continuity (Eq. 4.1). The vorticity equation is obtained by eliminating the pressure between the two momentum equations, i.e., by taking the y-derivative of Eq. (4.2) and subtracting from it the x-derivative of Eq. (4.3). This gives:

$$\frac{\partial\psi}{\partial y}\frac{\partial\omega}{\partial x}-\frac{\partial\psi}{\partial x}\frac{\partial\omega}{\partial y}=\upsilon_{\mathrm{nf}}\left(\frac{\partial^2\omega}{\partial x^2}+\frac{\partial^2\omega}{\partial y^2}\right)+\beta_{\mathrm{nf}}g\left(\frac{\partial T}{\partial x}\right)$$
$$+\frac{\sigma_{\mathrm{nf}}B^2}{\rho_{\mathrm{nf}}}\left(-\frac{\delta v}{\delta y}\sin\lambda\cos\lambda+\frac{\delta u}{\delta y}\sin^2\lambda+\frac{\delta u}{\delta x}\sin\lambda\cos\lambda-\frac{\delta v}{\delta x}\cos^2\lambda\right) \tag{4.6}$$

$$\frac{\partial\psi}{\partial y}\frac{\partial T}{\partial x}-\frac{\partial\psi}{\partial x}\frac{\partial T}{\partial y}=\alpha_{\mathrm{nf}}\left(\frac{\partial^2 T}{\partial x^2}+\frac{\partial^2 T}{\partial y^2}\right) \tag{4.7}$$

$$\frac{\partial^2\psi}{\partial x^2}+\frac{\partial^2\psi}{\partial y^2}=-\omega \tag{4.8}$$

By introducing the following nondimensional variables:

$$X=\frac{x}{L},\ Y=\frac{y}{L},\ \Omega=\frac{\omega L^2}{\alpha_{\mathrm{f}}},\ \Psi=\frac{\psi}{\alpha_{\mathrm{f}}},\ U=\frac{uL}{\alpha_{\mathrm{f}}},\ V=\frac{vL}{\alpha_{\mathrm{f}}},\ \Theta=\frac{T-T_{\mathrm{c}}}{\Delta T}\ \begin{cases}\Delta T=T_{\mathrm{h}}-T_{\mathrm{c}}\\ \Delta T=(q''L/k_{\mathrm{f}})\end{cases} \tag{4.9}$$

where in Eq. (4.9) $L=r_{\mathrm{out}}-r_{\mathrm{in}}=r_{\mathrm{in}}$. Also, $\Delta T=T_{\mathrm{h}}-T_{\mathrm{c}}$ and $\Delta T=(q''L/k_{\mathrm{f}})$ are used for constant temperature and constant heat flux boundary conditions, respectively. Using dimensionless parameters the equations now become:

$$\frac{\partial\Psi}{\partial Y}\frac{\partial\Omega}{\partial X}-\frac{\partial\Psi}{\partial X}\frac{\partial\Omega}{\partial Y}=Pr\frac{(\mu_{\mathrm{nf}}/\mu_{\mathrm{nf}})}{(\rho_{\mathrm{nf}}/\rho_{\mathrm{nf}})}\left(\frac{\partial^2\Omega}{\partial X^2}+\frac{\partial^2\Omega}{\partial Y^2}\right)$$
$$+RaPr\left[\frac{\beta_{\mathrm{nf}}}{\beta_{\mathrm{f}}}\right]\left(\frac{\partial\Theta}{\partial X}\right)+Ha^2Pr\frac{(\sigma_{\mathrm{nf}}/\sigma_{\mathrm{nf}})}{(\rho_{\mathrm{nf}}/\rho_{\mathrm{nf}})}\left(-\frac{\delta V}{\delta Y}\tan\lambda+\frac{\delta U}{\delta Y}\tan^2\lambda+\frac{\delta U}{\delta X}\tan\lambda-\frac{\delta V}{\delta X}\right) \tag{4.10}$$

$$\frac{\partial\Psi}{\partial Y}\frac{\partial\Theta}{\partial X}-\frac{\partial\Psi}{\partial X}\frac{\partial\Theta}{\partial Y}=\left[\frac{\dfrac{k_{\mathrm{nf}}}{k_{\mathrm{f}}}}{\dfrac{(\rho C_{\mathrm{p}})_{\mathrm{nf}}}{(\rho C_{\mathrm{p}})_{\mathrm{f}}}}\right]\left(\frac{\partial^2\Theta}{\partial X^2}+\frac{\partial^2\Theta}{\partial Y^2}\right) \tag{4.11}$$

$$\frac{\partial^2\Psi}{\partial X^2}+\frac{\partial^2\Psi}{\partial Y^2}=-\Omega \tag{4.12}$$

where the Rayleigh number, Hartmann number, and Prandtl number are defined as follows:

$$Ra=g\beta_{\mathrm{f}}L^3\Delta T/(\alpha_{\mathrm{f}}\upsilon_{\mathrm{f}}),\ \ Ha=LB_x\sqrt{\sigma_{\mathrm{f}}/\mu_{\mathrm{f}}},\ \ Pr=\upsilon_{\mathrm{f}}/\alpha_{\mathrm{f}} \tag{4.13}$$

4.2.1.2 Mixed convection

As described in the previous section, the governing equations of nanofluid heat transfer and fluid flow can be obtained in dimensional form:

$$\frac{\partial u}{\partial x}+\frac{\partial v}{\partial y}=0 \tag{4.14}$$

$$u\frac{\partial u}{\partial x}+v\frac{\partial u}{\partial y}=-\frac{1}{\rho_{\mathrm{nf}}}\frac{\partial P}{\partial x}+\upsilon_{\mathrm{nf}}\left(\frac{\partial^2 u}{\partial x^2}+\frac{\partial^2 u}{\partial y^2}\right)+\beta_{\mathrm{nf}}g\left(\frac{\partial T}{\partial x}\right)+\frac{\sigma_{\mathrm{nf}}B^2}{\rho_{\mathrm{nf}}}\left(v\sin\lambda\cos\lambda-u\sin^2\lambda\right) \tag{4.15}$$

$$u\frac{\partial v}{\partial x}+v\frac{\partial v}{\partial y}=-\frac{1}{\rho_{\mathrm{nf}}}\frac{\partial P}{\partial y}+\upsilon_{\mathrm{nf}}\left(\frac{\partial^2 v}{\partial x^2}+\frac{\partial^2 v}{\partial y^2}\right)+\frac{\sigma_{\mathrm{nf}}B^2}{\rho_{\mathrm{nf}}}\left(u\sin\lambda\cos\lambda-v\cos^2\lambda\right) \tag{4.16}$$

$$u\frac{\partial T}{\partial x}+v\frac{\partial T}{\partial y}=\alpha_{\mathrm{nf}}\left(\frac{\partial^2 T}{\partial x^2}+\frac{\partial^2 T}{\partial y^2}\right) \tag{4.17}$$

By introducing the following nondimensional variables:

$$X=\frac{x}{L},\quad Y=\frac{y}{L},\quad \Omega=\frac{\omega L}{u_{\mathrm{r}}},\quad \Psi=\frac{\psi}{u_{\mathrm{r}}L},\quad U=\frac{u}{u_{\mathrm{r}}},\quad V=\frac{v}{u_{\mathrm{r}}},\quad \Theta=\frac{T-T_{\mathrm{c}}}{\Delta T}\quad \begin{cases}\Delta T=T_{\mathrm{h}}-T_{\mathrm{c}}\\ \Delta T=(q''L/k_{\mathrm{f}})\end{cases} \tag{4.18}$$

where in Eq. (4.18) $L=r_{\mathrm{out}}-r_{\mathrm{in}}=r_{\mathrm{in}}$. Also, $\Delta T=T_{\mathrm{h}}-T_{\mathrm{c}}$ and $\Delta T=(q''L/k_{\mathrm{f}})$ are used for constant temperature and constant heat flux boundary conditions, respectively. The vorticity equation is obtained by eliminating the pressure between the two momentum equations, i.e., by taking the y-derivative of Eq. (4.15) and subtracting from it the x-derivative of Eq. (4.16). Using dimensionless parameters the equations now become:

$$\frac{\partial\Psi}{\partial Y}\frac{\partial\Omega}{\partial X}-\frac{\partial\Psi}{\partial X}\frac{\partial\Omega}{\partial Y}=\frac{1}{Re}\frac{(\mu_{\mathrm{nf}}/\mu_{\mathrm{nf}})}{(\rho_{\mathrm{nf}}/\rho_{\mathrm{nf}})}\left(\frac{\partial^2\Omega}{\partial X^2}+\frac{\partial^2\Omega}{\partial Y^2}\right)+\frac{Gr}{Re^2}\frac{\beta_{\mathrm{nf}}}{\beta_{\mathrm{f}}}\frac{\partial\Theta}{\partial X}+\frac{Ha^2}{Re}\frac{(\sigma_{\mathrm{nf}}/\sigma_{\mathrm{nf}})}{(\rho_{\mathrm{nf}}/\rho_{\mathrm{nf}})}\left(-\frac{\delta V}{\delta Y}\tan\lambda+\frac{\delta U}{\delta Y}\tan^2\lambda+\frac{\delta U}{\delta X}\tan\lambda-\frac{\delta V}{\delta X}\right) \tag{4.19}$$

$$\frac{\partial\Psi}{\partial Y}\frac{\partial\Theta}{\partial X}-\frac{\partial\Psi}{\partial X}\frac{\partial\Theta}{\partial Y}=\frac{1}{RePr}\left[\frac{\frac{k_{\mathrm{nf}}}{k_{\mathrm{f}}}}{\frac{(\rho C_{\mathrm{p}})_{\mathrm{nf}}}{(\rho C_{\mathrm{p}})_{\mathrm{f}}}}\right]\left(\frac{\partial^2\Theta}{\partial X^2}+\frac{\partial^2\Theta}{\partial Y^2}\right) \tag{4.20}$$

$$\frac{\partial^2\Psi}{\partial X^2}+\frac{\partial^2\Psi}{\partial Y^2}=-\Omega \tag{4.21}$$

where $Re = \frac{\rho_f u_r L}{\mu_f}$, $Gr = g\beta \Delta T L^3/\upsilon^2$, and $Ri = \frac{Gr}{Re^2}$ are the Reynolds number, Grashof number, and Richardson number, respectively.

4.2.2 MATHEMATICAL MODELING FOR A TWO-PHASE MODEL

4.2.2.1 Natural convection

A nanofluid's density ρ is

$$\begin{aligned}\rho &= \phi\rho_p + (1-\phi)\rho_f \\ &\cong \phi\rho_p + (1-\phi)\{\rho_{f_0}(1-\beta(T-T_c)\}\end{aligned} \tag{4.22}$$

where ρ_f is the base fluid's density, T_c is a reference temperature, ρ_{f_0} is the base fluid's density at the reference temperature, and β is the volumetric coefficient of expansion. Taking the density of the base fluid as that of the nanofluid, the density ρ in Eq. (4.22) thus becomes

$$\rho \cong \phi\rho_p + (1-\phi)\{\rho_0(1-\beta(T-T_c)\} \tag{4.23}$$

ρ_0 is the nanofluid's density at the reference temperature.

The continuity, momentum under Boussinesq approximation, and energy equations for the laminar and steady-state natural convection in a two-dimensional enclosure can be written in dimensional form:

$$\frac{\partial u}{\partial x} + \frac{\partial v}{\partial y} = 0 \tag{4.24}$$

$$\rho_f\left\{u\frac{\partial u}{\partial x} + v\frac{\partial u}{\partial y}\right\} = -\frac{\partial P}{\partial x} + \mu\left(\frac{\partial^2 u}{\partial x^2} + \frac{\partial^2 u}{\partial y^2}\right) + \sigma B^2\left(v\sin\lambda\cos\lambda - u\sin^2\lambda\right) \tag{4.25}$$

$$\begin{aligned}\rho_f\left\{u\frac{\partial v}{\partial x} + v\frac{\partial v}{\partial y}\right\} &= -\frac{\partial P}{\partial y} + \mu\left(\frac{\partial^2 v}{\partial x^2} + \frac{\partial^2 v}{\partial y^2}\right) \\ &\quad - (\phi-\phi_c)\left(\rho_p - \rho_{f_0}\right)g + (1-\phi_c)\rho_{f_0}(T-T_c)g + \sigma B^2\left(u\sin\lambda\cos\lambda - v\cos^2\lambda\right)\end{aligned} \tag{4.26}$$

$$u\frac{\partial T}{\partial x} + v\frac{\partial T}{\partial y} = \alpha\left(\frac{\partial^2 T}{\partial x^2} + \frac{\partial^2 T}{\partial y^2}\right) + \frac{(\rho c)_p}{(\rho c)_f}\left[D_B\left\{\frac{\partial \phi}{\partial x}\cdot\frac{\partial T}{\partial x} + \frac{\partial \phi}{\partial y}\cdot\frac{\partial T}{\partial y}\right\} + (D_T/T_c)\left\{\left(\frac{\partial T}{\partial x}\right)^2 + \left(\frac{\partial T}{\partial y}\right)^2\right\}\right] \tag{4.27}$$

$$u\frac{\partial \phi}{\partial x} + v\frac{\partial \phi}{\partial y} = D_B\left\{\frac{\partial^2 \phi}{\partial x^2} + \frac{\partial^2 \phi}{\partial y^2}\right\} + \left(\frac{D_T}{T_c}\right)\left\{\frac{\partial^2 T}{\partial x^2} + \frac{\partial^2 T}{\partial y^2}\right\} \tag{4.28}$$

The stream function and vorticity are defined as follows:

$$u = \frac{\partial \psi}{\partial y}, \quad v = -\frac{\partial \psi}{\partial x}, \quad \omega = \frac{\partial v}{\partial x} - \frac{\partial u}{\partial y} \tag{4.29}$$

The stream function satisfies the continuity (Eq. 4.24). The vorticity equation is obtained by eliminating the pressure between the two momentum equations, i.e., by taking thc y-derivative of Eq. (4.25) and subtracting from it the x-derivative of Eq. (4.26). Also, the following nondimensional variables should be introduced:

$$X=\frac{x}{L},\quad Y=\frac{y}{L},\quad \Omega=\frac{\omega L^2}{\alpha},\quad \Psi=\frac{\psi}{\alpha},\quad \Theta=\frac{T-T_c}{T_h-T_c},\quad \Phi=\frac{\phi-\phi_c}{\phi_h-\phi_c},\quad U=\frac{uL}{\alpha},\quad V=\frac{vL}{\alpha} \tag{4.30}$$

By using these dimensionless parameters the equations become:

$$\left[\frac{\partial\Psi}{\partial Y}\frac{\partial\Omega}{\partial X}-\frac{\partial\Psi}{\partial X}\frac{\partial\Omega}{\partial Y}\right]=Pr\left(\frac{\partial^2\Omega}{\partial X^2}+\frac{\partial^2\Omega}{\partial Y^2}\right)+PrRa\left(\frac{\partial\Theta}{\partial X}-Nr\frac{\partial\Theta}{\partial X}\right)$$
$$+Ha^2Pr\left(-\frac{\delta V}{\delta Y}\tan\lambda+\frac{\delta U}{\delta Y}\tan^2\lambda+\frac{\delta U}{\delta X}\tan\lambda-\frac{\delta V}{\delta X}\right) \tag{4.31}$$

$$\frac{\partial\Psi}{\partial Y}\frac{\partial\Theta}{\partial X}-\frac{\partial\Psi}{\partial X}\frac{\partial\Theta}{\partial Y}=\left(\frac{\partial^2\Theta}{\partial X^2}+\frac{\partial^2\Theta}{\partial Y^2}\right)+Nb\left(\frac{\partial\Phi}{\partial X}\frac{\partial\Theta}{\partial X}+\frac{\partial\Phi}{\partial Y}\frac{\partial\Theta}{\partial Y}\right)+Nt\left(\left(\frac{\partial\Theta}{\partial X}\right)^2+\left(\frac{\partial\Theta}{\partial Y}\right)^2\right) \tag{4.32}$$

$$\frac{\partial\Psi}{\partial Y}\frac{\partial\Phi}{\partial X}-\frac{\partial\Psi}{\partial X}\frac{\partial\Phi}{\partial Y}=\frac{1}{Le}\left(\frac{\partial^2\Phi}{\partial X^2}+\frac{\partial^2\Phi}{\partial Y^2}\right)+\frac{Nt}{Nb\,Le}\left(\frac{\partial^2\Theta}{\partial X^2}+\frac{\partial^2\Theta}{\partial Y^2}\right) \tag{4.33}$$

$$\frac{\partial^2\Psi}{\partial X^2}+\frac{\partial^2\Psi}{\partial Y^2}=-\Omega \tag{4.34}$$

where the thermal Rayleigh number, the buoyancy ratio, the Prandtl number, the Brownian motion parameter, the thermophoretic parameter, the Lewis number, and the Hartmann number of a nanofluid are defined as follows:

$$\begin{aligned}
&Ra=(1-\phi_c)\rho_{f_0}g\beta L^3(T_h-T_c)/(\mu\alpha),\\
&Nr=\left(\rho_p-\rho_0\right)(\phi_h-\phi_c)/\left[(1-\phi_c)\rho_{f_0}\beta L(T_h-T_c)\right],\\
&Pr=\mu/\rho_f\alpha,\\
&Nb=(\rho c)_p D_B(\phi_h-\phi_c)/((\rho c)_f\alpha),\\
&Nt=(\rho c)_p D_T(T_h-T_c)/\left[(\rho c)_f\alpha T_c\right],\\
&Le=\alpha/D_B,\\
&Ha=LB_x\sqrt{\sigma/\mu}
\end{aligned} \tag{4.35}$$

4.2.2.2 Mixed convection

To obtain the governing equations for force convection, the following nondimensional variables should be introduced:

$$X=\frac{x}{L},\quad Y=\frac{y}{L},\quad \Omega=\frac{\omega L}{u_r},\quad \Psi=\frac{\psi}{u_rL},\quad U=\frac{u}{u_r},\quad V=\frac{v}{u_r},\quad \Theta=\frac{T-T_c}{T_h-T_c},\quad \Phi=\frac{\phi-\phi_c}{\phi_h-\phi_c} \tag{4.36}$$

So, the governing equations are:

$$\left[\frac{\partial\Psi}{\partial Y}\frac{\partial\Omega}{\partial X}-\frac{\partial\Psi}{\partial X}\frac{\partial\Omega}{\partial Y}\right]=\frac{1}{Re}\left(\frac{\partial^2\Omega}{\partial X^2}+\frac{\partial^2\Omega}{\partial Y^2}\right)+\frac{Gr}{Re^2}\left(\frac{\partial\Theta}{\partial X}-Nr\frac{\partial\Theta}{\partial X}\right)$$
$$+\frac{Ha^2}{Re}\left(-\frac{\delta V}{\delta Y}\tan\lambda+\frac{\delta U}{\delta Y}\tan^2\lambda+\frac{\delta U}{\delta X}\tan\lambda-\frac{\delta V}{\delta X}\right) \tag{4.37}$$

$$\frac{\partial\Psi}{\partial Y}\frac{\partial\Theta}{\partial X}-\frac{\partial\Psi}{\partial X}\frac{\partial\Theta}{\partial Y}=\frac{1}{RePr}\left(\frac{\partial^2\Theta}{\partial X^2}+\frac{\partial^2\Theta}{\partial Y^2}\right)$$
$$+\frac{Nb}{Re}\left(\frac{\partial\Phi}{\partial X}\frac{\partial\Theta}{\partial X}+\frac{\partial\Phi}{\partial Y}\frac{\partial\Theta}{\partial Y}\right)+\frac{Nt}{Re}\left(\left(\frac{\partial\Theta}{\partial X}\right)^2+\left(\frac{\partial\Theta}{\partial Y}\right)^2\right) \tag{4.38}$$

$$\frac{\partial\Psi}{\partial Y}\frac{\partial\Phi}{\partial X}-\frac{\partial\Psi}{\partial X}\frac{\partial\Phi}{\partial Y}=\frac{1}{LeRe}\left(\frac{\partial^2\Phi}{\partial X^2}+\frac{\partial^2\Phi}{\partial Y^2}\right)+\frac{Nt}{NbLeRe}\left(\frac{\partial^2\Theta}{\partial X^2}+\frac{\partial^2\Theta}{\partial Y^2}\right) \tag{4.39}$$

$$\frac{\partial^2\Psi}{\partial X^2}+\frac{\partial^2\Psi}{\partial Y^2}=-\Omega \tag{4.40}$$

where the thermal Reynolds number, buoyancy ratio, Prandtl number, Brownian motion parameter, thermophoretic parameter, Lewis number, Hartmann number, Grashof number, and Richardson number of the nanofluid are defined as:

$$\begin{aligned}
&Re=\frac{\rho_f L u_r}{\mu},\\
&Nr=\left(\rho_p-\rho_0\right)(\phi_h-\phi_c)/\left[(1-\phi_c)\rho_{f_0}\beta L(T_h-T_c)\right],\\
&Pr=\mu/\rho_f\alpha,\\
&Nb=(\rho c)_p D_B(\phi_h-\phi_c)/((\rho c)_f\upsilon),\\
&Nt=(\rho c)_p D_T(T_h-T_c)/\left[(\rho c)_f\upsilon T_c\right],\\
&Le=\upsilon/D_B,\\
&Ha=LB_x\sqrt{\sigma/\mu},\ Gr=(1-\phi_c)\rho_{f_0}g\beta L^3\rho_f(T_h-T_c)/(\mu^2),\ Ri=\frac{Gr}{Re^2}
\end{aligned} \tag{4.41}$$

4.2.3 APPLICATION OF THE CVFEM FOR MHD NANOFLUID FLOW AND HEAT TRANSFER

4.2.3.1 Effects of MHD on copper-water nanofluid flow and heat transfer in an enclosure with an inclined elliptic hot cylinder

4.2.3.1.1 Problem definition

The schematic diagram and the mesh of the semiannulus enclosure used in the present CVFEM program are shown in Fig. 4.1 [30]. The system consists of a circular enclosure with radius r_{out}, within which an inclined elliptic cylinder is located and rotates from $\gamma=0°$ to 90°. T_h and T_c are the constant temperatures of the inner and

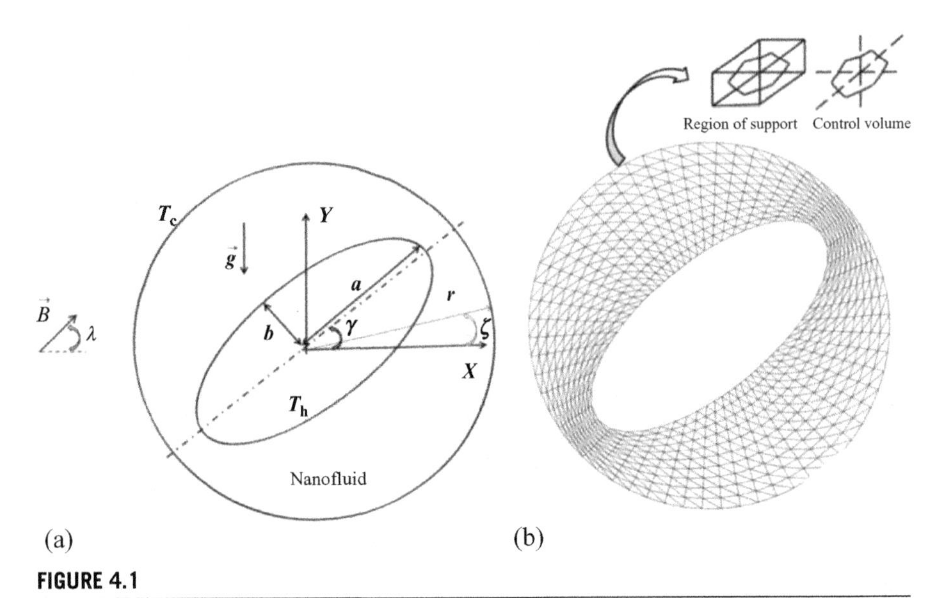

FIGURE 4.1

Geometry and the boundary conditions (a) with the mesh of the enclosure (b) considered in this work.

outer cylinders, respectively ($T_h > T_c$). Setting a as the major axis and b as the minor axis of the elliptic cylinder, the eccentricity (ε) of the inner cylinder is defined as

$$\varepsilon = \sqrt{a^2 - b^2}/a \text{ or } b = \sqrt{1-\varepsilon^2} \cdot a \quad (4.42)$$

For the inner ellipse in this study, the eccentricity and the major axis are 0.9 and 0.8L, respectively. Also, it is also assumed that the uniform magnetic field ($\vec{B} = B_x \vec{e}_x + B_y \vec{e}_y$) with a constant magnitude $B = \sqrt{B_x^2 + B_y^2}$ is applied, where $\vec{e}_x$ and $\vec{e}_y$ are unit vectors in the Cartesian coordinate system. The orientation of the magnetic field form an angle λ with a horizontal axis such that $\lambda = B_x/B_y$. The electric current J and the electromagnetic force F are defined by $J = \sigma\left(\vec{V} \times \vec{B}\right)$ and $F = \sigma\left(\vec{V} \times \vec{B}\right) \times \vec{B}$, respectively. In this study λ is equal to zero.

4.2.3.1.2 Effects of active parameters

MHD natural convection heat transfer between a circular enclosure and an elliptic cylinder filled with nanofluid is investigated numerically using the CVFEM. Constant eccentricity ($\varepsilon = 0.9$), major axis ($a = 0.8L$), and Prandtl ($Pr = 6.2$) at different Rayleigh numbers ($Ra = 10^3$, 10^4, 10^5), Hartmann numbers ($Ha = 0$, 20, 60, and 100), inclined angle of the inner cylinder ($\gamma = 0°$, 30°, 60°, 90°), and volume fraction of the nanoparticles ($\phi = 0\%$, 2%, 4%, and 6%) are calculated.

Isotherms and streamlines for different values of Ra, γ, and Ha are shown in Figs. 4.2 and 4.3. At $Ra = 10^3$, the isotherms are parallel to each other and take the form of the inner and outer walls; the stream function magnitude is relatively small, which indicates the domination of the conduction heat transfer mechanism. Increasing the inclination angle leads to an increase in the absolute value of the maximum stream function ($|\Psi_{max}|$) at this Rayleigh number. At $\gamma = 0°$ the streamlines and isotherms are symmetrical with respect to the vertical center line of the enclosure. Each pair has two cells. The top vortex is stronger because in this area the hot

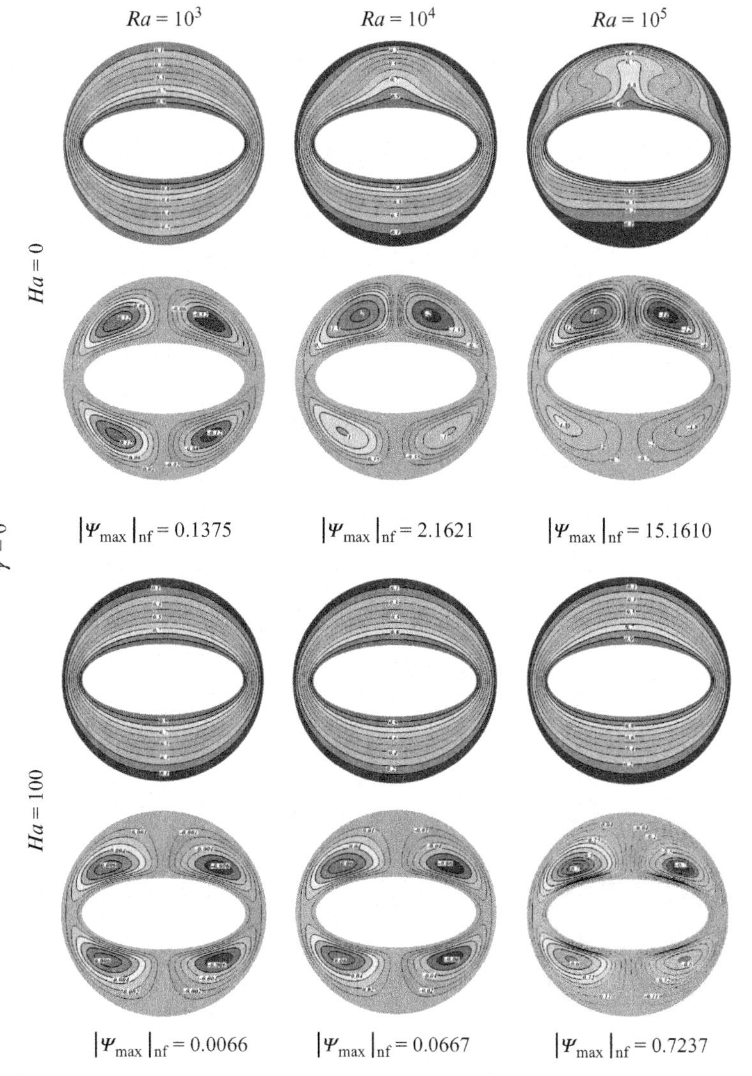

FIGURE 4.2

Isotherm (up) and streamline (down) contours for different values of Ra; $\gamma = 0°$, $30°$, and $45°$; and Ha at $\phi = 0.06$.

Continued

surface is located beneath the cold one, which helps the flow circulate, whereas the arrangement of the cold wall under the hot cylinder at the bottom of the enclosure resists flow circulation. As $\gamma = 0°$ increases, these two pairs of cells merge together and form two single cells at different locations inside the enclosure. At $\gamma = 90°$, again the streamline pattern becomes symmetrical with respect to the vertical centerline of

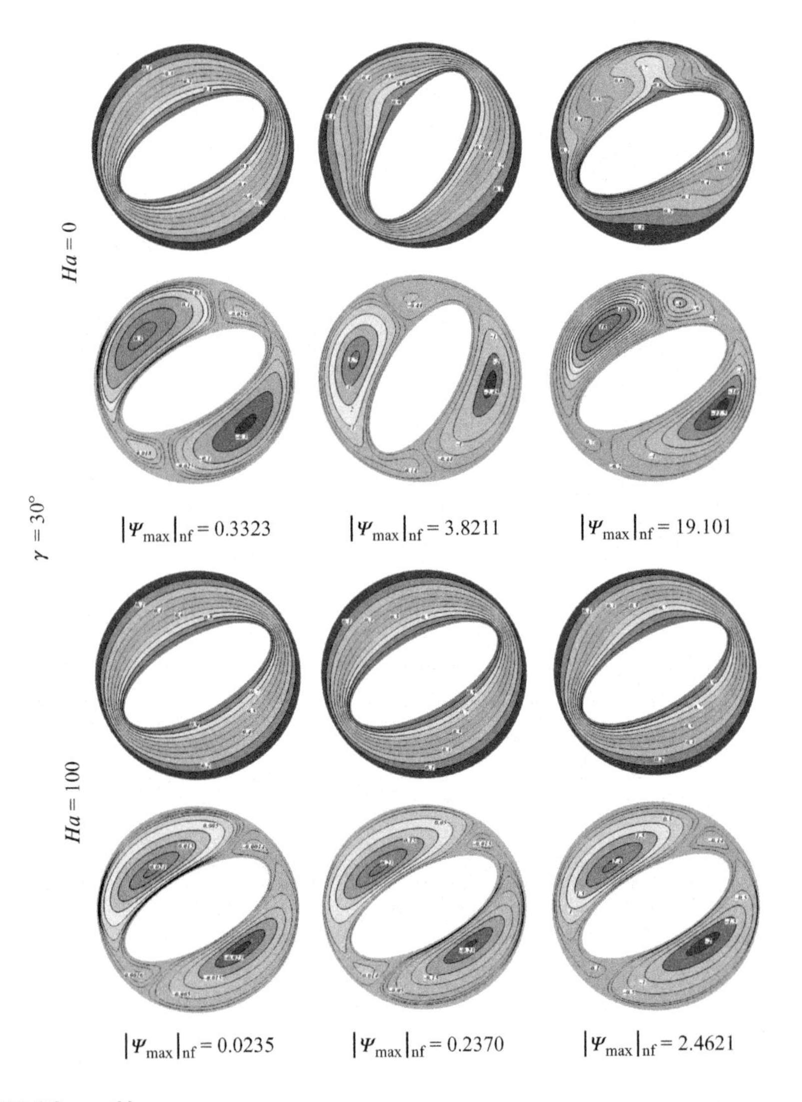

FIGURE 4.2, cont'd

Continued

the enclosure. At $Ra = 10^4$ a thermal plume starts to form over the hot elliptic cylinder. In addition, the stream function values start to grow, which shows that the convection heat transfer mechanism becomes comparable with the conduction mechanism. As $\gamma = 0°$ increases, the pattern of the streamline is similar to that of $Ra = 10^3$, but the size of the vortices at the bottom of the enclosure. In addition, the temperature contour becomes stratified beneath the hot cylinder. With an increase in γ to 30°, a secondary vortex appears at the top of the enclosure and

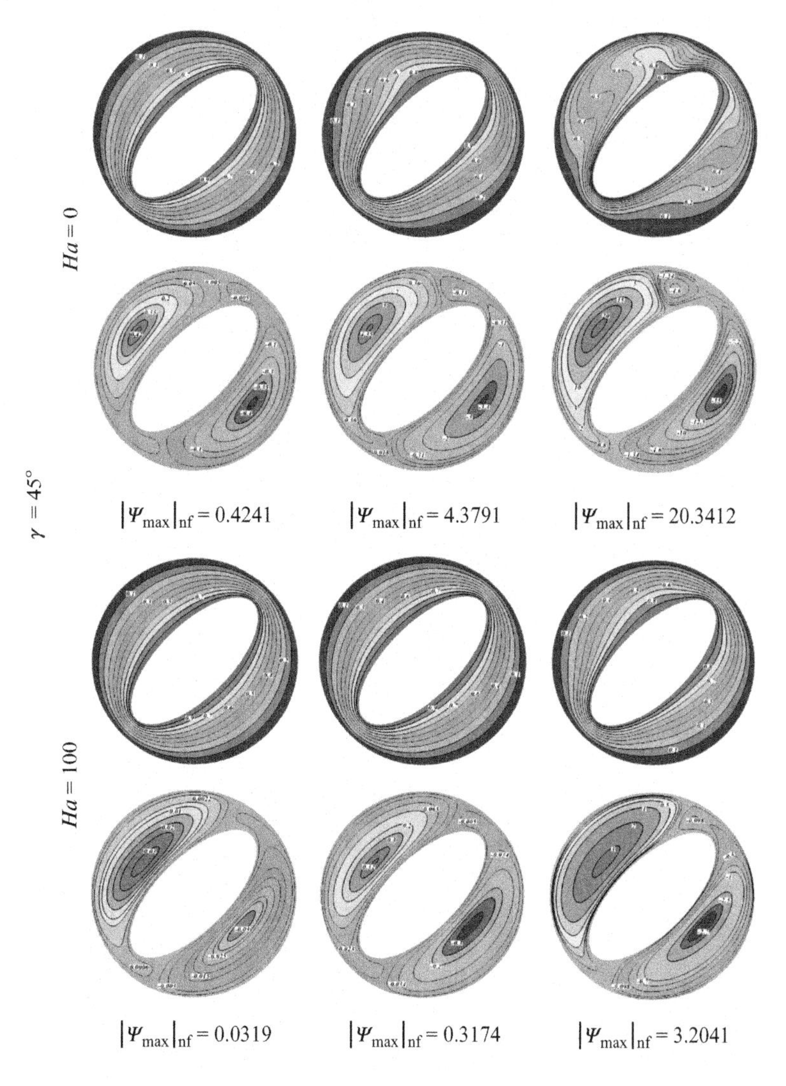

FIGURE 4.2, cont'd

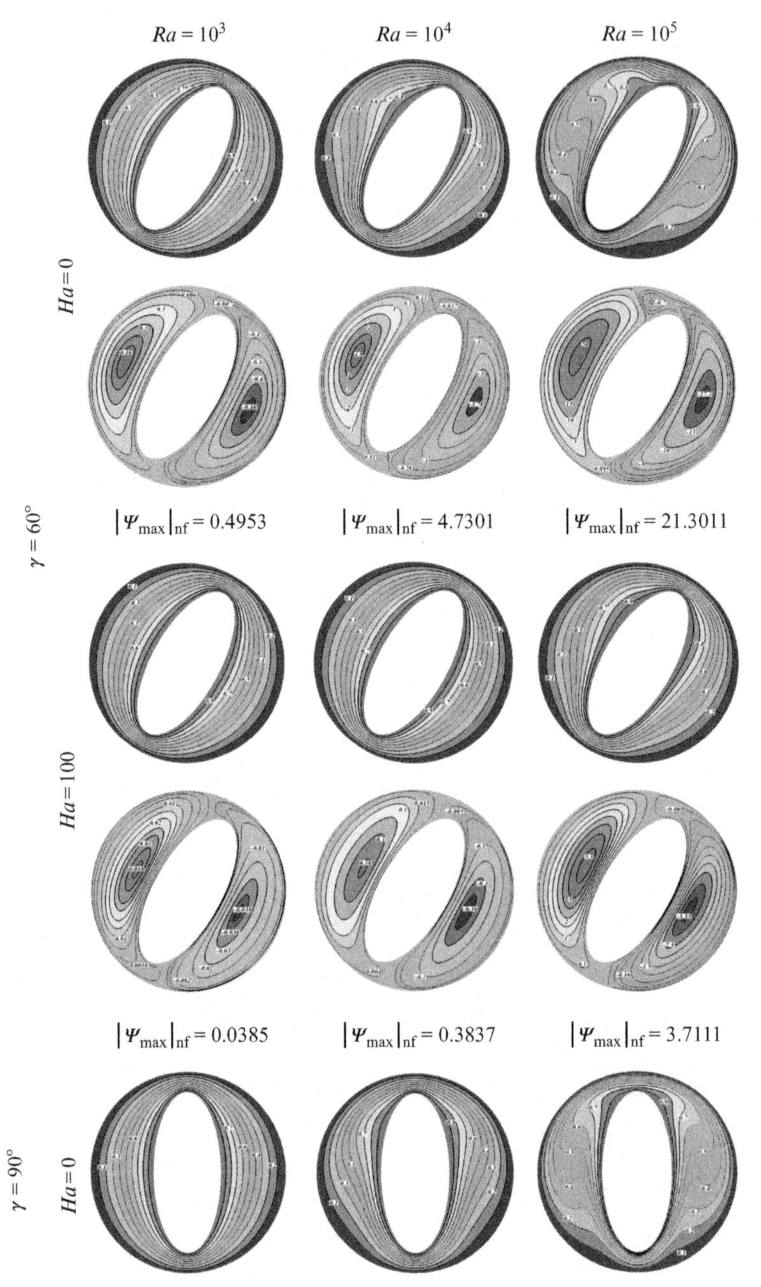

FIGURE 4.3

Isotherm (up) and streamline (down) contours for different values of Ra; $\gamma = 60°$ and $90°$; and Ha at $\phi = 0.06$.

Continued

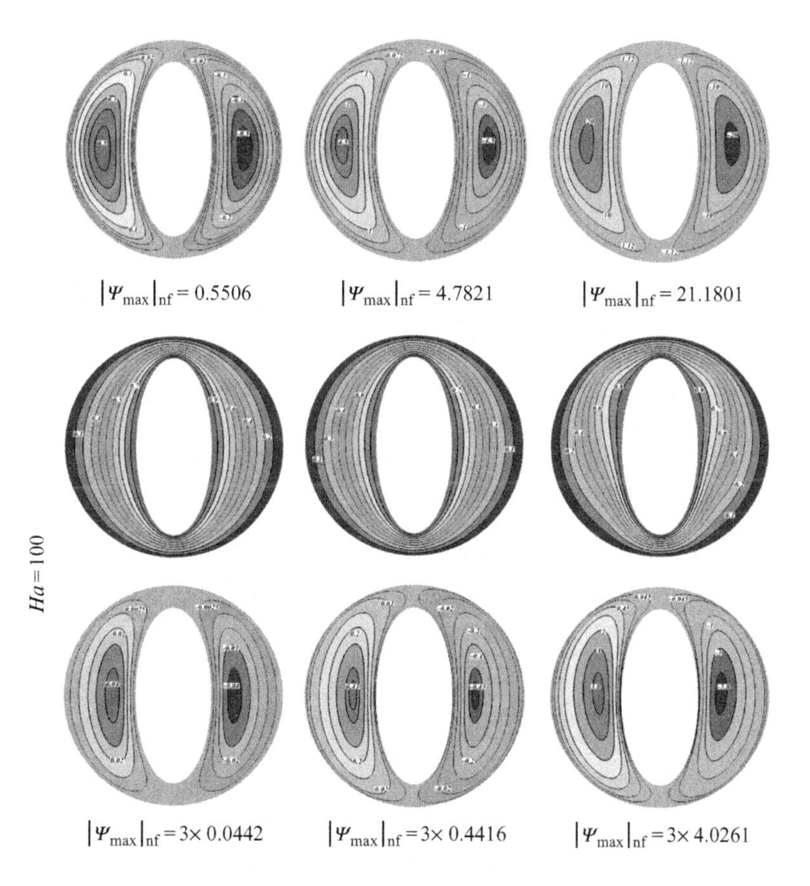

FIGURE 4.3, cont'd

the thermal plume slants to the left because of more available space in this area. As the inclination angle of the inner cylinder increases further, this secondary vortex disappears and the streamlines show two main vortices in the enclosure. Also, it the effect of an increasing inclination angle on $|\Psi_{max}|$ becomes less pronounced at $\gamma > 30°$. When the Rayleigh number increases to $Ra = 10^5$, isotherms are totally distorted at the top of the enclosure but stratified at the bottom of the enclosure, which shows the heat transfer mechanism is dominated by convection. Thermal boundary layer thickness over the cold wall becomes more thinner due to existence of thermal plume. The secondary vortex exists at $\gamma = 30°$ and $60°$,which can make the thermal plume slant to the left. Also, the maximum value of $|\Psi_{max}|$ occurs at $\gamma = 60°$.

It is worthwhile to mention that the effect of a magnetic field decreases the value of the velocity magnitude throughout the enclosure because the presence of the magnetic field introduces a force called the Lorentz force, which acts against the flow if the magnetic field is applied in the normal direction. This type of resisting force slows the fluid velocity. Increasing the Hartmann number makes the core of vortices move toward the horizontal center line. Also, the magnetic field makes the thermal plume to disappear and makes the isotherms parallel to each other through the domination of the conduction mode of heat transfer.

Figure 4.4 shows the distribution of local Nusselt numbers along the surface of the outer circular wall at different inclination angles, Rayleigh numbers, and Hartmann number. As the Rayleigh number increases the local Nusselt number increases because of an enlarged convection effect. At $Ra = 10^3$, the Nu_{loc} profile is nearly symmetrical with respect to the horizontal center line. As the Rayleigh number increases (e.g., $Ra = 10^4$ and 10^5), the Nu_{loc} profile is no longer symmetrical and the local Nusselt number is considerably small along the bottom wall of the enclosure. Increasing the Hartmann number causes the local Nusselt number to decrease. These local Nusselt number profiles are more complex because of the presence of a thermal plume in the vicinity of the top wall of the enclosure.

The corresponding polynomial representation of such a Nusselt number model is as follows:

$$\begin{aligned} Nu &= a_{13} + a_{23}Y_1 + a_{33}Y_2 + a_{43}{Y_1}^2 + a_{53}{Y_2}^2 + a_{63}Y_1Y_2 \\ Y_1 &= a_{11} + a_{21}Ra^* + a_{31}Ha^* + a_{41}Ra^{*2} + a_{51}Ha^{*2} + a_{61}Ha^*Ra^* \\ Y_2 &= a_{12} + a_{22}\gamma + a_{32}\phi + a_{42}\gamma^2 + a_{52}\phi^2 + a_{62}\gamma\phi \end{aligned} \tag{4.43}$$

Also, values of a_{ij} can be found in Table 4.1. For example, a_{21} equals -1.52352.

Effects of the volume fraction of nanoparticles, inclination angle, Hartmann number, and Rayleigh number on the average Nusselt number are shown in Figs. 4.5 and 4.6. Adding nanoparticles leads to an increase in the thermal boundary layer thickness; the Nusselt number increases because it is the Nusselt number is calculated by multiplying the temperature gradient and the thermal conductivity ratio. A reduction in the temperature gradient due to the presence of nanoparticles is much smaller than the thermal conductivity ratio. An increase in the Rayleigh number is associated with an increase in the heat transfer and the Nusselt number. This is caused by stronger convective heat transfer at higher Rayleigh numbers. Increasing the Hartmann number causes the Lorenz force to increase and substantially suppresses the convection. Therefore the Nusselt number has a reveres relationship with the Hartmann number. These figures also show that the inclination angle has a direct relationship with the average Nusselt number. As seen in Fig. 4.6, the inclination angle has no significant effect on the average Nusselt number at high Hartmann numbers.

To estimate the enhanced heat transfer between $\phi = 0.06$ and a pure fluid (base fluid), the heat transfer enhancement is defined as:

$$En = \frac{Nu(\phi = 0.06) - Nu(\text{ basefluid})}{Nu(\text{ basefluid})} \times 100 \tag{4.44}$$

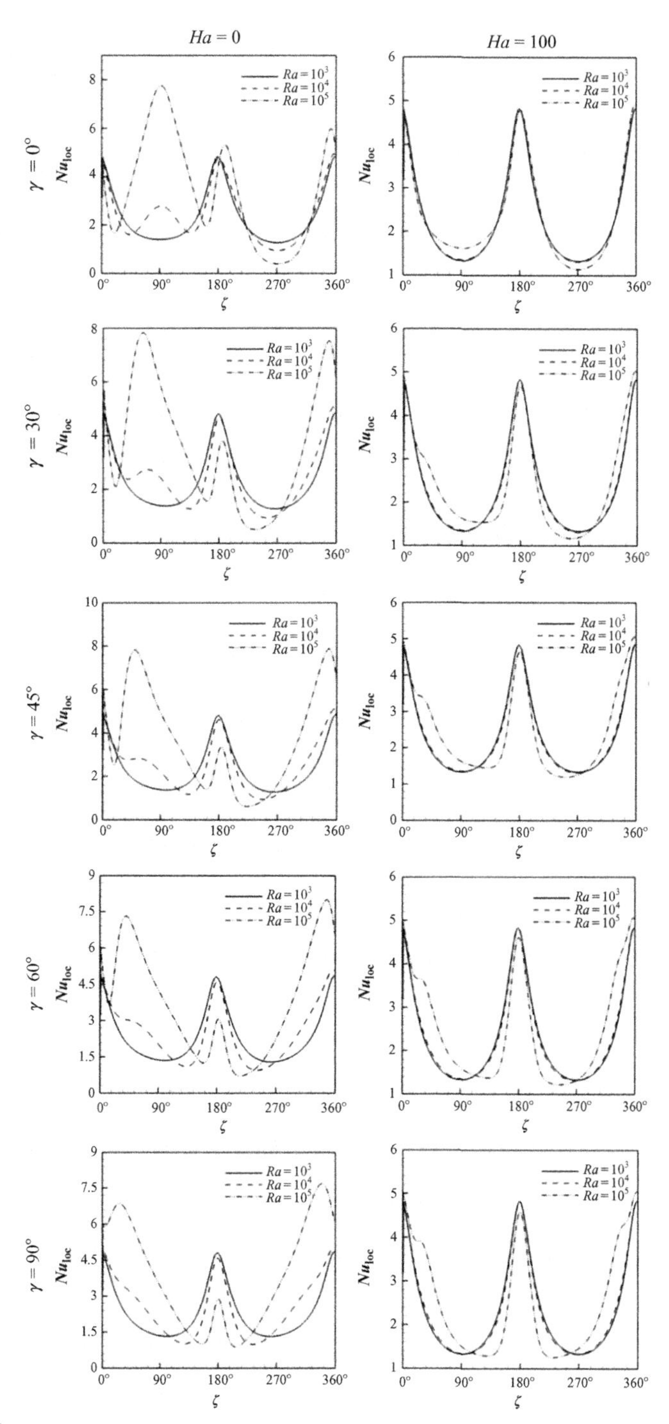

FIGURE 4.4

Effects of the inclination angle, Hartmann number, and Rayleigh number for copper-water ($\phi = 0.06$) nanofluids on the local Nusselt number.

Table 4.1 Constant Coefficients (a_{ij}) for Using Eq. (4.43)

a_{ij}	$i=1$	$i=2$	$i=3$	$i=4$	$i=5$	$i=6$
$j=1$	4.302581	−1.52352	1.51045	0.267076	0.380695	−0.59522
$j=2$	2.170725	−0.05626	6.043703	0.071014	−2.55729	−0.09138
$j=3$	3.084162	−2.27497	−0.06891	0.556024	0.155194	0.131637

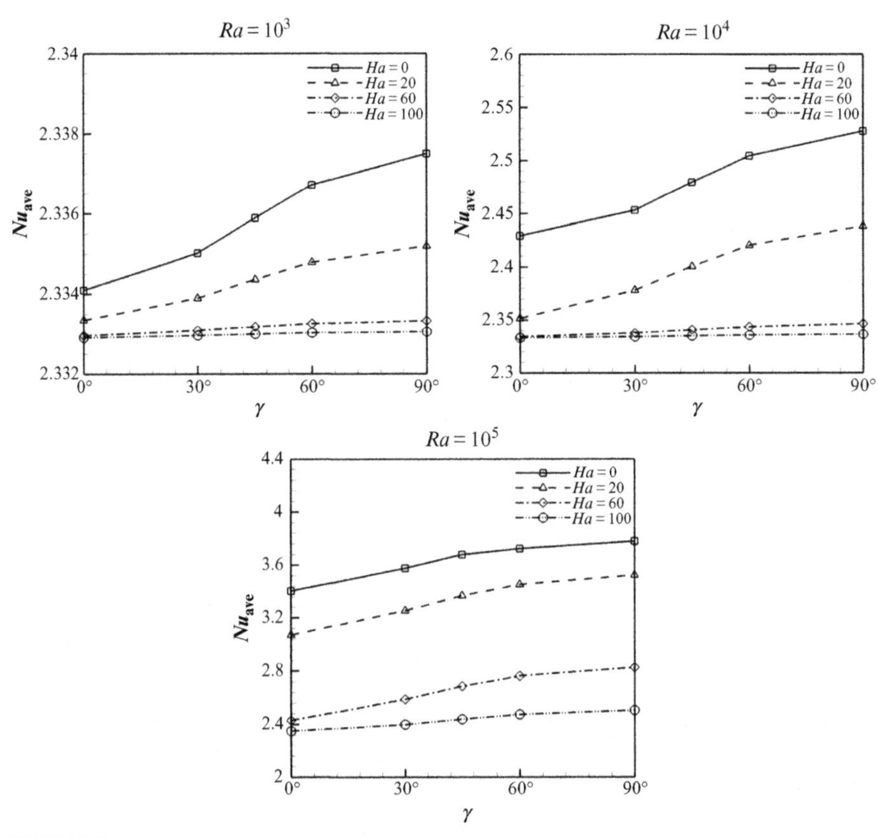

FIGURE 4.5

Effects of the inclination angle, Hartmann number, and Rayleigh number for copper-water ($\phi = 0.06$) nanofluids on the average Nusselt number.

The heat transfer enhancement ratio with the addition of nanoparticles at different values of Ha, γ, and Ra is shown in Fig. 4.7. In general, increasing the Rayleigh number decreases the heat transfer because of the domination of the conduction mechanism at low Rayleigh numbers. Also, the Hartmann number is an increasing function of En. As the inclination angle increases, En decreases, except at $Ra = 10^4$.

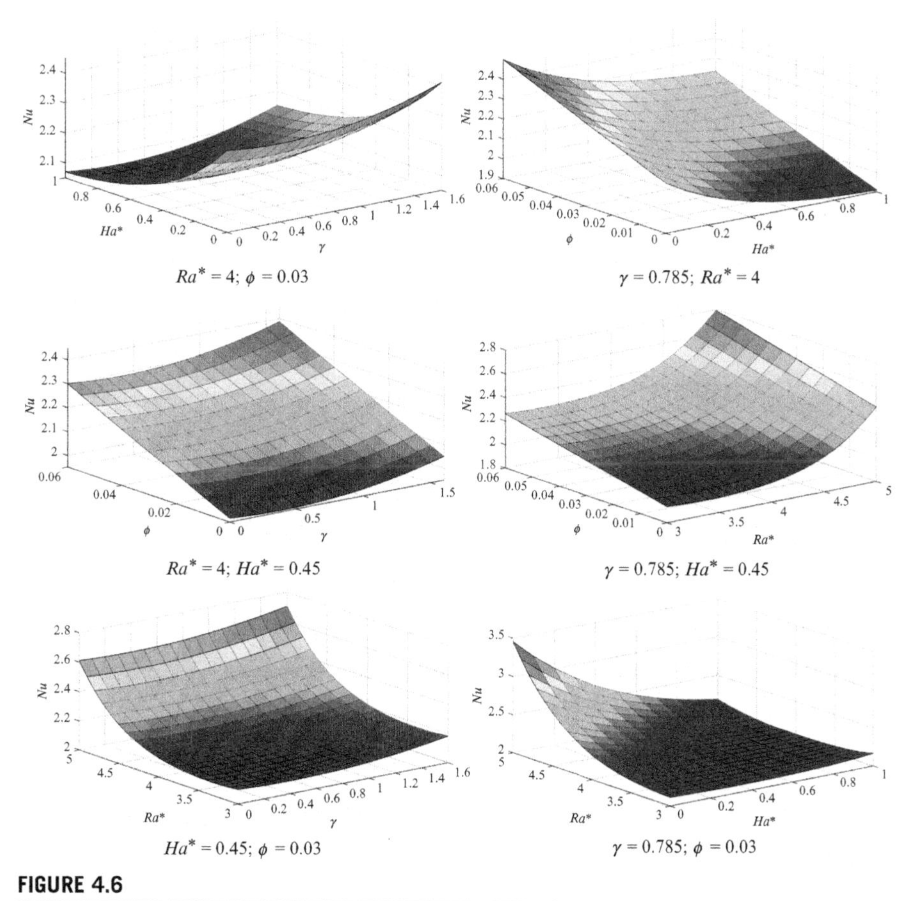

FIGURE 4.6

Variation of Nu_{ave} for various input parameters.

At $Ra = 10^4$, maximum values of En occur at $\gamma = 30°$ and $45°$ for $Ha = 0$ and 100, respectively.

4.2.3.2 Natural convection heat transfer in a cavity with a sinusoidal wall filled with CuO-water nanofluid in the presence of a magnetic field

4.2.3.2.1 Problem definition

A schematic of the problem and the related boundary conditions, as well as the mesh of the enclosure, that are used in the present CVFEM program are shown in Fig. 4.8 [31]. The enclosure has a width-to-height aspect ratio of 2. The shape of the left sinusoidal wall profile is assumed to follow the pattern:

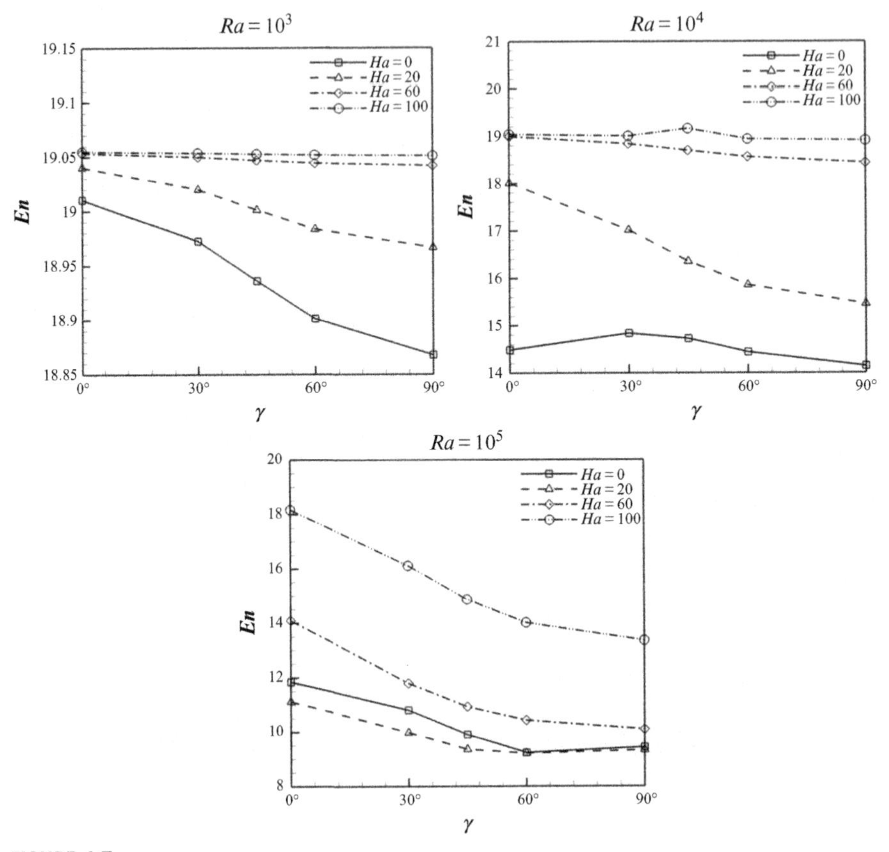

FIGURE 4.7

Effects of the inclination angle, Hartmann number, and Rayleigh number on the enhancement of the heat transfer ratio due to the addition of nanoparticles when $Pr = 6.2$ (in a copper-water fluid).

$$\frac{x}{H} = 1 - \left\{ a\left(1 + \sin\left(\pi\frac{y}{H} - \pi/2\right)\right)\right\} \tag{4.45}$$

where a is the dimensionless amplitude of the sinusoidal wall. The left sinusoidal wall is under constant heat flux (q''), the right flat wall is maintained at a constant temperature (T_c), and the two other walls are thermally insulated. It is also assumed that a uniform magnetic field ($\vec{B} = B_x\vec{e}_x + B_y\vec{e}_y$) of constant magnitude $B = \sqrt{B_x^2 + B_y^2}$ is applied, where $\vec{e}_x$ and $\vec{e}_y$ are unit vectors in the Cartesian coordinate system. The orientation of the magnetic field forms an angle γ with the horizontal axis such that $\gamma = B_x/B_y$. In this study γ equals zero. The electric current J and the electromagnetic force F are defined by $J = \sigma\left(\vec{V} \times \vec{B}\right)$ and $F = \sigma\left(\vec{V} \times \vec{B}\right) \times \vec{B}$, respectively.

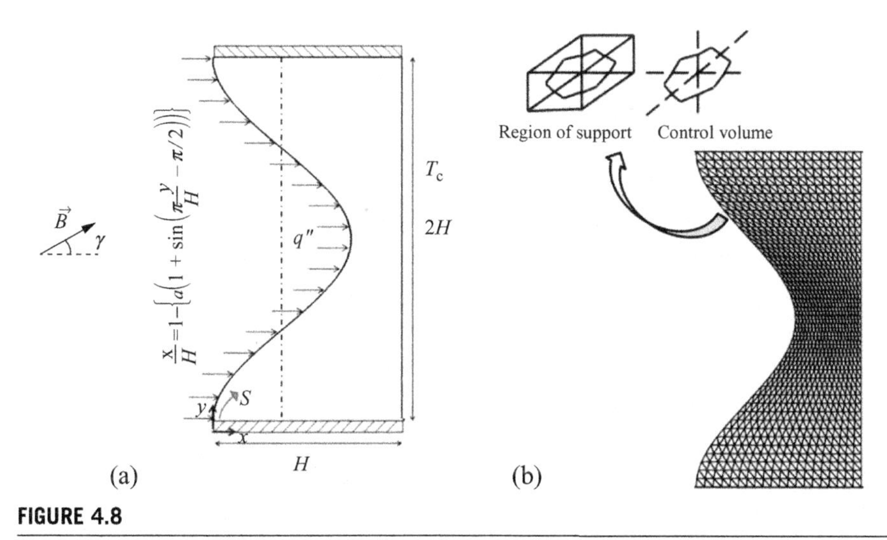

FIGURE 4.8

Geometry and the boundary conditions (a) with the mesh of enclosure (b) considered in this work. (c) A sample triangular element and its corresponding control volume.

4.2.3.2.2 Effects of active parameters

Numerical simulations of natural convection in an enclosure filled with a CuO-water nanofluid were performed using the CVFEM. Various nanoparticle volume fractions ($\phi = 0 - 4\%$), Rayleigh numbers ($Ra = 10^3$, 10^4, and 10^5), dimensionless amplitudes of the sinusoidal wall ($a = 0.1, 0.2$, and 0.3), and Hartmann numbers ($Ha = 0\text{-}100$) at a constant Prandtl number ($Pr = 6.2$) were calculated.

Figure 4.9 shows the effects of the dimensionless amplitude of the sinusoidal wall (a) and Rayleigh number (Ra) in the absence of a magnetic field. At $Ra = 10^3$, for all amplitudes of the sinusoidal wall, the isotherms are nearly smooth curves and nearly parallel to each other; this pattern is characteristic of a conduction-dominant mechanism of heat transfer at low Rayleigh numbers. The space in which to accelerate flow inside the cavity decreases as the amplitude of the sinusoidal wall increases. At $a = 0.3$, one counterclockwise and two clockwise eddies are observed. By increasing a to 0.3, another tiny clockwise vorticity appears. In general, as the Rayleigh number increases, the buoyancy-driven circulations inside the enclosure become stronger, as is clear from the additional distortion that appears in the isotherms.

Isotherm (up) and streamline (down) contours for different values of a and Ra in the presence of a magnetic field are shown in Fig. 4.10. When the magnetic field is imposed on the enclosure, the velocity field is suppressed because of the retarding effect of the Lorenz force. Therefore the intensity of the convection weakens significantly. Increasing the Hartmann number leads to stretch eddies along a vertical line. The magnetic field also makes the isotherms align parallel to each other because of the domination of the conduction mode of heat transfer.

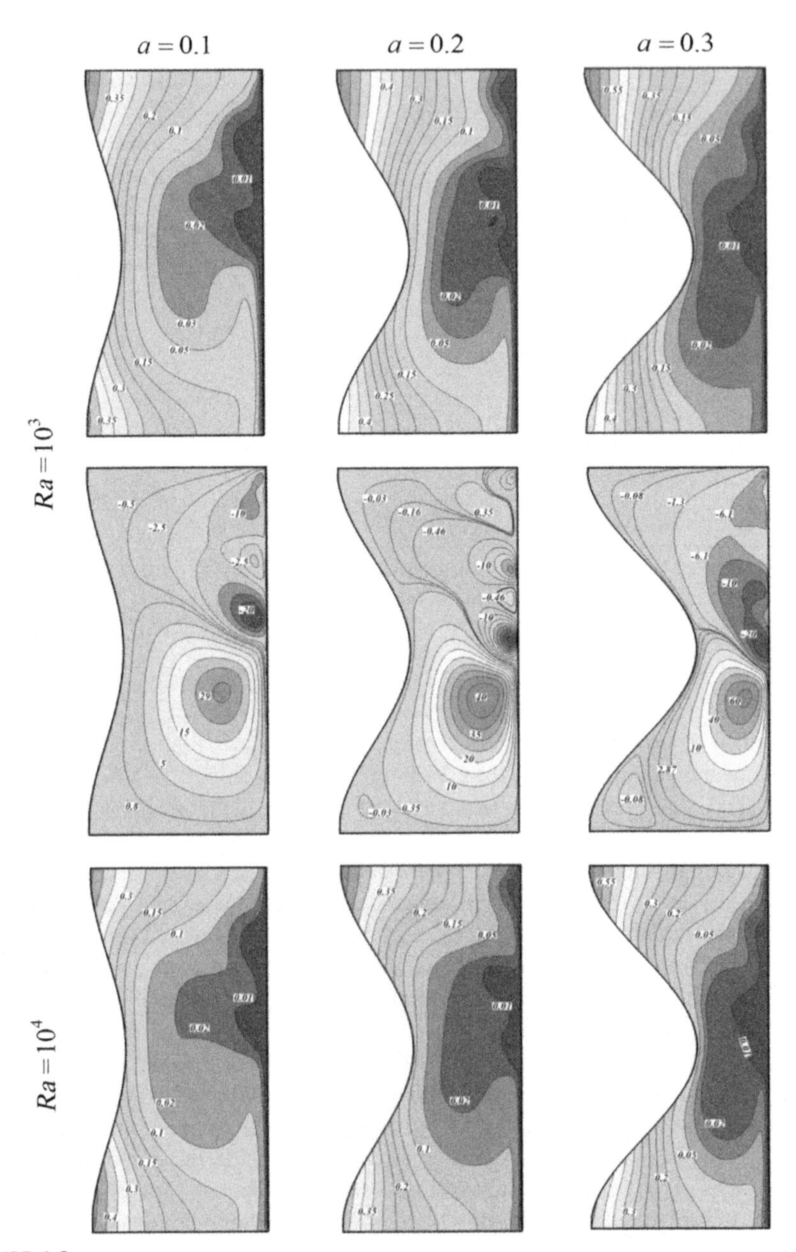

FIGURE 4.9

Isotherm (up) and streamline (down) contours for different values of a and Ra at $Ha=0$, $\phi=0.04$, and $Pr=6.2$.

Continued

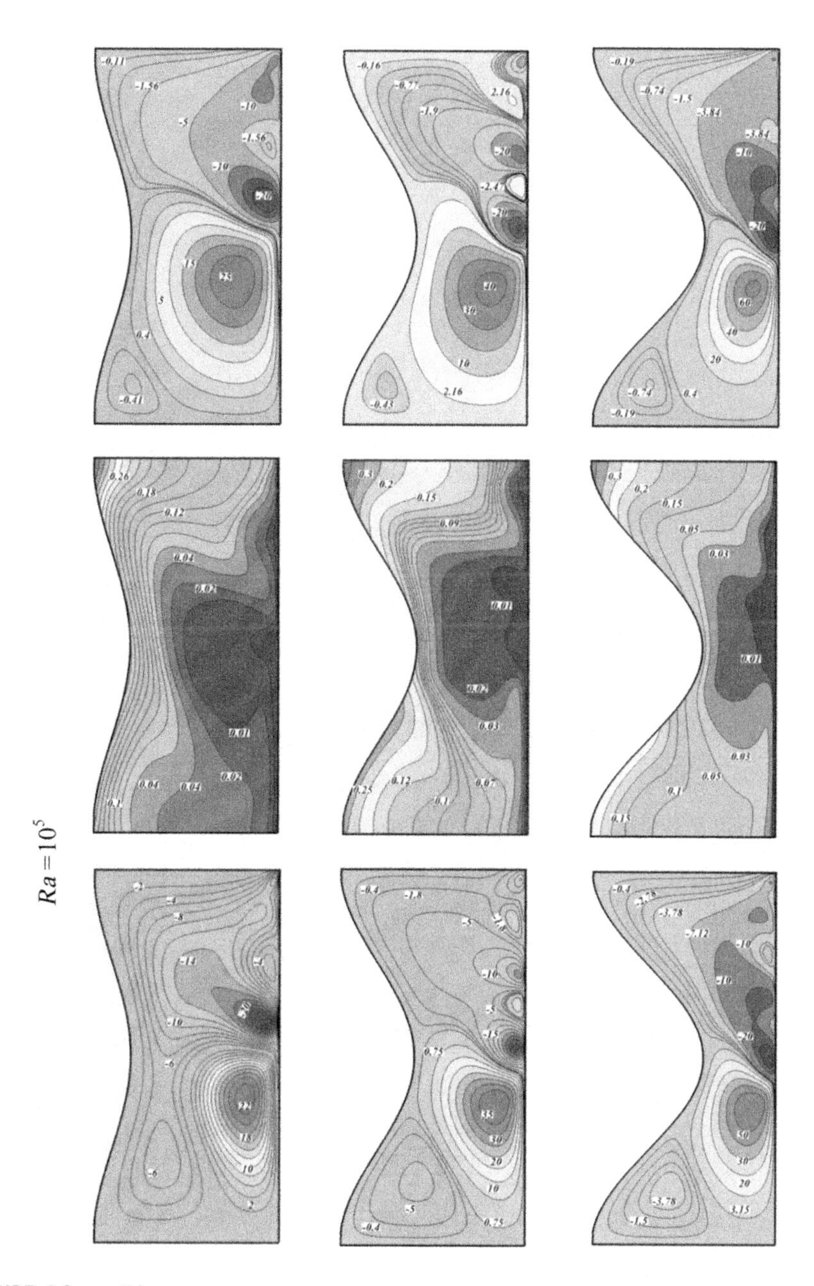

FIGURE 4.9, cont'd

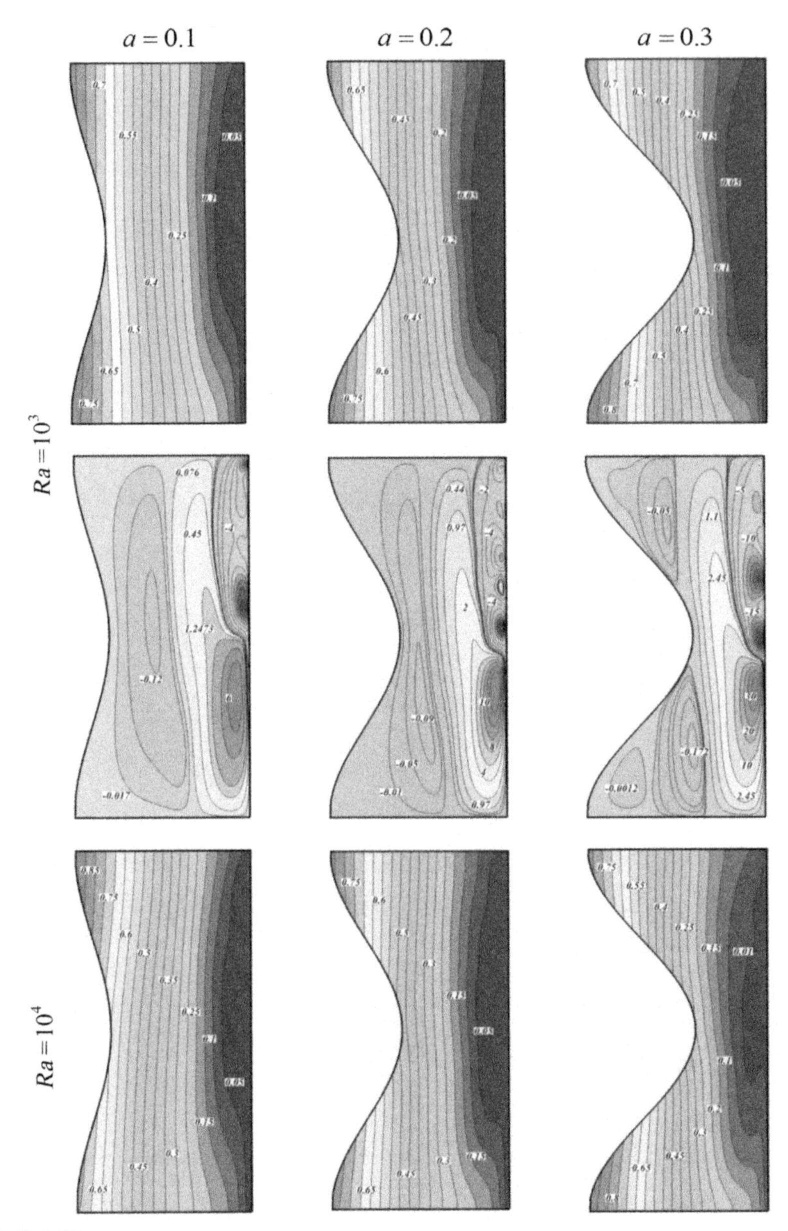

FIGURE 4.10

Isotherm (up) and streamline (down) contours for different values of a and Ra at $Ha=100$, $\phi=0.04$, and $Pr=6.2$.

Continued

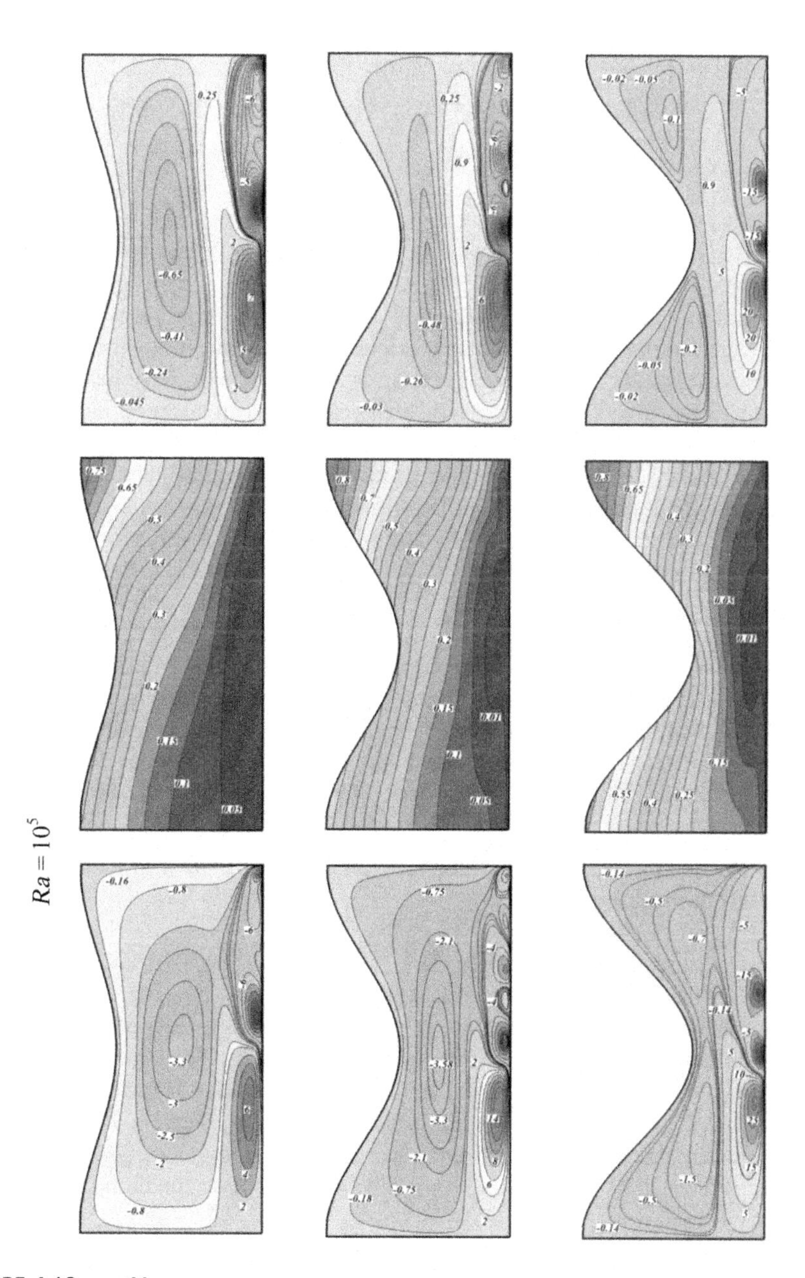

FIGURE 4.10, cont'd

The corresponding polynomial representation of the Nusselt number is:

$$\begin{aligned} Nu &= a_{13} + a_{23}Y_1 + a_{33}Y_2 + a_{43}Y_1^2 + a_{53}Y_2^2 + a_{63}Y_1Y_2 \\ Y_1 &= a_{11} + a_{21}Ha^* + a_{31}\phi + a_{41}Ha^{*2} + a_{51}\phi^2 + a_{61}Ha^*\phi \\ Y_2 &= a_{12} + a_{22}Ra^* + a_{32}a + a_{42}Ra^{*2} + a_{52}a^2 + a_{62}Ra^*a \end{aligned} \tag{4.46}$$

where $Ha^* = Ha/100$ and $Ra^* = \log(Ra)$. Also, values for a_{ij} are listed in Table 4.2. For example, a_{21}, equals −6.49034.

Figures 4.11 and 4.12 show the effects of active parameters on the average Nusselt number. Increasing the Rayleigh number is associated with an increase in the heat transfer and the Nusselt number. This is because there is a stronger convective heat transfer at higher Rayleigh numbers. Although the thermal boundary layer thickness increases with an increase in the volume fraction of nanoparticles, the Nusselt number increases as ϕ is augmented. As the volume fraction of nanoparticles increases, thermal diffusivity increases. So, the high values of thermal diffusivity cause the boundary thickness to increase and, accordingly, decrease $1/\theta$. The Nusselt number is a function of $1/\theta$ and k_{nf}/k_f. Because the reduction in $1/\theta$ due to the presence of nanoparticles is much smaller than the thermal conductivity ratio; therefore the Nusselt number is augmented by increasing the nanoparticle volume fraction. Increasing the Hartmann number causes the Lorenz force to increase and leads to a substantial suppression of the convection. So, the Nusselt number has a reverse relationship with the Hartmann number. As the dimensionless amplitude of the sinusoidal wall increases, space in which to accelerate the flow inside the cavity decreases. Therefore the thermal boundary layer thickness decreases and, in turn, $1/\theta$ increases. This is the reason the Nusselt number increases as a is enhanced.

4.2.3.3 MHD effect on natural convection heat transfer in an inclined, L-shaped enclosure

4.2.3.3.1 Problem definition

The physical model, along with important geometrical parameters and the mesh of the enclosure used in the present CVFEM program, are shown in Fig. 4.13 [32]. The width and height of the enclosure are H. The right and top walls of the enclosure are maintained at a constant cold temperature T_c, whereas the inner circular wall is maintained at a constant hot temperature T_h and the two bottom and left walls, with a length of $H/2$, are thermally isolated. Under all instances, the $T_h > T_c$ condition is maintained. To assess the shape of the inner circular and outer rectangular boundaries, which consist of the right and top walls, a superelliptic function can be used:

Table 4.2 Constant Coefficients (a_{ij}) for Using Eq. (4.46)

a_{ij}	$i=1$	$i=2$	$i=3$	$i=4$	$i=5$	$i=6$
$j=1$	4.707421	−6.49034	15.71694	3.436246	0.628678	−6.63023
$j=2$	5.70953	−2.38534	0.866596	0.402887	28.59651	−1.4327
$j=3$	−0.45209	−0.09317	0.376291	0.017439	−0.05352	0.300014

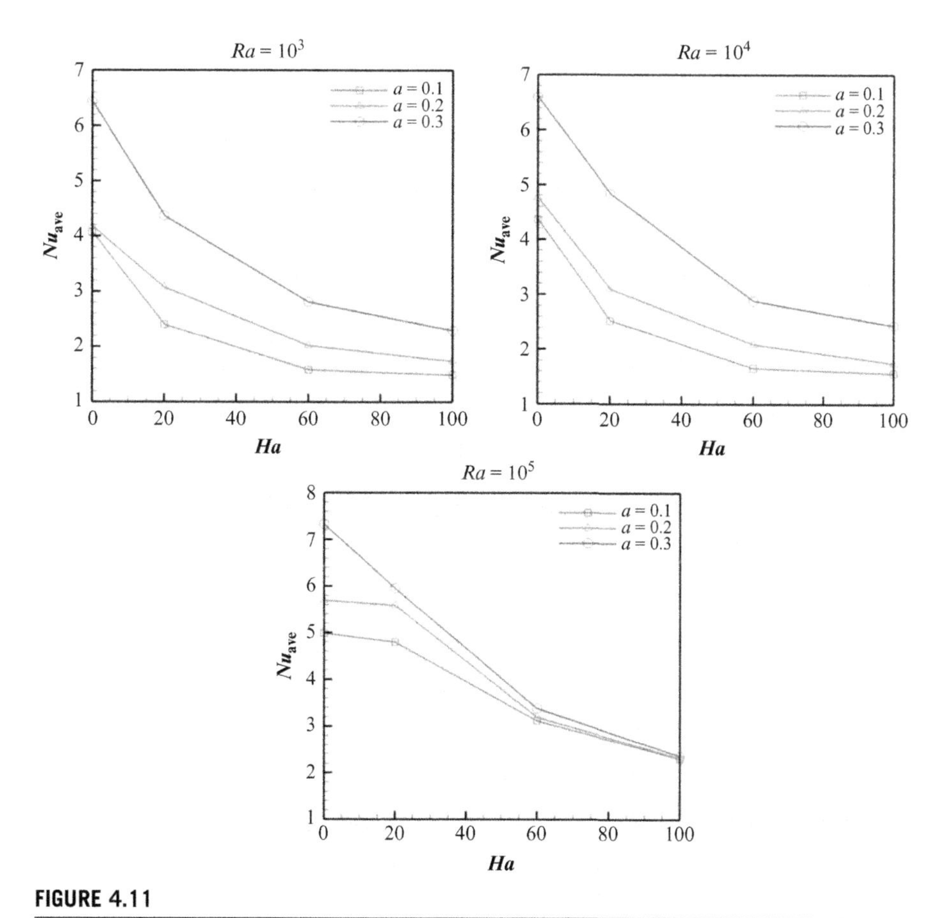

FIGURE 4.11

Variations in the average Nusselt number along the wall with a constant heat flux for different values of a and Ra at $Pr=6.2$ and $\phi=0.04$.

$$\left(\frac{X}{a}\right)^{2n}+\left(\frac{Y}{b}\right)^{2n}=1 \tag{4.47}$$

When $a=b$ and $n=1$, the geometry becomes a circle. As n increases from 1 the geometry approaches a rectangle for $a\neq b$ and a square for $a=b$. In this study $r_{in}/r_{out}=0.75$. It is also assumed that the uniform magnetic field ($\vec{B}=B_x\vec{e}_x+B_y\vec{e}_y$) of the constant magnitude $B=\sqrt{B_x^2+B_y^2}$ is applied, where $\vec{e}_x$ and $\vec{e}_y$ are unit vectors in the Cartesian coordinate system. The orientation of the magnetic field forms an angle λ with the horizontal axis, such that $\lambda=B_x/B_y$. In this study λ equals zero. The electric current J and the electromagnetic force F are defined by $J=\sigma\left(\vec{V}\times\vec{B}\right)$ and $F=\sigma\left(\vec{V}\times\vec{B}\right)\times\vec{B}$, respectively.

The heatlines are adequate tools for visualizing and analyzing two-dimensional convection heat transfer through an extension of the heat flux

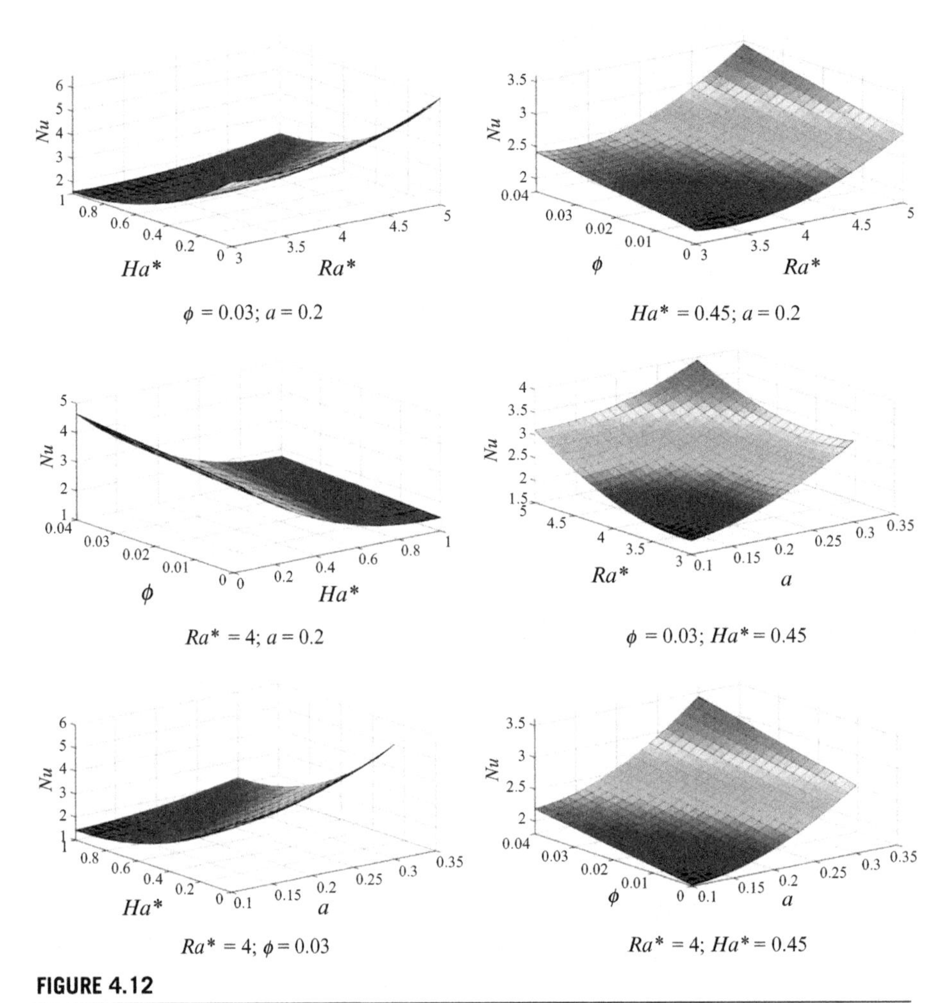

FIGURE 4.12

Variations in *Nu* for various active parameters.

line concept to include advection terms. The heat function (H) is defined in terms of the energy equation:

$$\frac{\partial H}{\partial Y} = U\Theta - \frac{\partial \Theta}{\partial X} \tag{4.48}$$

$$-\frac{\partial H}{\partial X} = V\Theta - \frac{\partial \Theta}{\partial Y} \tag{4.49}$$

4.2.3.3.2 Effects of active parameters

In this study the effect of MHD on natural convection heat transfer in an L-shaped, inclined enclosure filled with a nanofluid is investigated numerically using the CVFEM. The fluid in the enclosure is an aluminum oxide-water nanofluid. Various values for the

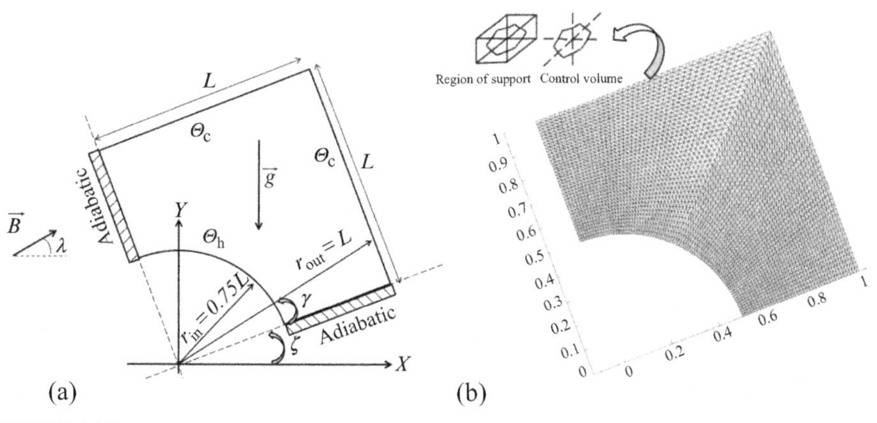

FIGURE 4.13

Geometry and the boundary conditions (a) with the mesh of enclosure (b) considered in this work.

Hartmann number ($Ha=0$, 20, 60, and 100), the volume fraction of nanoparticles ($\phi=0\%$ and 4%), the Rayleigh number ($Ra=10^3$, 10^4, and 10^5), and the inclination angle ($\gamma=-90°$, $-60°$, $-30°$, and $0°$) and a constant Prandtl number ($Pr=6.2$).

The isotherms, streamlines, and heatlines at different Hartmann numbers, inclined angles, and Rayleigh numbers are compared in Figs. 4.14 and 4.15.

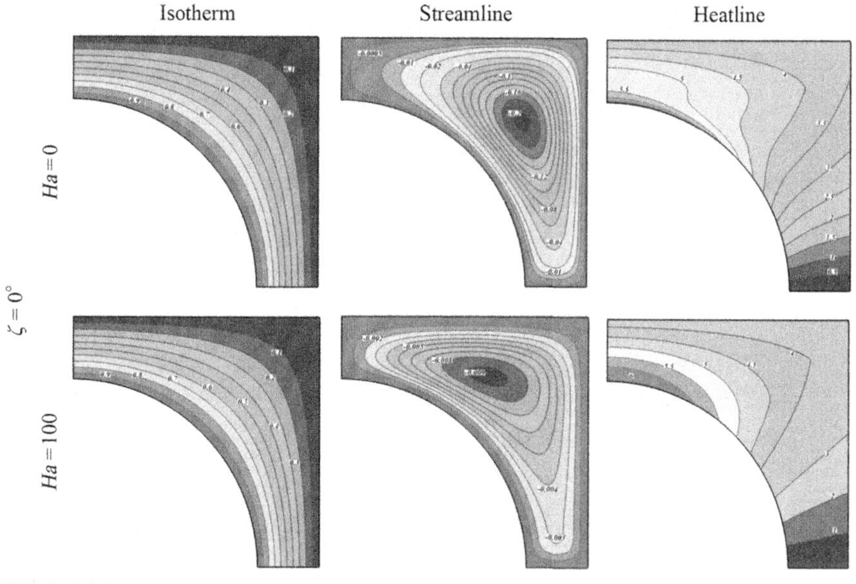

FIGURE 4.14

Comparison of the isotherms, streamlines, and heatlines for different Hartmann numbers and inclination angles at $Ra=10^3$, $\phi=0.04$, and $Pr=6.2$.

Continued

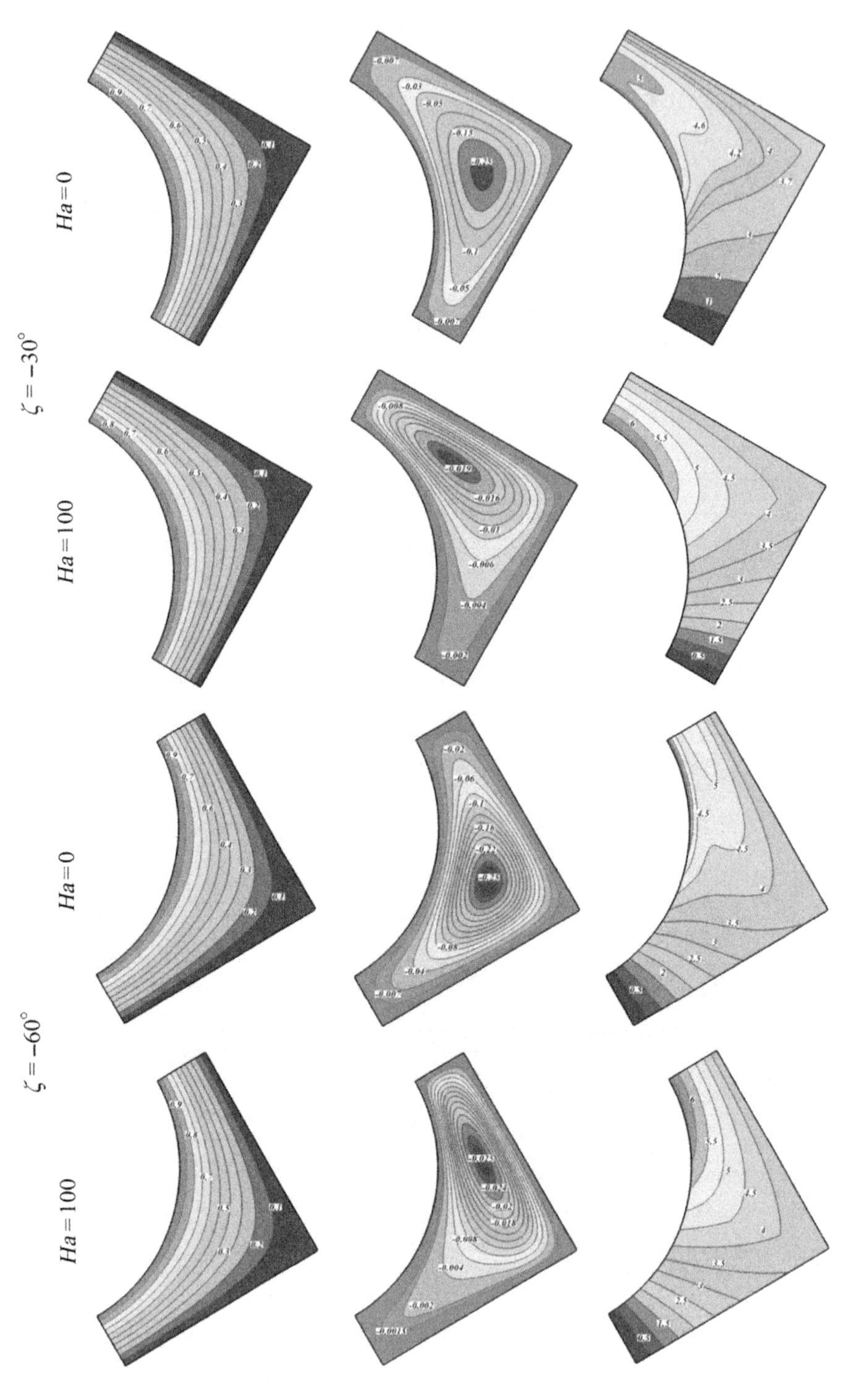

FIGURE 4.14, cont'd

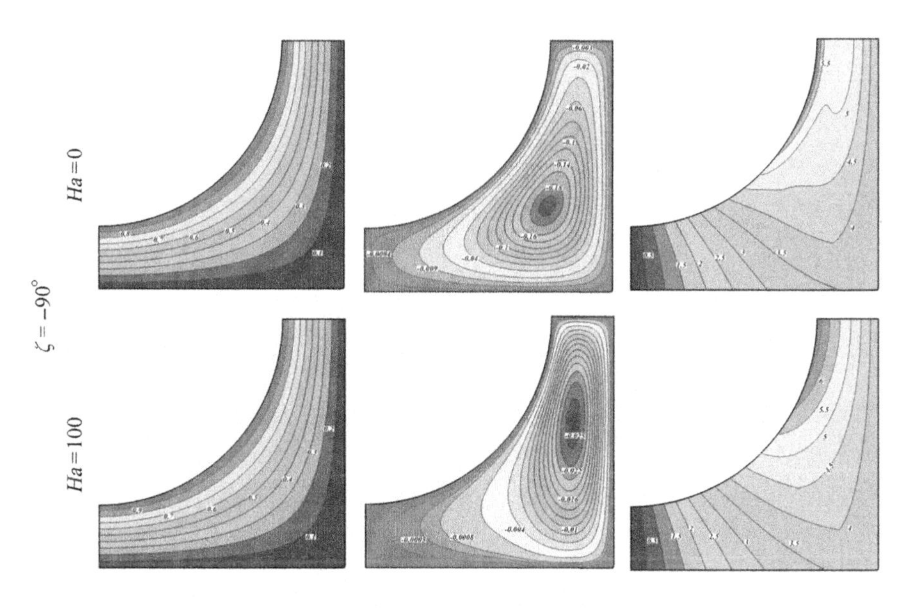

FIGURE 4.14, cont'd

At $Ra = 10^3$ the isotherms are parallel to each other and take the shape of the enclosure, which is the main characteristic of the conduction heat transfer mechanism. As the Rayleigh number increases the isotherms become more distorted and the stream function values are enhanced, which is due to the domination of the convective heat transfer mechanism at higher Rayleigh numbers. At $Ra = 10^5$, a thermal plume appears on the hot circular wall and three vortices exist in the streamline. By increasing $|\zeta|$, these vortices merge into one eddy; as $|\zeta|$ increases, the hot wall locates on the cold wall. Therefore convective heat transfer weakens and, in turn, the Nusselt number decreases as $|\zeta|$ increases.

When a magnetic field is imposed on the enclosure, the velocity field is suppressed by the retarding effect of the Lorenz force. Therefore the intensity of the convection weakens significantly. The braking effect of the magnetic field is based on the maximum stream function value. The core vortex shifts upward vertically as the Hartmann number increases. Imposing a magnetic field also removes the thermal plume over the inner wall. At high Hartmann numbers the conduction heat transfer mechanism is more pronounced. For this reason the isotherms are parallel to each other. The increase in Ra causes the heatlines to cluster from the hot to the cold wall and generates an area of passive heat transfer in which heat is rotated without having a significant effect on heat transfer between the walls.

Figure 4.16 shows the distribution of local Nusselt numbers along the surface of the inner circular wall at different inclination angles, Rayleigh numbers, and Hartmann numbers. Increasing the Rayleigh number leads to an increase in the local

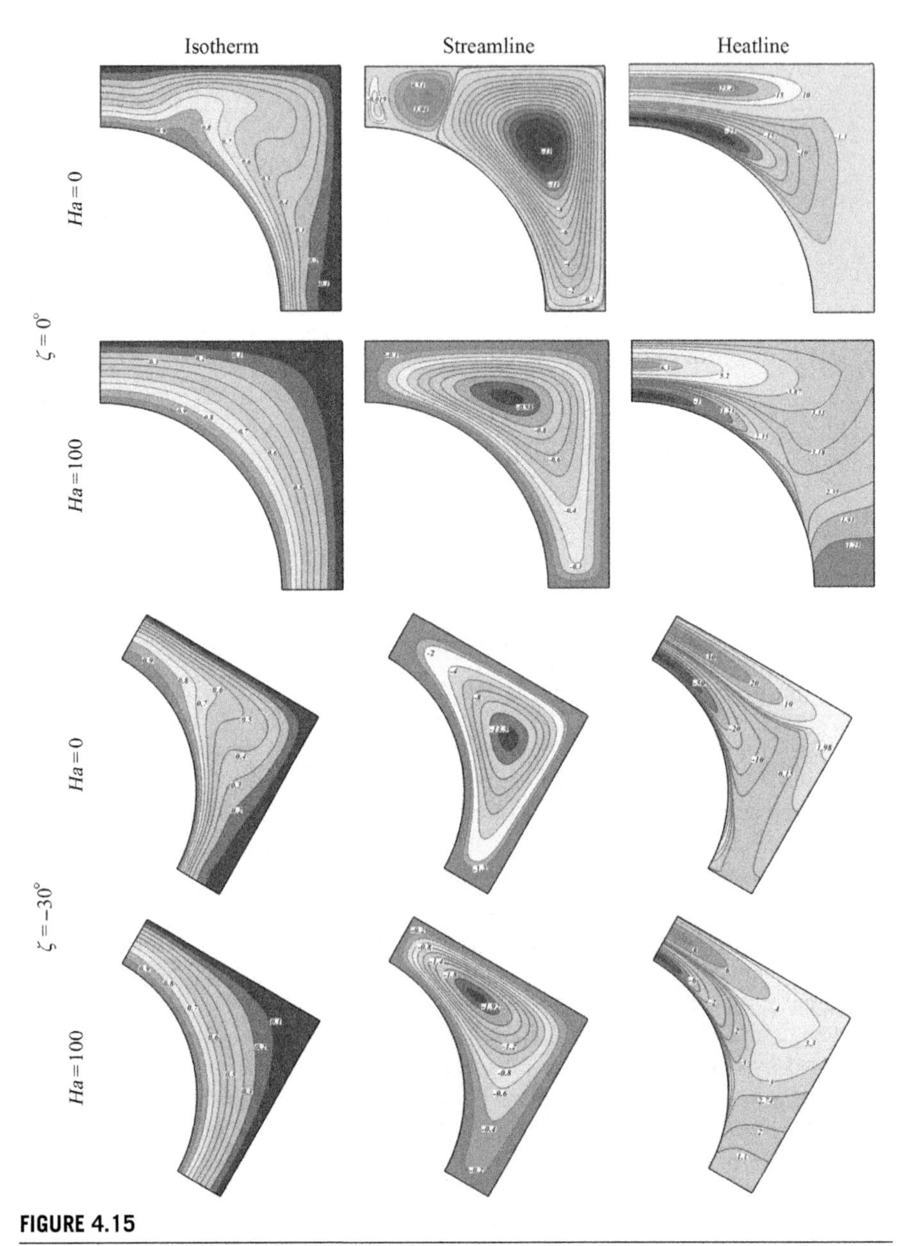

FIGURE 4.15

Comparison of the isotherms, streamlines, and heatlines for different Hartmann numbers and inclination angles at $Ra=10^5$, $\phi=0.04$, and $Pr=6.2$.

Continued

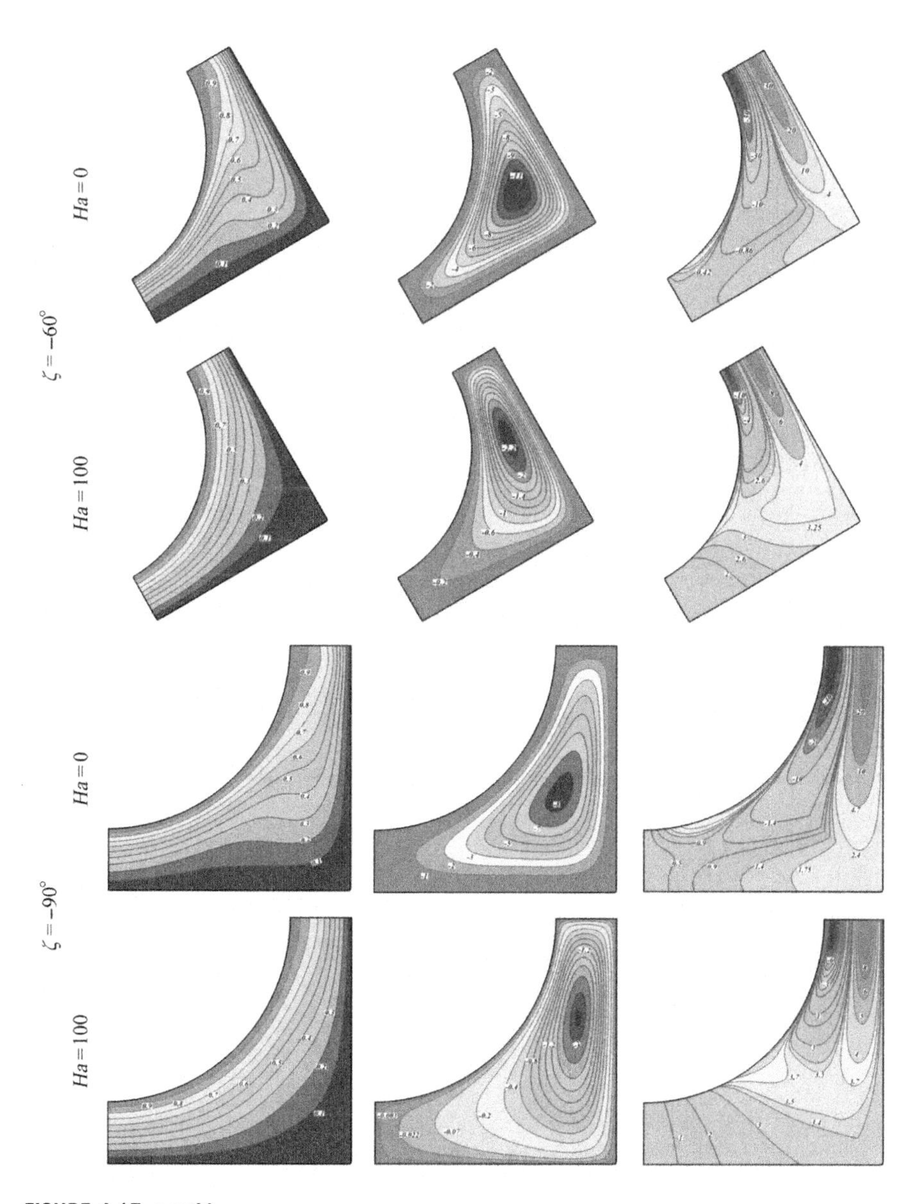

FIGURE 4.15, cont'd

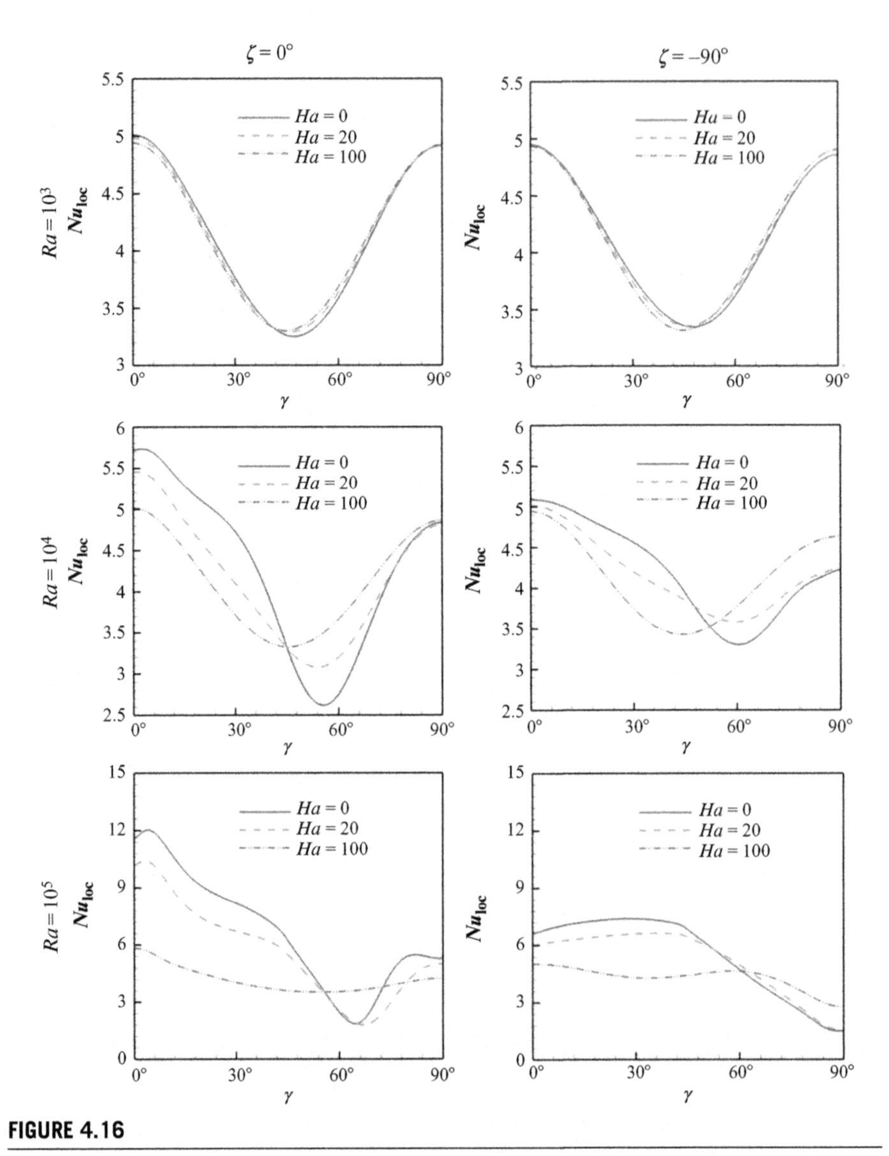

FIGURE 4.16

Effects of the Hartmann number, Rayleigh number, and inclination angle for a copper-water nanofluid on the local Nusselt number.

Nusselt number, but increasing the Hartmann number and inclination angle causes the local Nusselt number to decrease. At $Ra = 10^3$, because of the dominating conduction heat transfer mechanism, the distribution of the local Nusselt numbers along the surface of the inner circular wall shows the symmetrical shape. It is interesting to note that at high Rayleigh numbers the local Nusselt number profiles are more complex because of the presence of a thermal plume. In all cases—expect for $Ra = 10^5$, $\gamma = -90°$—one minimum point exists in the local Nusselt number profile, which occurs for lower values of $|\gamma|$ with an increase in the Hartmann number. Effects of the Rayleigh number, Hartmann number, and inclination angle on the average Nusselt number are shown in Fig. 4.17. The Nusselt number is an increasing function of the Rayleigh number, but it is a decreasing function of the Hartmann number and inclined angle. The effect of the Hartmann number on the Nusselt number is more pronounced at $\zeta = 0°$.

The heat transfer enhancement ratio as affected by the addition of nanoparticles for different values of ζ, Ha, and Ra is shown in Fig. 4.18. In general, the effect of nanoparticles is more pronounced at low Rayleigh numbers and high Hartmann numbers because of the greater enhancement rate. This observation can be explained by noting that at low Rayleigh numbers the heat transfer is dominated by conduction. Therefore, the addition of nanoparticles with high thermal conductivity increases conduction and makes the enhancement more effective. The inclination angle has no significant effect on the rate of enhancement when $Ra = 10^3$ and 10^4, whereas E increases with an increase in $|\zeta|$ when $Ra = 10^5$. Finally, the corresponding polynomial representation of such a model for each Nusselt number and rate of enhancement are presented as follows:

$$\begin{aligned} Nu &= a_{13} + a_{23}Y_1 + a_{33}Y_2 + a_{43}{Y_1}^2 + a_{53}{Y_2}^2 + a_{63}Y_1Y_2 \\ Y_1 &= a_{11} + a_{21}\zeta + a_{31}\phi + a_{41}\zeta^2 + a_{51}\phi^2 + a_{61}\zeta\phi \\ Y_2 &= a_{12} + a_{22}Ra^* + a_{32}Ha^* + a_{42}Ra^{*2} + a_{52}Ha^{*2} + a_{62}Ra^*Ha^* \end{aligned} \tag{4.50}$$

$$\begin{aligned} E &= b_{13} + b_{23}Y_1 + b_{33}Y_2 + b_{43}{Y_1}^2 + b_{53}{Y_2}^2 + b_{63}Y_1Y_2 \\ Y_1 &= b_{11} + b_{21}\zeta + b_{31}Ha^* + b_{41}\zeta^2 + b_{51}Ha^{*2} + b_{61}\zeta Ha^* \\ Y_2 &= b_{12} + b_{22}Ra^* + b_{32}\zeta + b_{42}Ra^{*2} + b_{52}\zeta^2 + b_{62}Ra^*\zeta \end{aligned} \tag{4.51}$$

where $Ra^* = \log(Ra)$ and $Ha^* = Ha/100$. Also, values for a_{ij} and b_{ij} can be found in Tables 4.3 and 4.4, respectively.

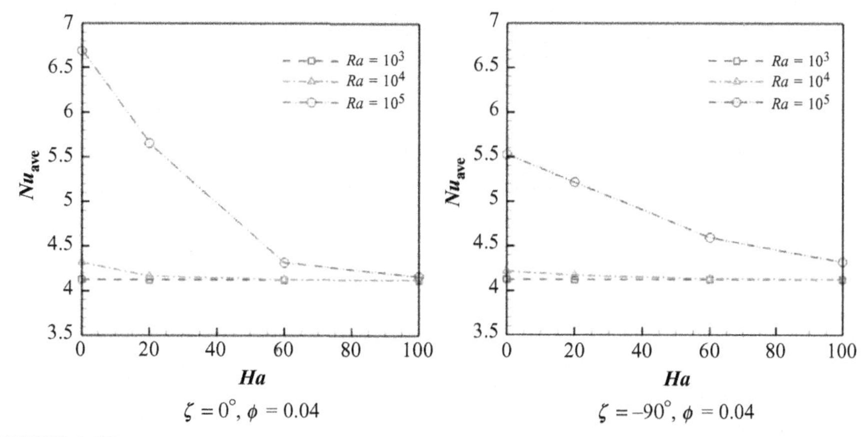

FIGURE 4.17

Effects of the Rayleigh number, Hartmann number, and inclination angle on Nu_{ave}.

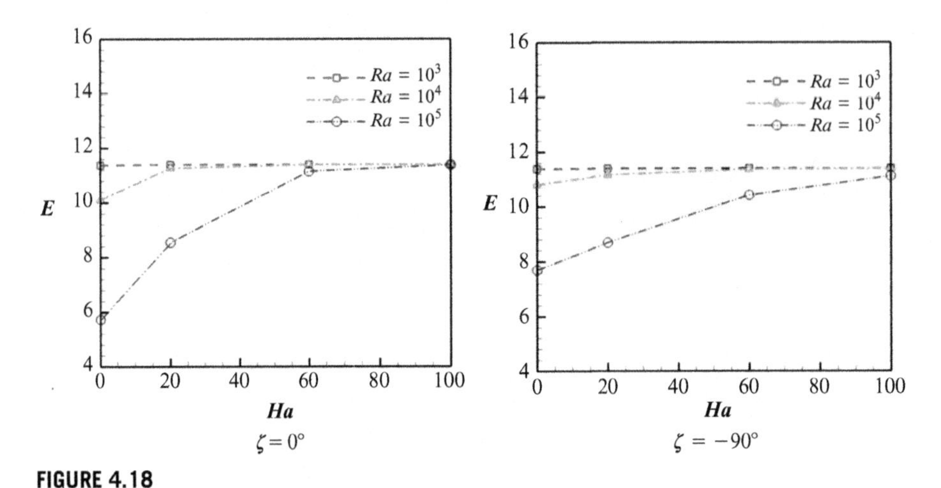

FIGURE 4.18

Effects of the Rayleigh number, Hartmann number, and inclination angle on the ratio of heat transfer enhancement due to the addition of nanoparticles.

4.2.3.4 Heat flux boundary condition for a nanofluid-filled enclosure in the presence of a magnetic field

4.2.3.4.1 Problem definition

The physical model, along with important geometrical parameters, are shown in Fig. 4.19 [33]. The width and height of the enclosure are L. The outer cylinder is maintained at a constant cold temperature T_c, whereas the inner circular wall is under constant heat flux. To assess the shape of the inner circular and outer rectangular

Table 4.3 Constant Coefficients (a_{ij}) for Using Eq. (4.50)

a_{ij}	$i=1$	$i=2$	$i=3$	$i=4$	$i=5$	$i=6$
$j=1$	4.080931	−0.10849	10.44936	−0.10891	0.417974	−0.03188
$j=2$	7.809771	−.67057	2.414376	0.452514	0.790997	−0.97843
$j=3$	8.465516	0.484181	−4.13987	−0.01217	0.4827	0.145098

Table 4.4 Constant Coefficients (b_{ij}) for Using Eq. (4.51)

b_{ij}	$i=1$	$i=2$	$i=3$	$i=4$	$i=5$	$i=6$
$j=1$	9.333563	−0.00506	4.537482	0.195157	−2.33844	0.508573
$j=2$	3.204931	5.085197	0.530967	−0.77932	0.195157	−0.07679
$j=3$	−134.848	14.51322	11.03078	−0.10825	0.060356	−1.0671

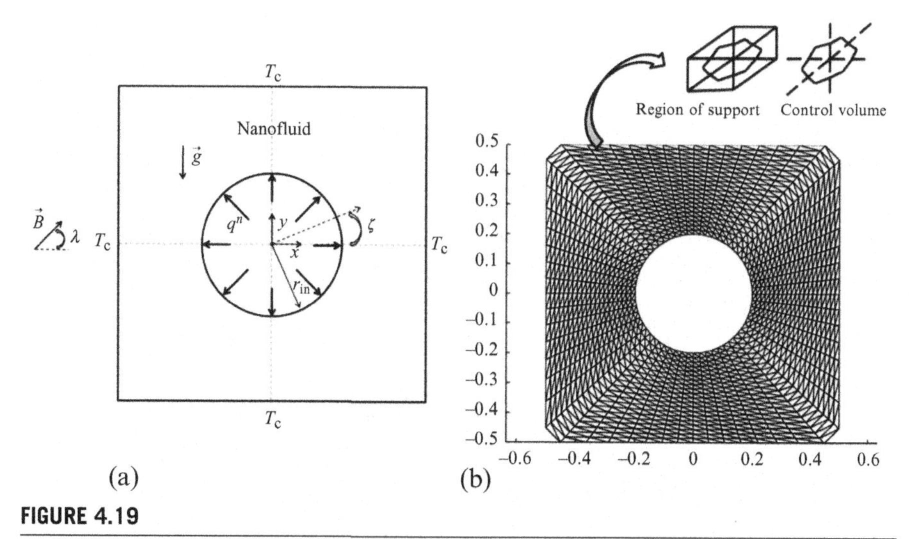

FIGURE 4.19

Geometry and the boundary conditions (a) with the mesh of the geometry (b) considered in this work.

boundaries, which consist of the right and top walls, a superelliptic function can be used:

$$(X/a)^{2\hat{n}} + (Y/b)^{2\hat{n}} = 1 \tag{4.52}$$

When $a=b$ and $\hat{n}=1$, the geometry becomes a circle. As $\hat{n}$ increases from 1, the geometry approaches a rectangle for $a \neq b$ and a square for $a=b$. It is also assumed that the uniform magnetic field $(\vec{B}=B_x\vec{e}_x+B_y\vec{e}_y)$ of constant magnitude $B=\sqrt{B_x^2+B_y^2}$ is applied, where $\vec{e}_x$ and $\vec{e}_y$ are unit vectors in the Cartesian coordinate system. The orientation of the magnetic field forms an angle λ with the horizontal axis, such that $\lambda=B_x/B_y$. In this study λ equals zero. The electric current J and the electromagnetic force F are defined by $J=\sigma\left(\vec{V}\times\vec{B}\right)$ and $F=\sigma\left(\vec{V}\times\vec{B}\right)\times\vec{B}$, respectively.

4.2.3.4.2 Effects of active parameters

The CVFEM is applied to solve the problem of natural convection in an enclosure filled with an aluminum oxide-water nanofluid in the presence of a magnetic field. The effective thermal conductivity and viscosity of the nanofluid are calculated using the Koo-Kleinstreuer-Li correlation. Various nanoparticle volume fractions ($\phi=0$–4%), Rayleigh numbers ($Ra=10^3$, 10^4, and 10^5), aspect ratios ($r_{in}/L=0.2$, 0.3, and 0.4), and Hartmann numbers ($Ha=0$-100) at a constant Prandtl number ($Pr=6.2$) are calculated.

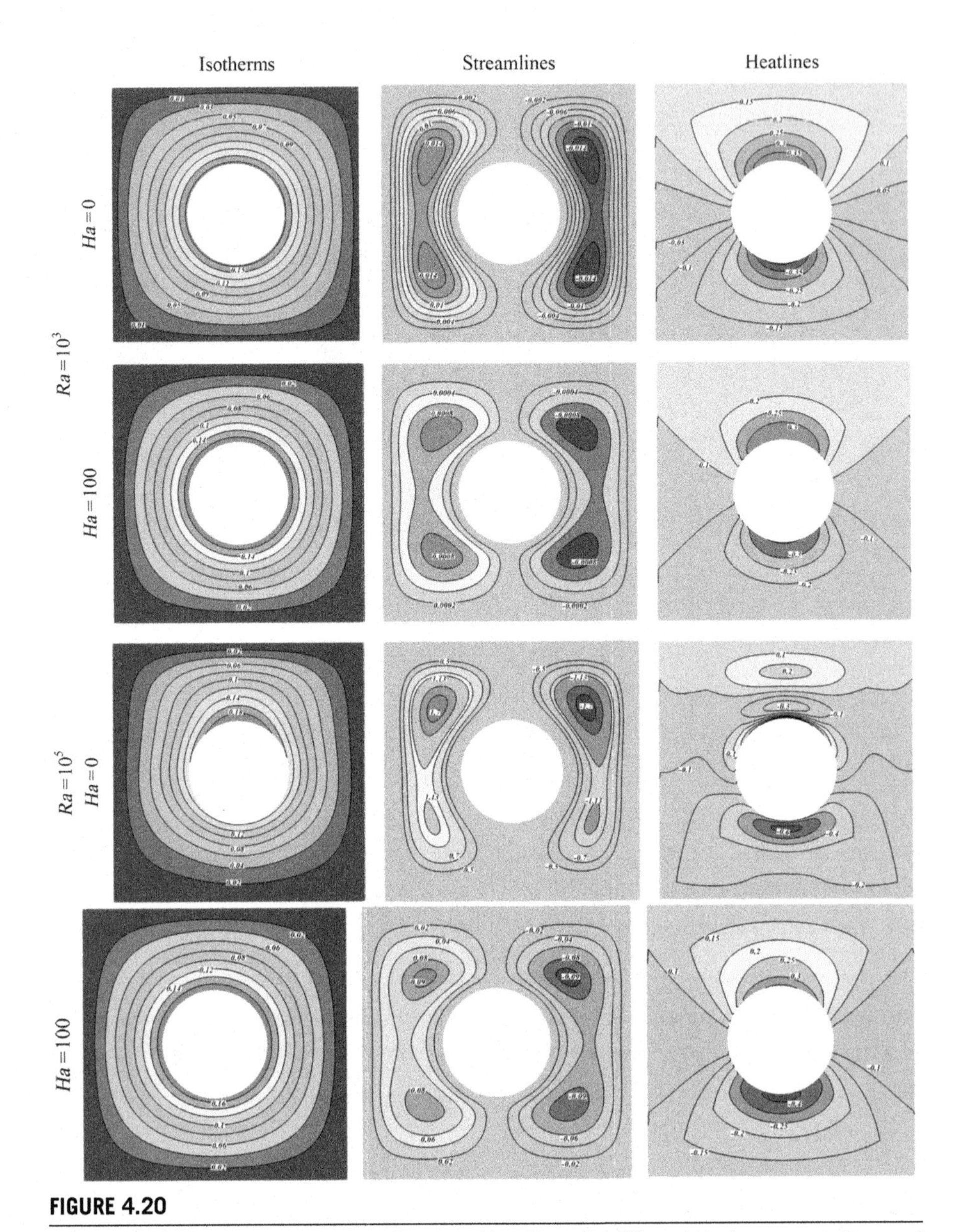

FIGURE 4.20

Comparison of the isotherms, streamlines, and heatlines for different Hartmann numbers and Raleigh numbers at $r_{in}/L = 0.2$, $\phi = 0.04$, and $Pr = 6.2$.

The isotherms, streamlines, and heatlines for different Hartmann numbers, aspect ratios, and Rayleigh numbers are compared in Figs. 4.20 and 4.21. By increasing the Rayleigh number the prominent heat transfer mechanism is changed from conduction to convection. Increasing the aspect ratio also leads to a decrease in the thermal boundary layer thickness and the intensity of convection because of the dominating

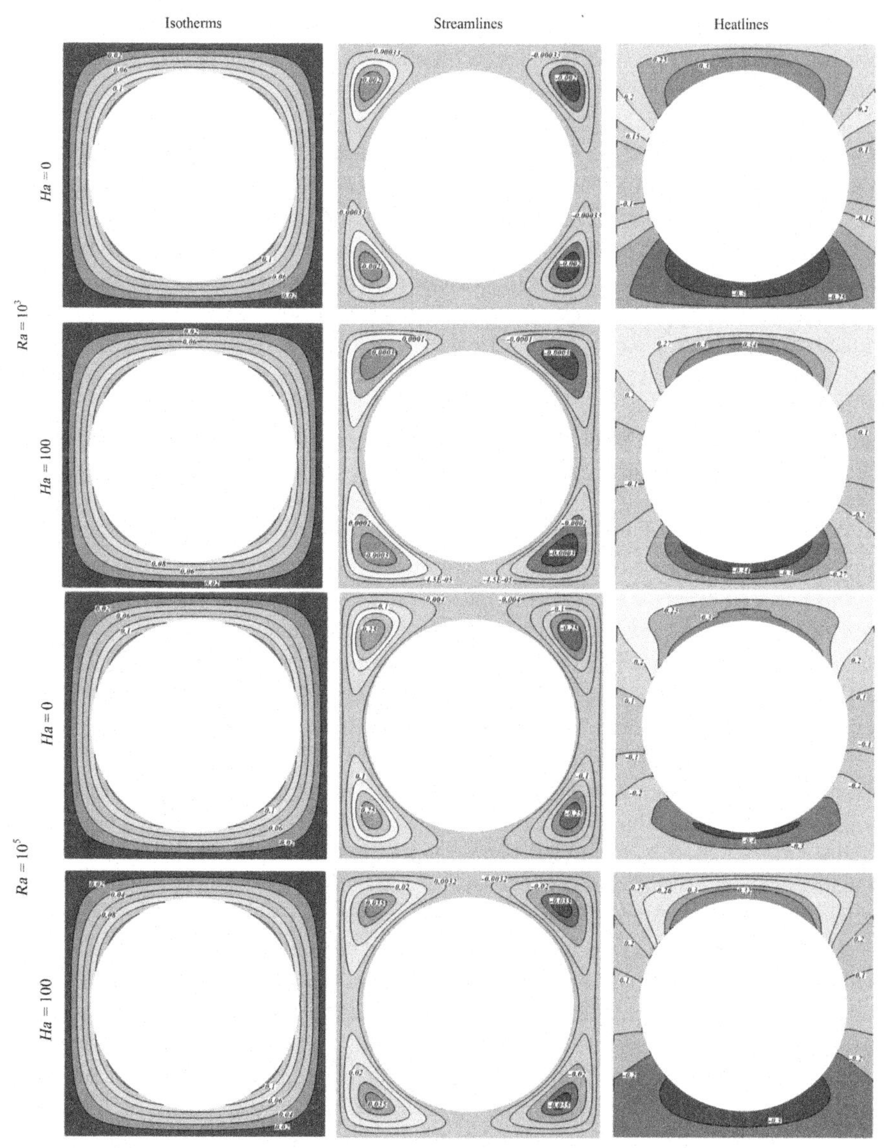

FIGURE 4.21

Comparison of the isotherms, streamlines, and heatlines for different Hartmann numbers and Raleigh numbers at $r_{in}/L = 0.4$, $\phi = 0.04$, and $Pr = 6.2$.

conduction heat transfer. When a magnetic field is imposed on the enclosure, the velocity field is suppressed because of the retarding effect of the Lorenz force. Therefore the intensity of the convection weakens significantly. The braking effect of the magnetic field is observed from the maximum stream function value. The core vortex shifts upward vertically as the Hartmann number increases. Also, imposing a magnetic field precludes the thermal plume over the inner wall. At high Hartmann numbers the conduction heat transfer mechanism is more pronounced. For this reason the isotherms are parallel to each other.

The heat flow within the enclosure is displayed using the heat function obtained from conductive heat fluxes ($\partial\Theta/\partial X$, $\partial\Theta/\partial Y$), as well as convective heat fluxes ($V\Theta$, $U\Theta$). Heatlines emanate from hot regimes and end on cold regimes, illustrating the path of heat flow. The domination of conduction heat transfer at low Rayleigh numbers and high Hartmann numbers can be observed from the heatline patterns because no passive area exists. The increase in Ra causes heatlines to move from the hot wall to cluster at the cold wall and generates a passive heat transfer area in which heat is rotated without having a significant effect on heat transfer between walls.

Distribution of local Nusselt numbers along the surface of the inner circular wall at different aspect ratios, Rayleigh number, and Hartmann numbers are shown in Fig. 4.22. Increasing the Rayleigh number and aspect ratio leads to an increase in the local Nusselt number, but increasing the Hartmann number causes the local Nusselt number to decrease.

Figure 4.23 shows the effects of the aspect ratio, Rayleigh number, and Hartmann number on the average Nusselt number. Increasing the Hartmann number causes the Lorenz force to increase and substantially suppresses the convection. Therefore the Nusselt number has a reverse relationship with the Hartmann number. As the aspect ratio increases, the space in which to accelerate the flow inside the cavity decreases. So the thermal boundary layer thickness decreases and, in turn, $1/\theta$ increases. As the nanoparticle volume fraction increases, thermal diffusivity increases. High values of thermal diffusivity cause the boundary thickness to increase and, accordingly, decrease $1/\theta$. The Nusselt number is a function of $1/\theta$ and k_{nf}/k_f. Because the reduction in $1/\theta$ is much smaller than the thermal conductivity ratio because of the presence of nanoparticles, the Nusselt number is augmented by increasing the volume fraction of nanoparticles. The distance between the cold and hot walls decreases with an increase in the aspect ratio. So, the Nusselt number decreases as r_{in}/L increases.

The heat transfer enhancement ratio as affected by the addition of nanoparticles at different values of r_{in}/L, Ha, and Ra is shown in Fig. 4.24. The heat transfer enhancement ratio has a direct relationship with the Hartmann number and aspect ratio but has a reverse relationship with the Rayleigh number. This occurs because of the domination of conduction heat transfer at low Rayleigh numbers and high Hartmann numbers or aspect ratios. Therefore the addition of nanoparticles with high thermal conductivity increases the conduction and makes the enhancement more effective.

4.2.3.5 Natural convection of nanofluids in an enclosure between a circular and a sinusoidal cylinder in the presence of a magnetic field

4.2.3.5.1 Problem definition

The physical model and the corresponding triangular elements used in the present CVFEM program are shown in Fig. 4.25 [11]. The inner and outer walls are maintained at constant temperatures T_h and T_c, respectively. The shape of the inner cylinder profile is assumed to mimic the following pattern:

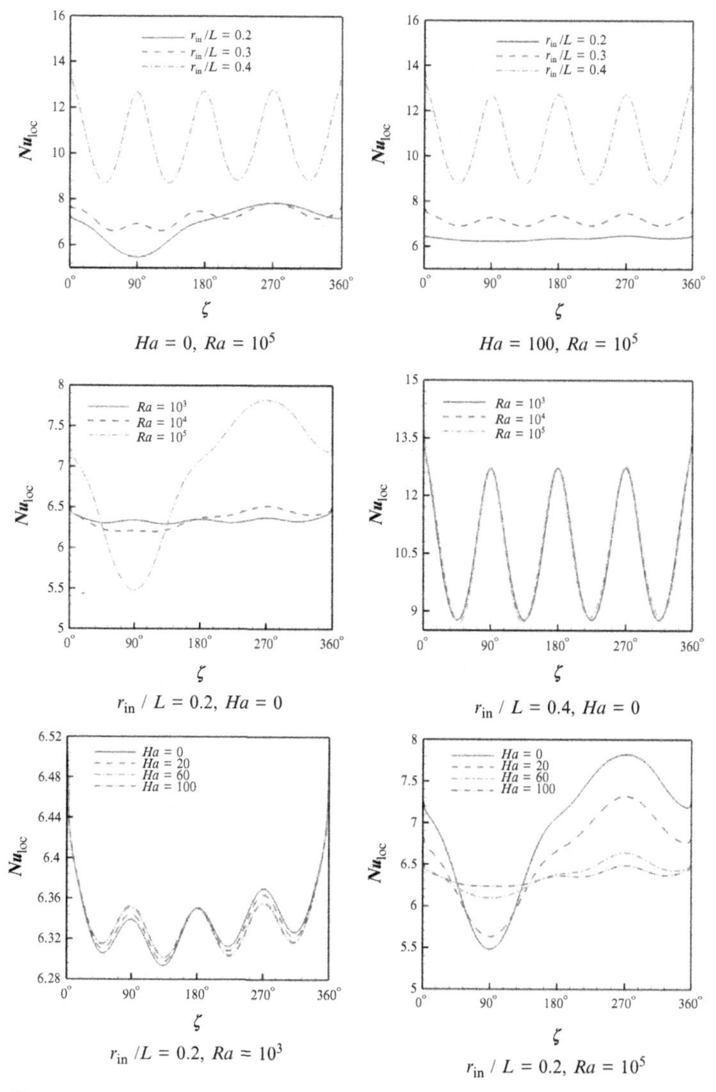

FIGURE 4.22

Effects of the aspect ratio, Rayleigh number, and Hartmann number on the local Nusselt number.

$$r = r_{in} + A\cos(N(\zeta - \zeta_0)) \tag{4.53}$$

in which r_{in} is the radius of the base circle, r_{out} is the radius of the outer cylinder, A and N are the amplitude and number of undulations, respectively, and ζ is the rotation angle. It is also assumed that the uniform magnetic field $\vec{B} = B_x\vec{e}_x + B_y\vec{e}_y$ of constant magnitude $B = \sqrt{B_x^2 + B_y^2}$ is applied horizontally. The electric current J and the

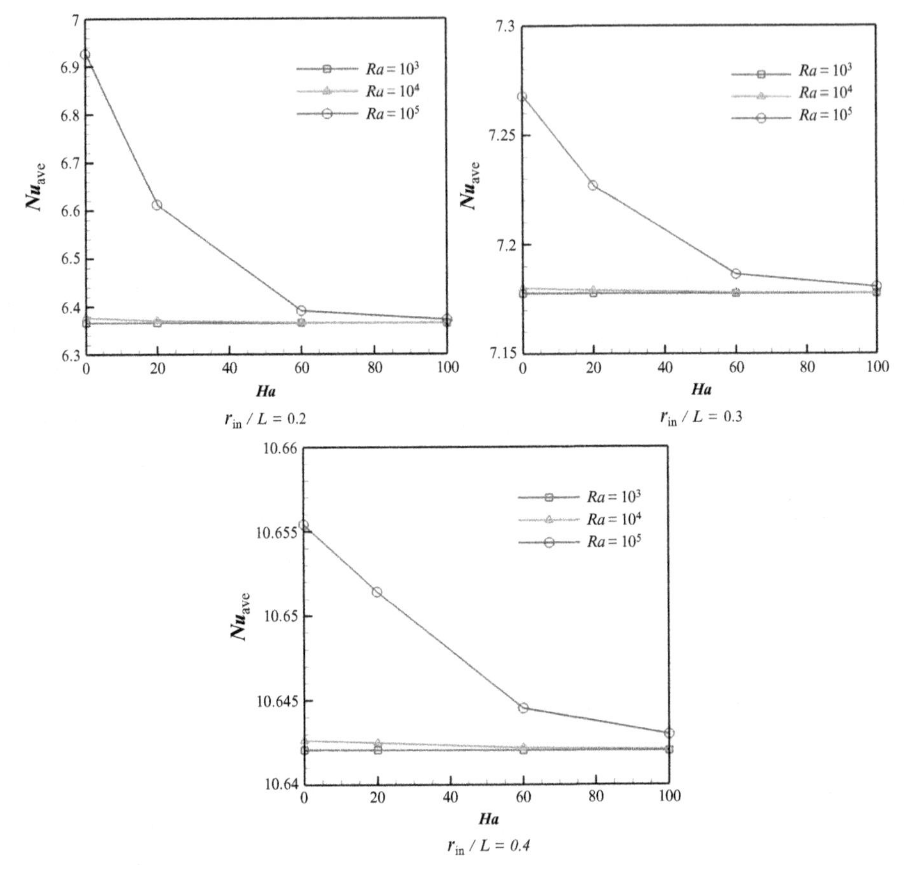

FIGURE 4.23

Effects of the aspect ratio, Rayleigh number, and Hartmann number on the average Nusselt number.

electromagnetic force F are defined by $J=\sigma\left(\vec{V}\times\vec{B}\right)$ and $F=\sigma\left(\vec{V}\times\vec{B}\right)\times\vec{B}$, respectively.

4.2.3.5.2 Effects of active parameters

Various Hartmann numbers ($Ha=0$, 20, 60, and 100), Rayleigh numbers ($Ra=10^3$, 10^4, and 10^5), volume fractions of nanoparticles ($\phi=0\%$, 2%, 4%, and 6%), and numbers of undulations ($N=3$ and 6) at a constant amplitude ($A=0.5$) and Prandtl number ($Pr=6.2$) are calculated.

The isotherm (left) and streamline (right) contours for different Hartmann numbers and Raleigh numbers for $N=3$ and $N=6$ are compared in Figs. 4.26 and 4.27, respectively. The absolute value of the stream function increases as the Rayleigh number increases, and it decreases as the Hartmann number increases. For both cases

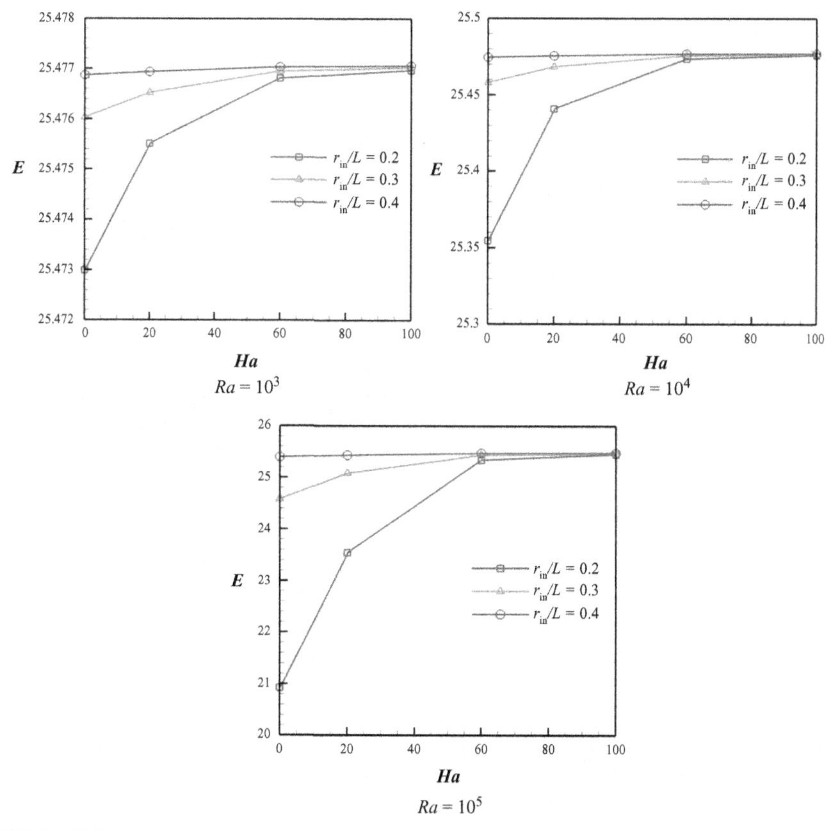

FIGURE 4.24

Effects of the aspect ratio, Rayleigh number, and Hartmann number on heat transfer enhancement.

at $Ra=10^3$, the isotherms are uniform and follow the shape of the inner and outer walls, and the values of the stream function are relatively small, which indicates the domination of the conduction heat transfer mechanism. For $N=3$, under the hot surface a small secondary eddy appears near the vertical center line, caused by the resistance of the crest against the flow circulating between the inner and outer cylinders. But for $N=6$, no small secondary eddy in the bottom half of the enclosure is observed because the lower crest of the inner surface is parallel to the flow. The isotherms start to be disturbed at $Ra=10^4$, and thermal plumes gradually appear on the hot surface of the inner cylinder, especially at the upper half of the enclosure, which indicates that the convective heat transfer becomes comparable with conduction in this Rayleigh number. In addition, the thermal boundary layer between the crests of the inner cylinder becomes thinner compared with those at $Ra=10^3$, which indicates that the unventilated region between the crests diminishes; hence the thermal boundary layer is suppressed toward the inner hot cylinder in these areas. At

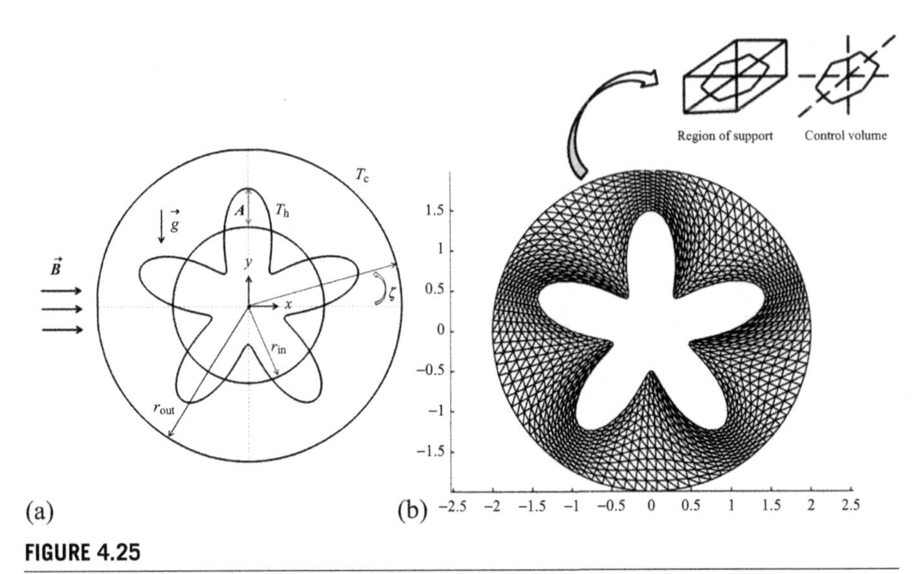

FIGURE 4.25

Geometry and the boundary conditions (a) with the mesh of the geometry (b) this work.

$Ra = 10^5$, the isotherms are crowed near the crests and depict the diminution of the thermal boundary layer thickness at the bottom of the enclosure. At the top of the inner cylinder ($\zeta = 90°$), a plume strongly drives the flow against the outer wall of the enclosure. It is worthwhile mentioning that the effect of a magnetic field decreases the value of the velocity magnitude throughout the enclosure because the presence of a magnetic field introduces the Lorentz force, which acts against the flow if the magnetic field is applied in the normal direction. This type of resisting force slows the fluid velocity. Increasing the Hartmann number makes the core of vortices move toward the horizontal center line. Also, a magnetic field causes the thermal plume to disappear and makes the isotherms parallel to each other because of the domination of the conduction mode of heat transfer.

Effects of the number of undulations, Rayleigh number, and Hartmann number on the local Nusselt number over the cold wall and between $\zeta = 90°$ and 270° are depicted in Fig. 4.28. The number of extrema in the local Nusselt number profile corresponds to the number of undulations of the inner hot surface. For $N = 3$ there is one local maximum and one local minimum value, whereas for $N = 6$ there are two maximum and three minimum values. In the presence of a magnetic field, the extremum points of local Nusselt number profiles occur at smaller ζ because of the change in the position of eddies. At $Ra = 10^5$ and $Ha = 0$, the local Nusselt number decreases as ζ increases from 90° and 270°; at the top of the inner cylinder ($\zeta = 90°$), the flow accompanying the cold fluid circulates against the upper cold surface, which results in a thicker thermal boundary layer over the enclosure wall, whereas when ζ approaches 270° the hot surface is located at the top of the cold wall and the conduction mechanism becomes more pronounced; hence the local Nusselt number decreases noticeably.

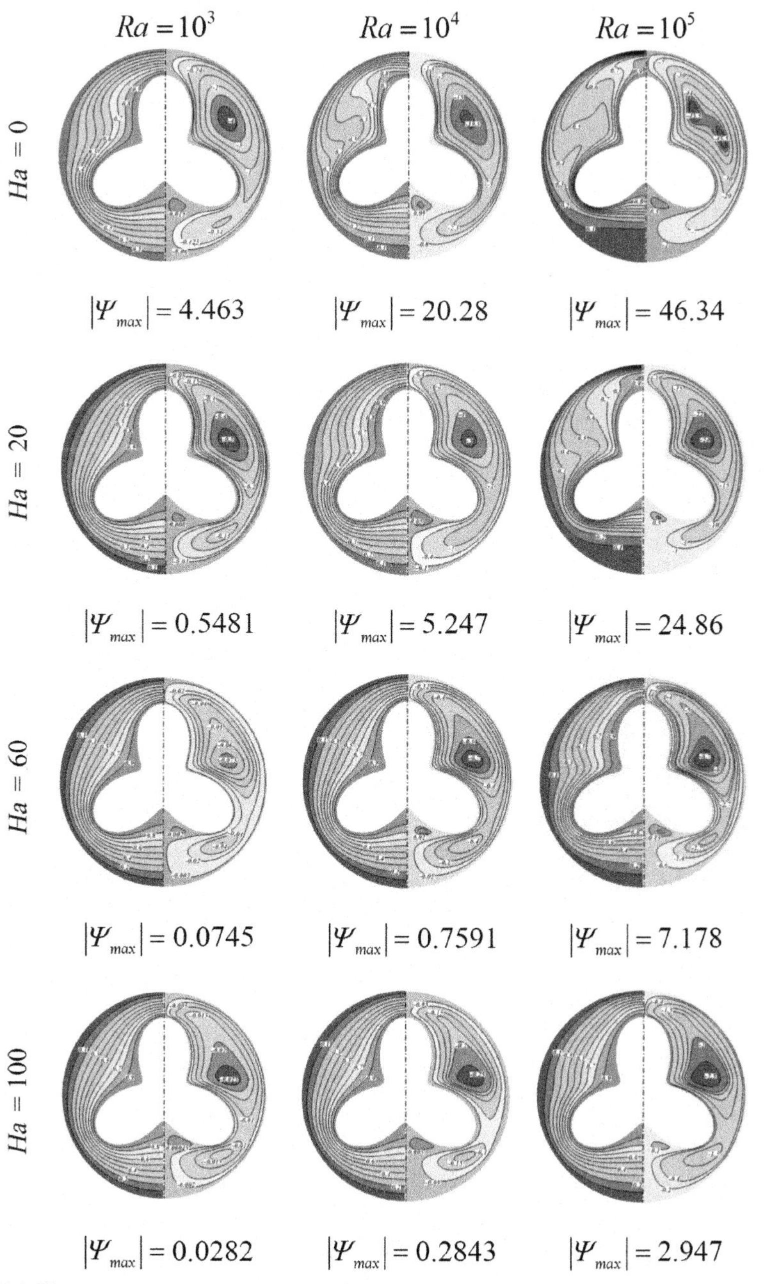

FIGURE 4.26

Comparison of the isotherm (left) and streamline (right) contours for different Hartmann numbers and Raleigh numbers at $A=0.5$ and $N=3$.

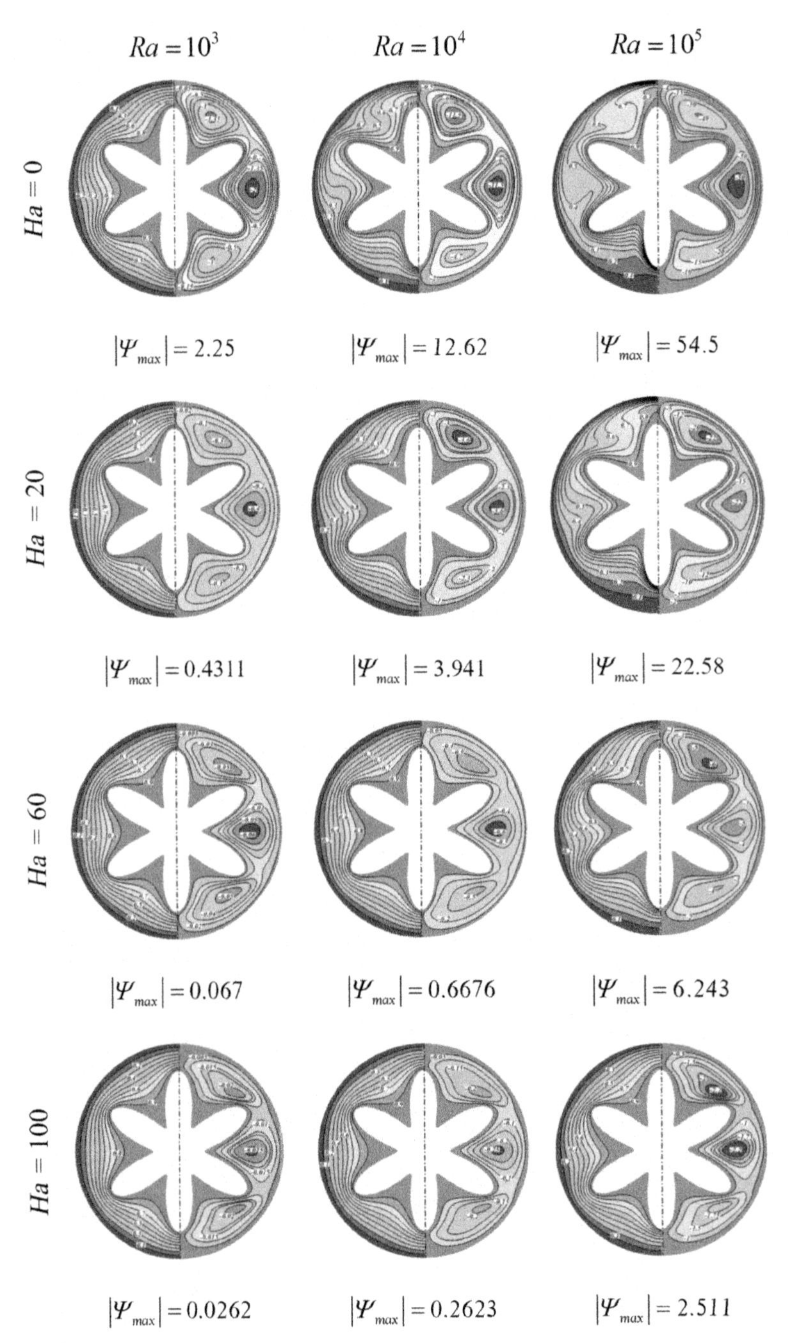

FIGURE 4.27

Comparison of the isotherm (left) and streamline (right) contours for different Hartmann numbers and Raleigh numbers at $A=0.5$ and $N=5$.

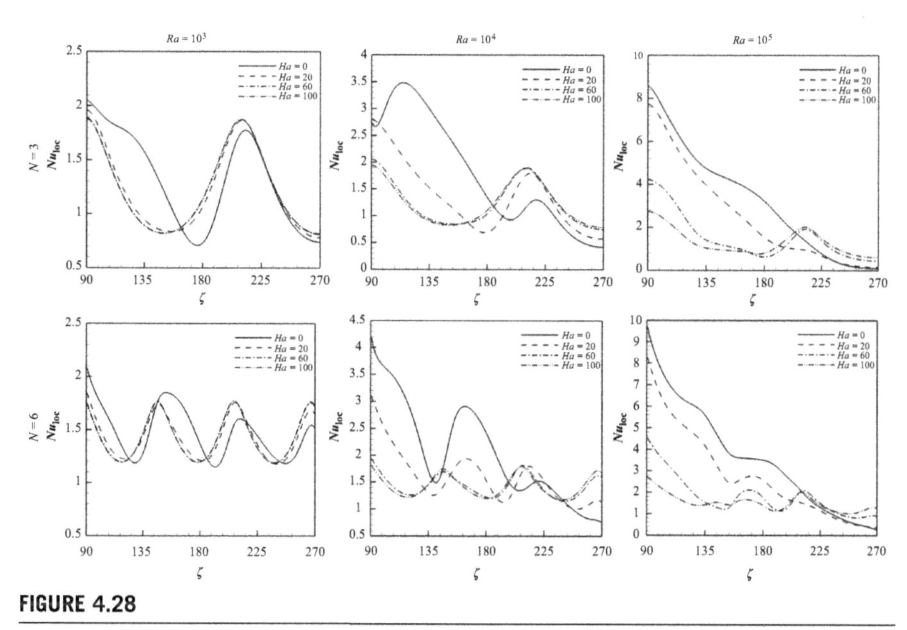

FIGURE 4.28

Effects of the number of undulations, Rayleigh number, and Hartmann number on the local Nusselt number at $\phi = 0.06$ and $A = 0.5$.

Figure 4.29 shows the effects of the number of undulations, nanoparticle volume fraction, Rayleigh number, and Hartmann number on the average Nusselt number. As a whole, the average Nusselt number increases as the Rayleigh number and nanoparticle volume fraction increase, whereas it decreases as the Hartmann number increases. Adding nanoparticles with high thermal conductivity leads to an increase in the conduction. Because conduction dominates at low Rayleigh numbers, the effect of the nanoparticle volume fraction on the average Nusselt number is more pronounced at low Rayleigh numbers than at high Rayleigh numbers. Also, Fig. 4.29 shows that an increasing the number of undulations causes the average Nusselt number to increase.To estimate the enhancement of heat transfer when $\phi = 0.06$ and a pure fluid (base fluid) is used, the enhancement is defined as:

$$E = \frac{Nu(0.06) - Nu(\text{basefluid})}{Nu(\text{basefluid})} \times 100 \tag{4.54}$$

Figure 4.30 shows the effect of the number of undulations, Hartmann number, and Rayleigh number on the heat transfer enhancement ratio. There is an insignificant change in enhancement at $Ra = 10^3$. In the absence of a magnetic field, the enhancement ratio decreases as the Rayleigh number increases, whereas the opposite behavior is observed in the presence of a magnetic field. The maximum values of enhancement for $Ra = 10^3$and 10^5 are observed at $Ha = 20$ and 60, respectively. It is an interesting observation that at $Ra = 10^5$ the enhancement in heat transfer for

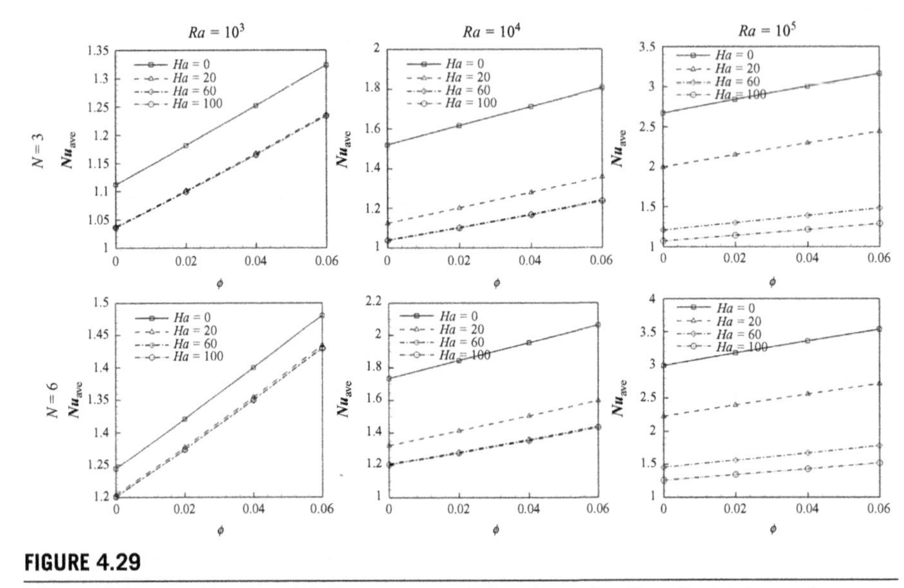

FIGURE 4.29

Effects of the number of undulations, nanoparticle volume fraction, Rayleigh number, and Hartmann number on the average Nusselt number at $A=0.5$.

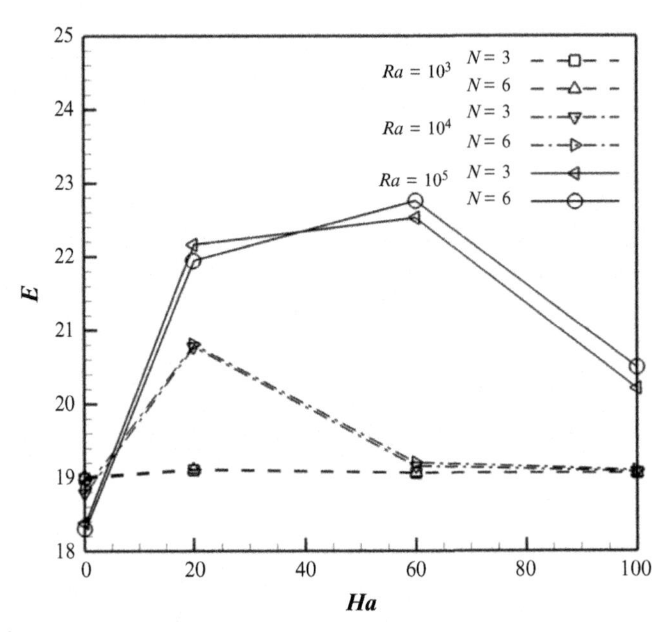

FIGURE 4.30

Effects of the number of undulations, Hartmann number, and Rayleigh number on the ratio of heat transfer enhancement due to the addition of nanoparticles.

$N=6$ is greater than those obtained for $N=3$ when $Ha>40$; for other Rayleigh numbers the number of undulations has no effect on enhancement.

4.2.3.6 Effect of a magnetic field on natural convection in an inclined half-annulus enclosure filled with a copper-water nanofluid

4.2.3.6.1 Problem definition

The physical model, along with important geometric parameters and the mesh of the half-annulus enclosure used in the present CVFEM program, are shown in Fig. 4.31 [9]. The inner wall is under constant heat flux (q''), the outer wall is maintained at a constant temperature (T_c); and the two other walls are thermally insulated. It is also assumed that the uniform magnetic field ($\vec{B}=B_x\vec{e}_x+B_y\vec{e}_y$) of a constant magnitude $B=\sqrt{B_x^2+B_y^2}$ is applied, where $\vec{e}_x$ and $\vec{e}_y$ are unit vectors in the Cartesian coordinate system. The orientation of the magnetic field forms an angle $\vec{e}_y$ with the horizontal axis such that $\lambda=B_x/B_y$. The electric current J and the electromagnetic force $F$are defined by $J=\sigma\left(\vec{V}\times\vec{B}\right)$ and $F=\sigma\left(\vec{V}\times\vec{B}\right)\times\vec{B}$, respectively.

4.2.3.6.2 Effects of active parameters

Various Hartmann numbers ($Ha=0$, 20, 60, and 100), Rayleigh numbers ($Ra=10^3$, 10^4, and 10^5), nanoparticle volume fractions ($\phi=0\%$, 2%, 4%, and 6%), and inclination angles ($\lambda=0°$, 45°, and 90°) at a constant Prandtl number ($Pr=6.2$) are calculated. Isotherm (down) and streamline (up) contours for different Rayleigh numbers and Hartmann numbers when $\lambda=0°$, 45°, and 90° are shown in Figs. 4.32, 4.33, and 4.34, respectively. In general, $|\Psi_{max}|_{nf}$ increases with an increase in the Rayleigh number and inclination angle, but it decreases with an increase in the Hartmann number. At $\lambda=0°$ and $Ra=10^3$, the heat transfer in the enclosure is mainly dominated by the conduction mode. The streamlines show

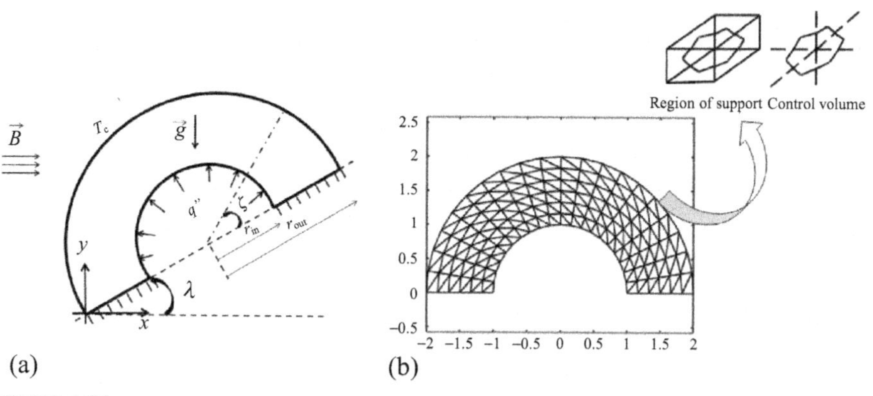

FIGURE 4.31

Geometry and the boundary conditions (a) with the mesh of half-annulus enclosure (b) considered in this work.

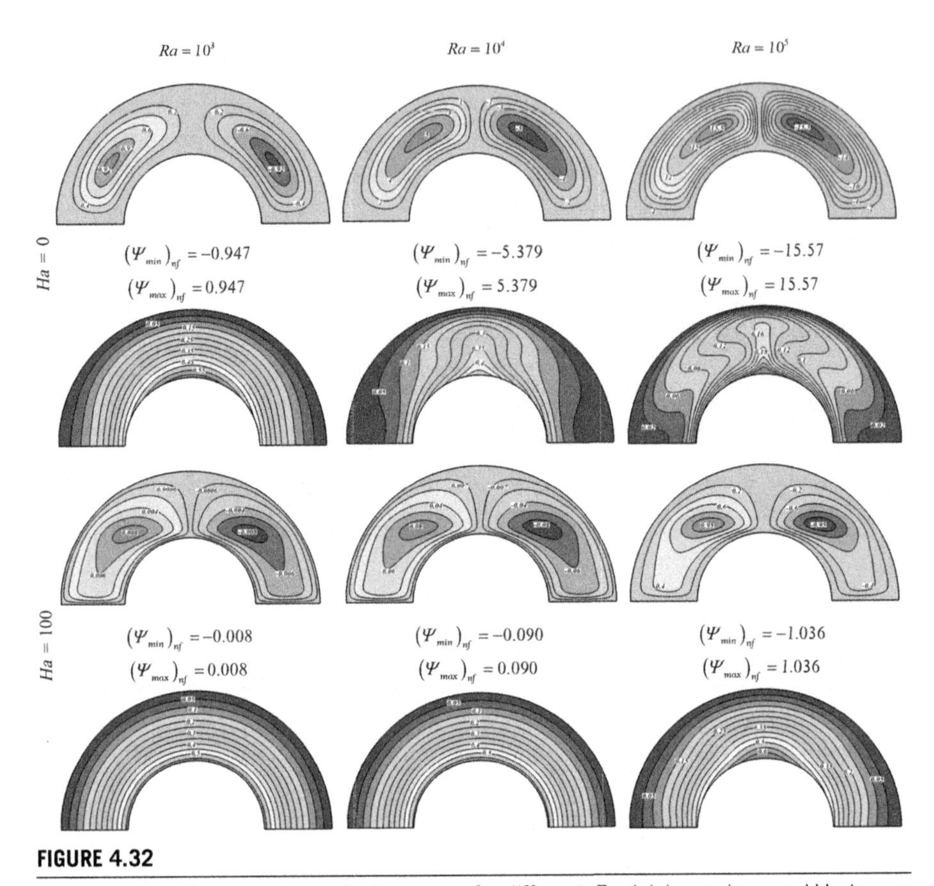

FIGURE 4.32

Isotherm (down) and streamline (up) contours for different. Rayleigh numbers and Hartmann numbers when $\lambda = 0°$ for a copper-water nanofluid ($\phi = 0.06$).

two overall rotating, symmetrical eddies. As the Rayleigh number increases up to 10^4, the role of convection in heat transfer becomes more significant, and consequently the thermal boundary layer on the surface of the inner wall becomes thinner. In addition, a plume starts to appear on the top of the inner circular wall. At $Ra = 10^5$, a strong plume forces the flow strongly against the top of the enclosure. In addition, the core of the primary eddies keeps moving upward because of the increasing convection velocity. At $\lambda = 45°$, two cells are formed, and they turn in opposite directions but are not symmetrical; the main cell, which turns in a counterclockwise direction, is dominant because more space is available for circulation. It is an interesting observation that for $Ra = 10^5$, as the Hartman number increases up to 100, the secondary eddy, which turns clockwise, is divided into two smaller eddies. At $\lambda = 90°$, the two main eddies merge into one counterclockwise eddy. The isotherms are distorted at the top of the enclosure at higher Rayleigh numbers because of the stronger convection effects. When a magnetic field is imposed on the enclosure, the velocity field is suppressed

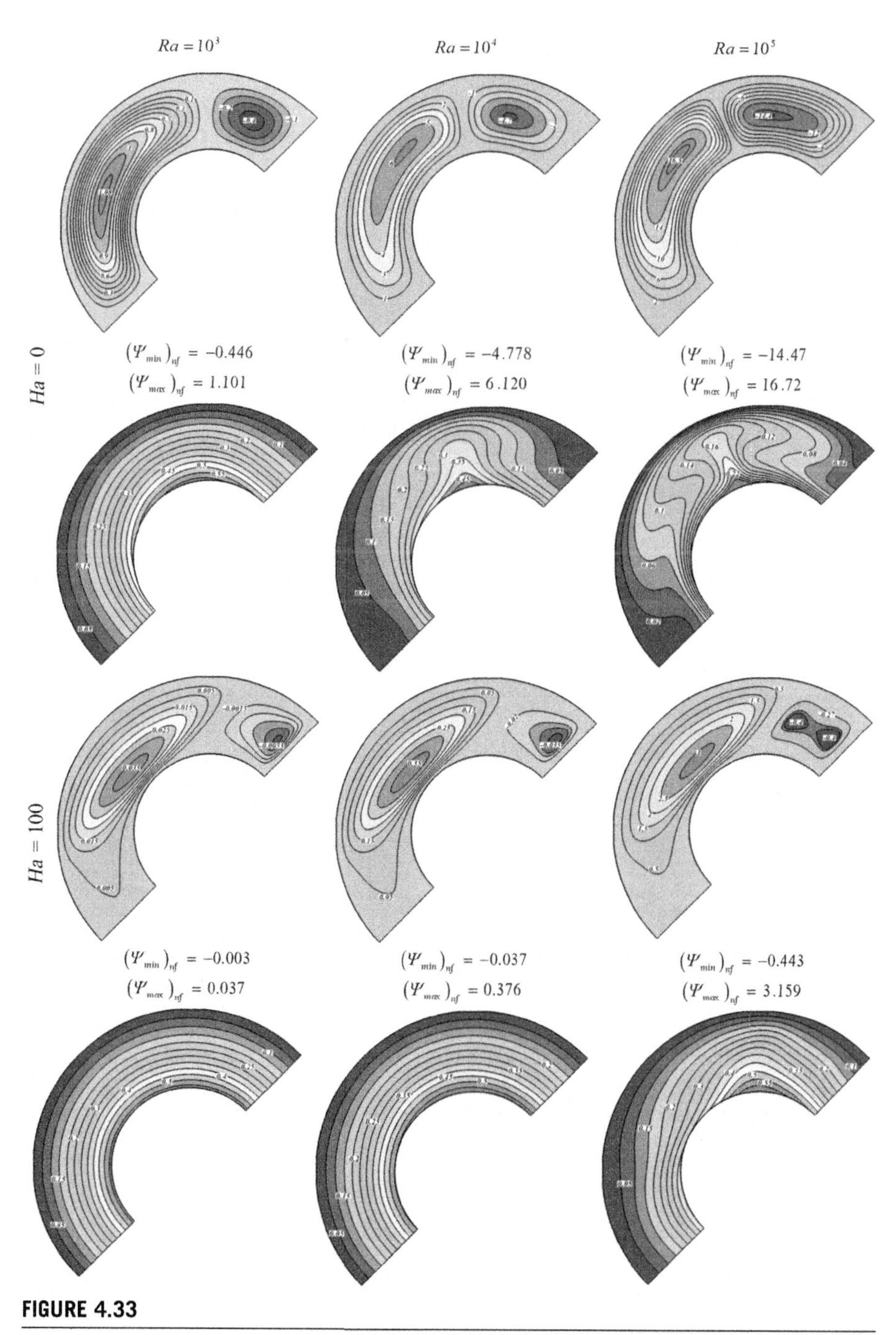

FIGURE 4.33

Isotherm (down) and streamline (up) contours for different Rayleigh numbers and Hartmann numbers when $\lambda = 45°$ for a copper-water nanofluid ($\phi = 0.06$).

because of the retarding effect of the Lorenz force. So, the intensity of the convection weakens significantly. The braking effect of the magnetic field is based on the maximum stream function value. The core vortex is shift vertically as the Hartmann number increases. Also, imposing a magnetic field avoids the creation of a thermal plume over the inner wall.

Figure 4.35 shows the effects of the Hartmann number, Rayleigh number, and inclination angle for copper-water ($\phi = 0.06$) nanofluids on the local Nusselt

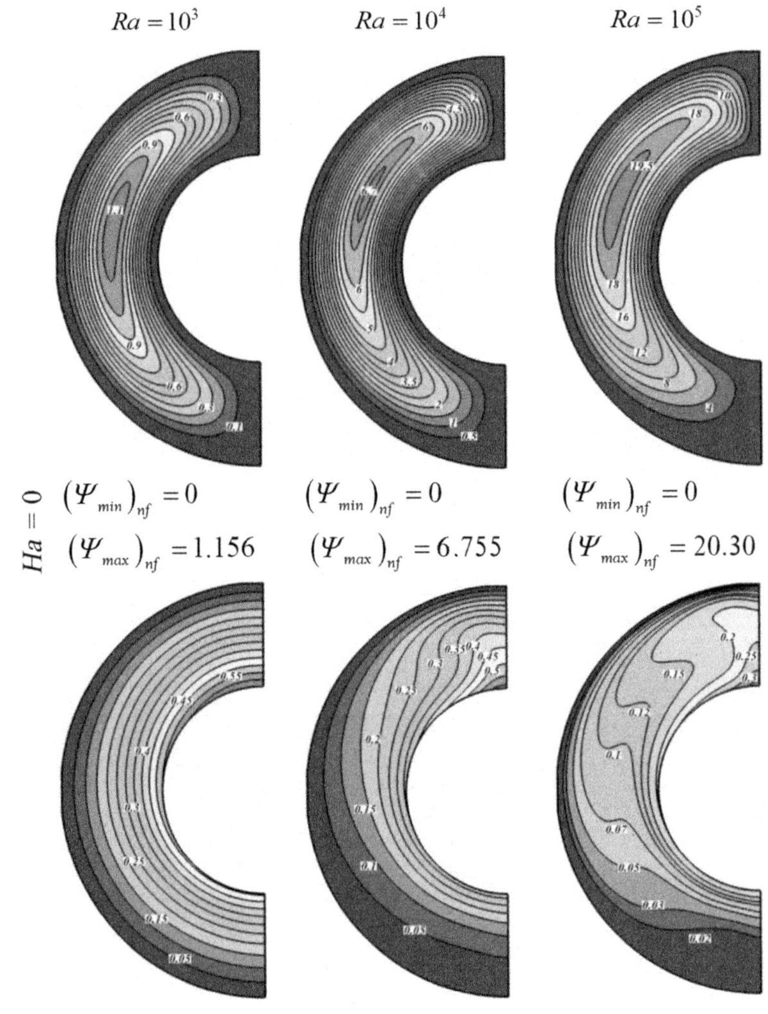

FIGURE 4.34

Isotherm (down) and streamline (up) contours for different Rayleigh numbers and Hartmann numbers when $\lambda = 90°$ for a copper-water nanofluid ($\phi = 0.06$)

Continued

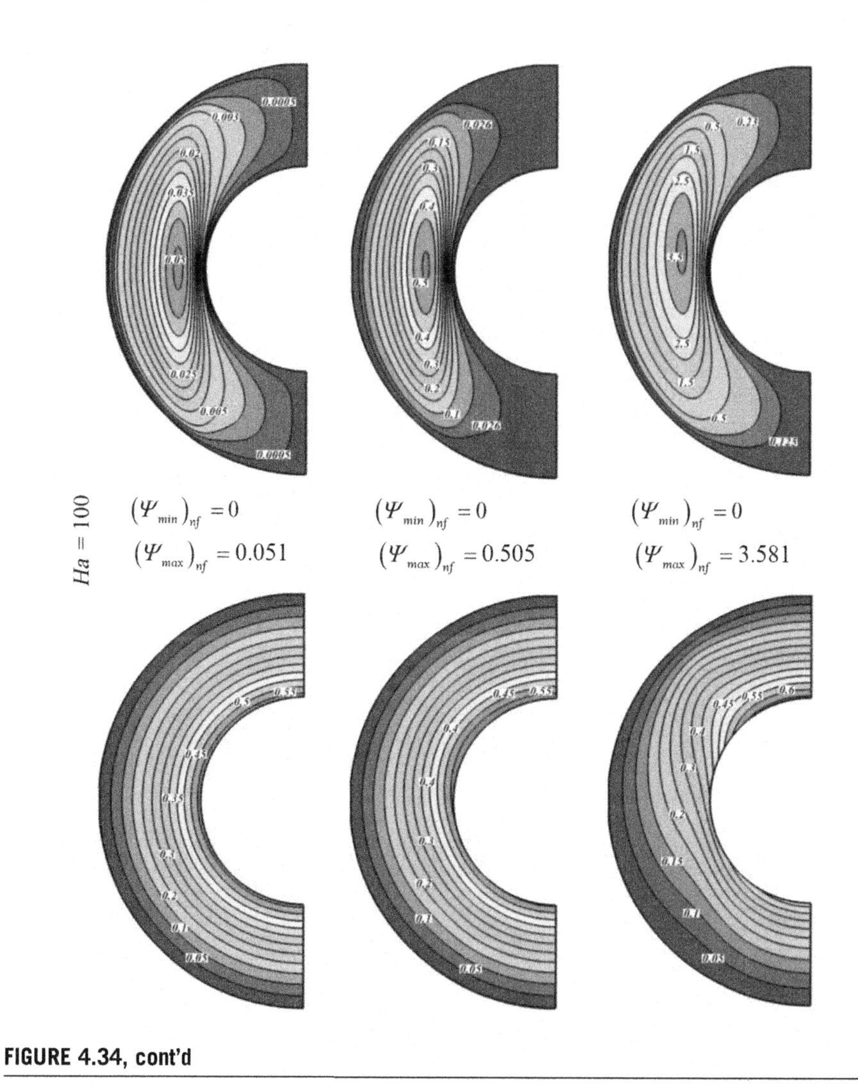

FIGURE 4.34, cont'd

number. At $\gamma = 0°$, the local Nusselt number profile is symmetrical with respect to the vertical center line of the enclosure because of the existence of symmetrical geometry and boundary conditions. But for other inclination angle it is not symmetric. In the presence of a magnetic field, because the conduction mechanism dominates, the maximum local Nusselt number occurs at the vertical center line. At $\gamma = 90°$, the local Nusselt number decreases with an increase in S, and increasing Hartmann number leads to a decrease in the Nusselt number. When $Ha = 0$, the number of extrema in the local Nusselt number profile corresponds to the existance of a thermal plume. The presence of a magnetic field causes the thermal plume over

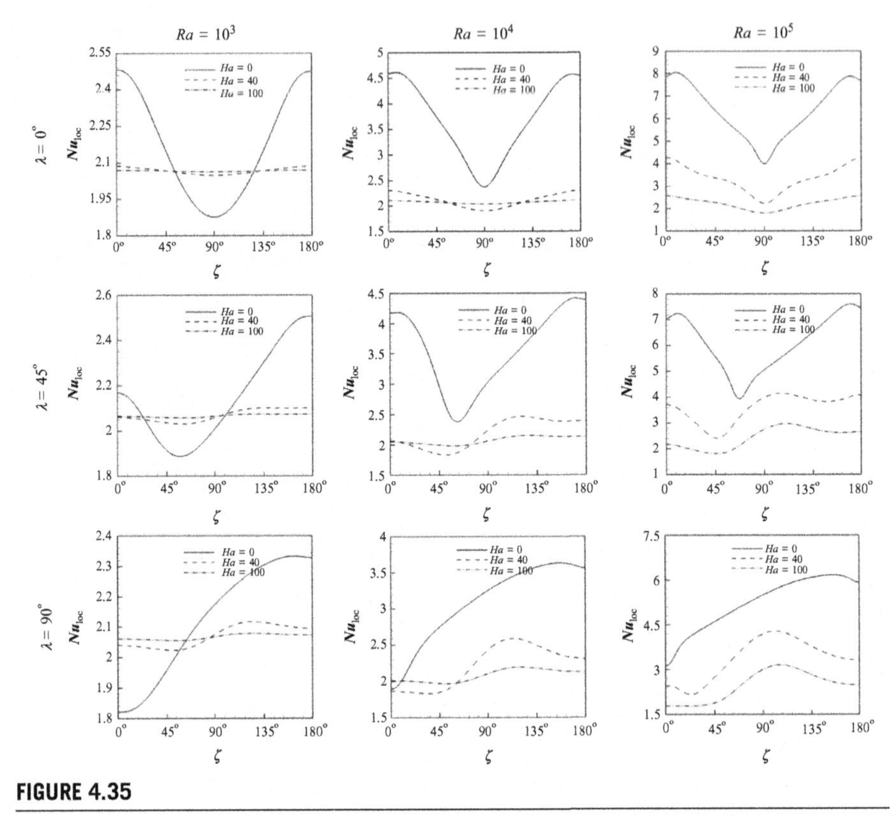

FIGURE 4.35

Effects of the Hartmann number, Rayleigh number, and inclination angle for copper-water nanofluids ($\phi = 0.06$) on the local Nusselt number.

the inner wall to disappear and makes the isotherms parallel to each other due to the domination of the conduction mode of heat transfer. So, the extremum points of the local Nusselt number profiles are omitted when the Hartmann number increases.

Effects of the Hartmann number, Rayleigh number, and inclination angle on the average Nusselt number are shown in Fig. 4.36. For all Rayleigh numbers, in the absence of a magnetic field, the maximum value of Nu_{ave} occurs at $\lambda = 0°$, but for high Hartmann numbers the maximum value of occurs at $\lambda = 90°$. The average Nusselt number increases as the Rayleigh number increases, whereas it decreases as the Hartmann number increases. At $Ra = 10^5$, in the absence of a magnetic field, the maximum average Nusselt number is obtained at $\gamma = 0°$, but for higher Hartmann numbers the maximum value of Nu_{ave} occurs at $\gamma = 90°$.

The effects of the Hartmann number, Rayleigh number, and inclination angle on the heat transfer enhancement ratio are shown in Fig. 4.37. In general, as the Rayleigh number increases the maximum values of enhancement decrease.

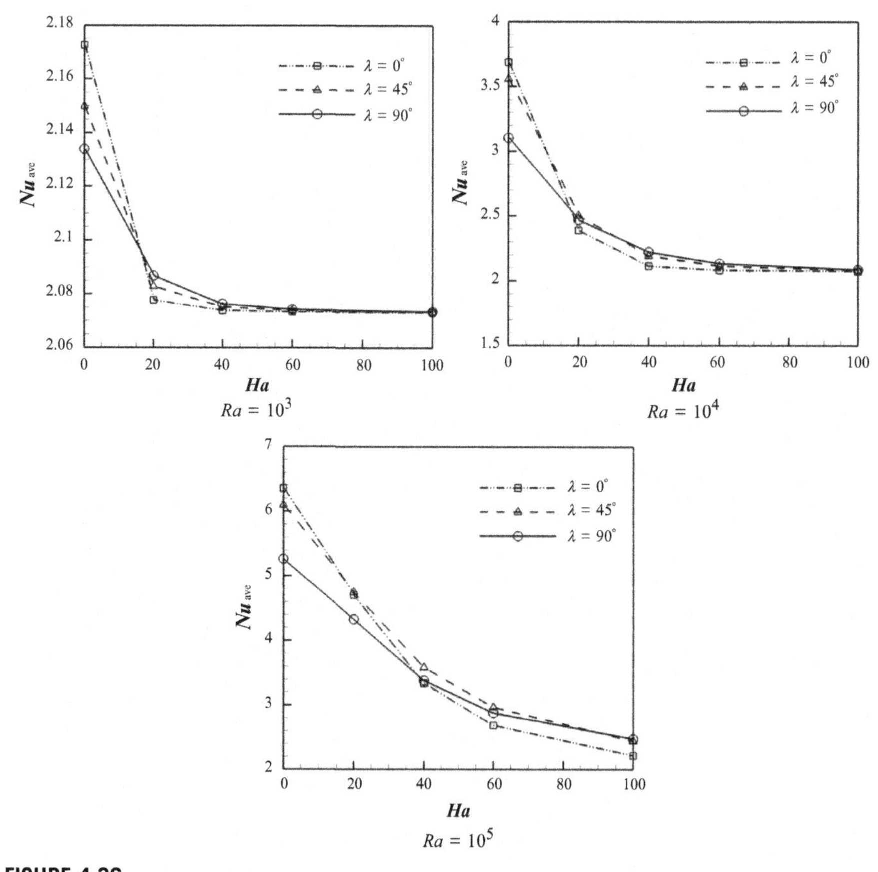

FIGURE 4.36

Effects of the Hartmann number, Rayleigh number, and inclination angle for copper-water nanofluids ($\phi = 0.06$) on the average Nusselt number.

At $Ra = 10^3$ and 10^4, it can be seen than maximum values of enhancement increases as Hartmann number increases, but for $Ra = 10^5$ minimum values of enhancement occurs at $Ha = 40$. An increase in heat transfer enhancement is related to the effect of nanoparticles on the fluid's thermal conductivity. At smaller Rayleigh numbers the dominant mechanism is conduction, and an increase in the Hartmann number enhances the conduction heat transfer by slowing the fluid's circulation. For $Ra = 10^5$, the largest proportion of heat transfer belongs to the convective heat transfer mechanism. Hence an increase in the Hartman number up to at $Ha = 40$ decreases the fluid velocity and convective heat transfer. For higher Hartman numbers the magnetic field slows the fluid flow; conduction again becomes the predominant mechanism of heat transfer and the value of enhancement increases.

At low Rayleigh numbers ($Ra = 10^3$ and 10^4), maximum values of enhancement for low Hartmann numbers are obtained at $\lambda = 90°$, whereas for high Hartmann

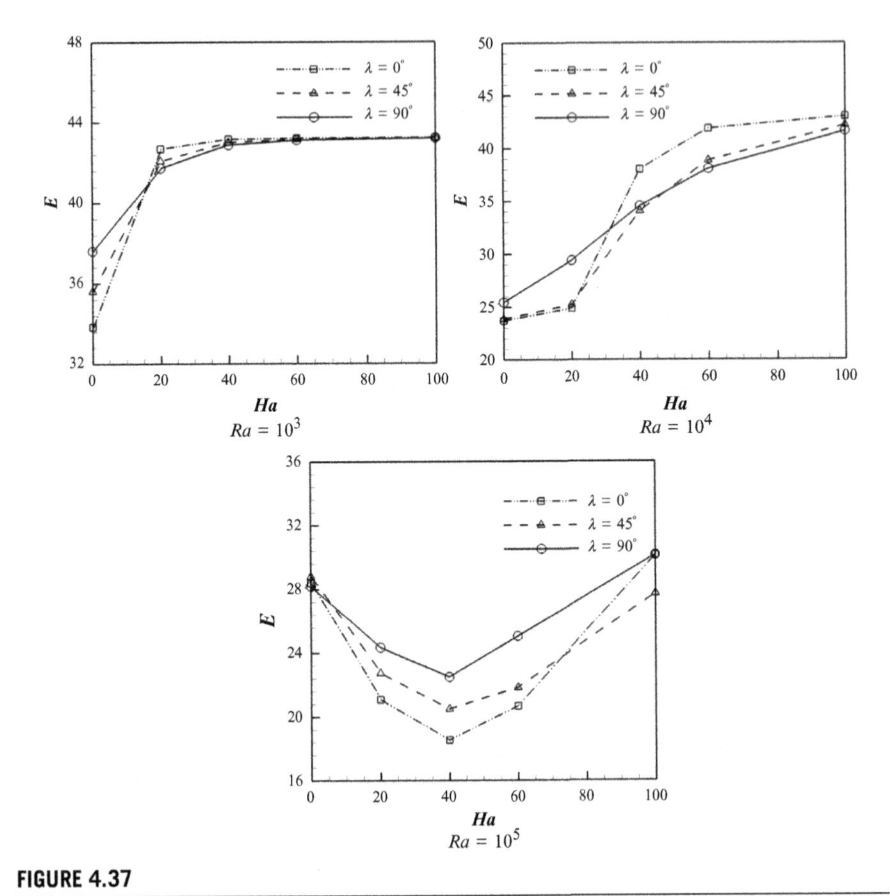

FIGURE 4.37

Effects of *Ha*, λ, and *Ra* on the heat transfer enhancement due to the addition of nanoparticles when $Pr = 6.2$ (a copper-water nanofluid).

numbers maximum values of enhancement occurred at $\lambda = 0°$. It is an interesting observation that at $Ra = 10^5$ the enhancement in heat transfer for case of $\lambda = 90°$ is greater than two other cases for all Hartmann numbers.

4.2.3.7 MHD free convection of an aluminum oxide-water nanofluid considering thermophoresis and Brownian motion effects

4.2.3.7.1 Problem definition

A schematic diagram and the mesh of the semiannulus enclosure used in the present CVFEM program are shown in Fig. 4.38 [34]. The inner and outer walls are maintained at constant temperatures T_h and T_c, respectively, while the two other walls are thermally insulated. Also, the boundary conditions of concentration are similar to temperature. It is assumed that the uniform magnetic field $(\vec{B} = B_x\vec{e}_x + B_y\vec{e}_y)$ of

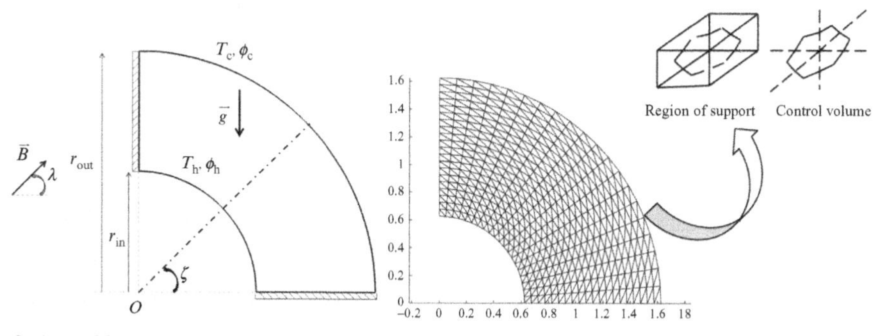

FIGURE 4.38

Geometry and the boundary conditions (a) and the mesh of the enclosure (b) considered in this work.

constant magnitude $B=\sqrt{B_x^2+B_y^2}$ is applied, where $\vec{e}_x$ and $\vec{e}_y$ are unit vectors in the Cartesian coordinate system. The orientation of the magnetic field forms an angle λ with the horizontal axis, such that $\lambda=B_x/B_y$. The electric current and the electromagnetic force are defined by $J=\sigma\left(\vec{V}\times\vec{B}\right)$ and $F=\sigma\left(\vec{V}\times\vec{B}\right)\times\vec{B}$, respectively.

The local Nusselt number on the hot circular wall can be expressed as:

$$Nu_{\text{loc}}=-\frac{\partial\Theta}{\partial n} \tag{4.55}$$

where n is the normal direction of the inner cylinder's surface. The average number on the hot circular wall is evaluated as:

$$Nu_{\text{ave}}=\frac{1}{0.5\pi}\int_0^{0.5\pi} Nu_{\text{loc}}(\zeta)\mathrm{d}\zeta \tag{4.56}$$

4.2.3.7.2 Effects of active parameters

The MHD effect on natural convection heat transfer in an enclosure filled with a nanofluid is investigated numerically using CVFEM. Effects of the Hartmann number ($Ha=0$, 30, 60, and 100), buoyancy ratio number ($Nr=0.1-4$), and Lewis number ($Le=2$, 4, 6, and 8) on flow and heat transfer characteristics are examined. The Brownian motion parameter of nanofluids ($Nb=0.5$), the thermophoretic parameter of nanofluids ($Nt=0.5$), the thermal Rayleigh number ($Ra=10^5$), and the Prandtl number ($Pr=10$) are fixed.

Effects of the Hartmann number, Lewis number, and buoyancy ratio on isotherms, streamlines, isoconcentrations, and heatline contours are shown in Figs. 4.39–4.42. When the magnetic field is imposed on the enclosure, the velocity field suppressed because of the retarding effect of the Lorenz force. So, the intensity of the convection weakens significantly. The braking effect of the magnetic field is

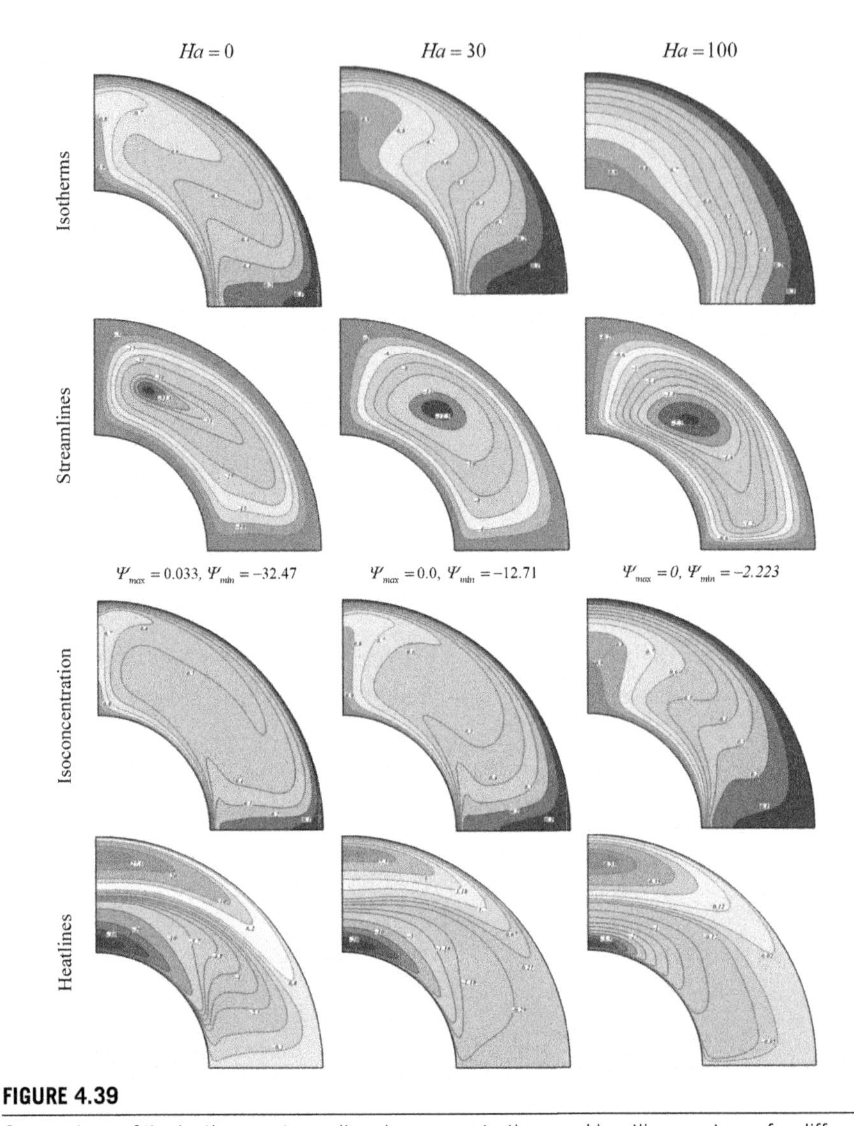

FIGURE 4.39

Comparison of the isotherm, streamline, isoconcentration, and heatline contours for different thermal Hartmann numbers when $Nr = 0.1$, $Le = 8$, $Nt = Nb = 0.5$, $Ra = 10^5$, and $Pr = 10$.

observed from the maximum stream function value. The core of the vortex is shift downward vertically as the Hartmann number increases. Imposing a magnetic field omits the creation of a thermal plume over the inner wall. At high Hartmann numbers the conduction heat transfer mechanism is more pronounced. For this reason the isotherms are parallel to each other. Increasing the Hartmann number causes the concentration boundary layer thickness near the inner wall to increase.

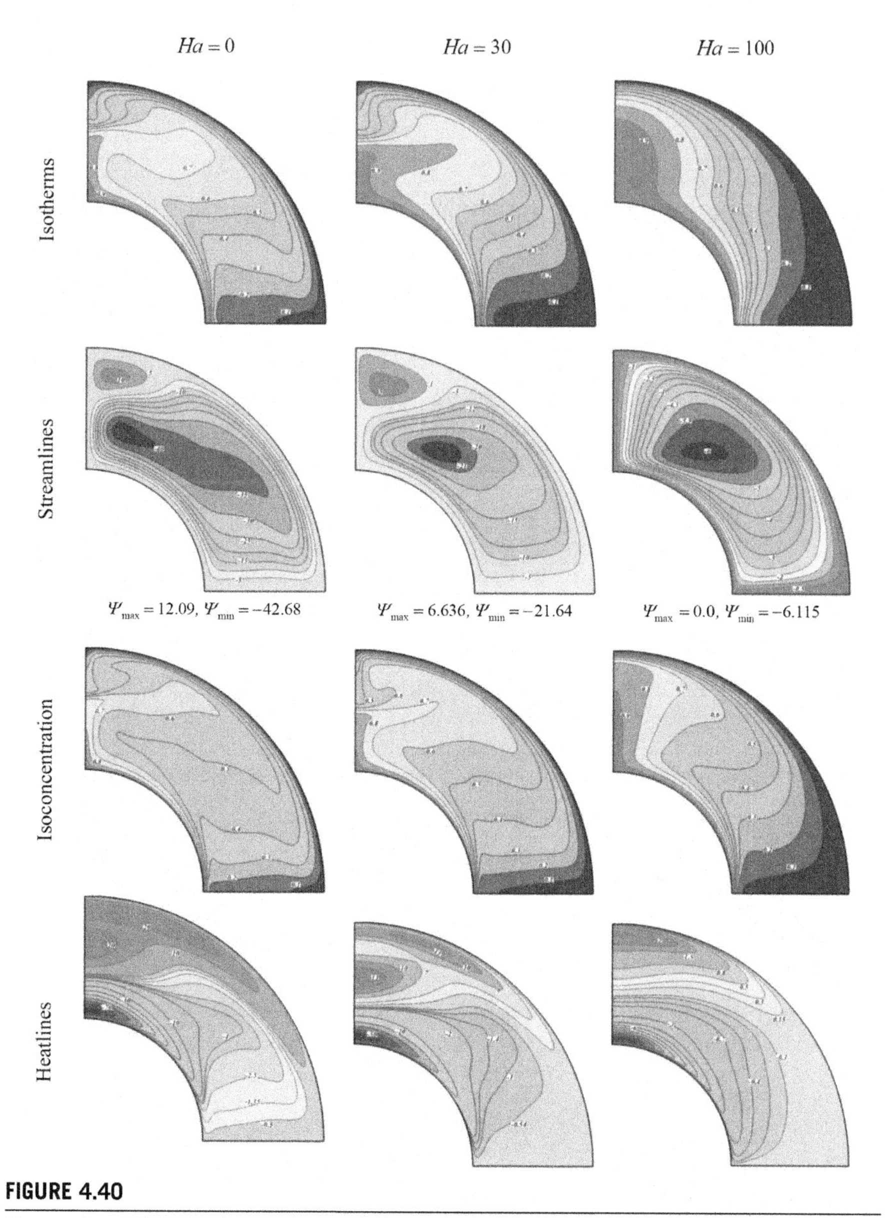

FIGURE 4.40

Comparison of the isotherm, streamline, isoconcentration, and heatline contours for different thermal Hartmann numbers when $Nr=4$, $Le=8$, $Nt=Nb=0.5$, $Ra=10^5$, and $Pr=10$.

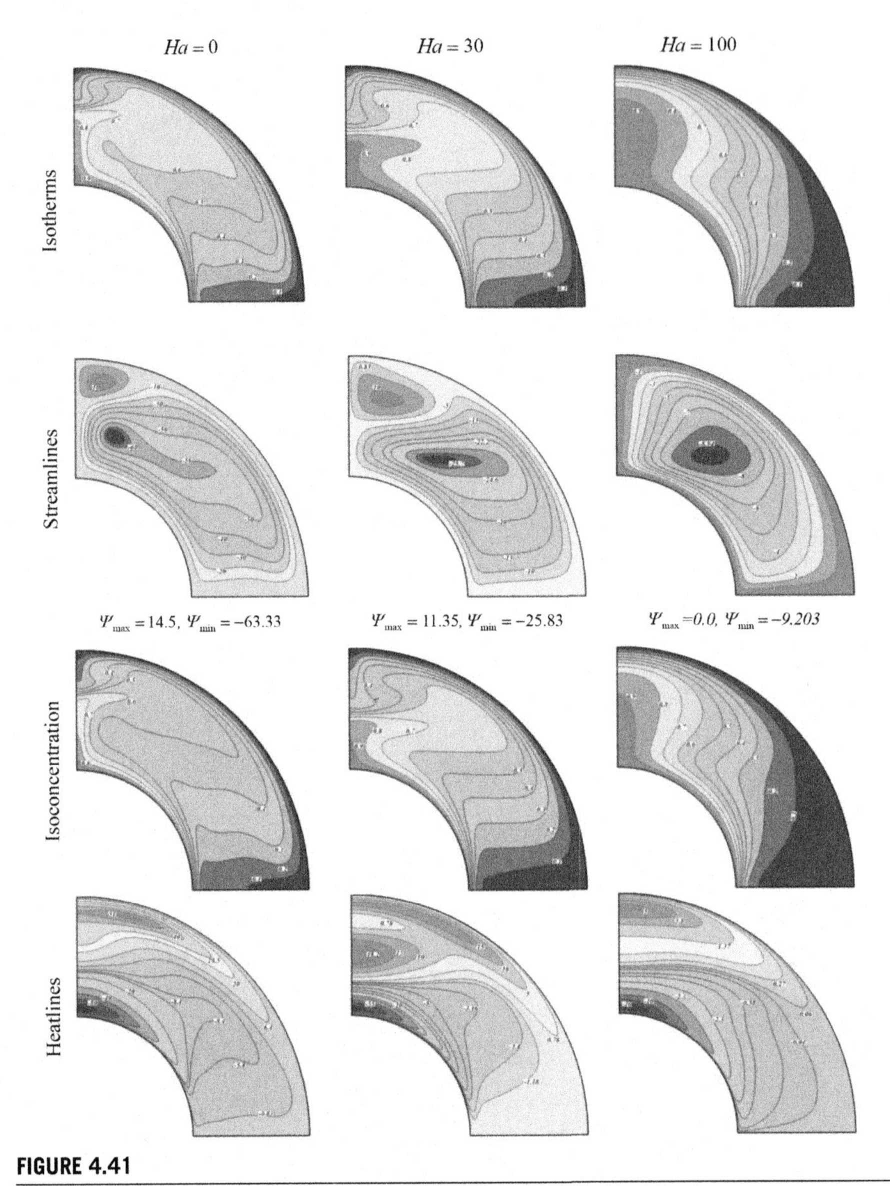

FIGURE 4.41

Comparison of the isotherm, streamline, isoconcentration, and heatline contours for different thermal Hartmann numbers when $Nr=4$, $Le=2$, $Nt=Nb=0.5$, $Ra=10^5$, and $Pr=10$.

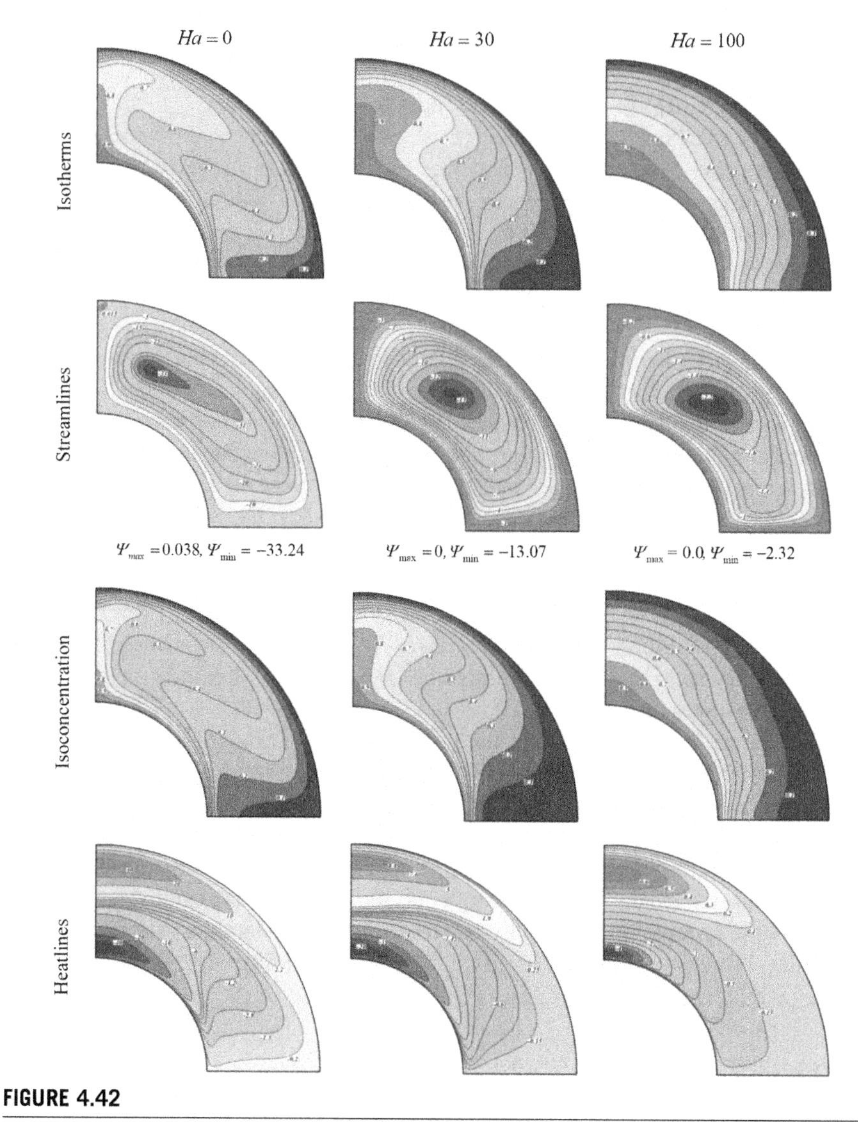

FIGURE 4.42

Comparison of the isotherm, streamline, isoconcentration, and heatline contours for different thermal Hartmann numbers when $Nr = 1$, $Le = 2$, $Nt = Nb = 0.5$, $Ra = 10^5$, and $Pr = 10$.

The heat flow within the enclosure is displayed using the heat function obtained from conductive heat fluxes ($\partial\Theta/\partial X$, $\partial\Theta/\partial Y$) as well as convective heat fluxes ($V\Theta$, $U\Theta$). Heatlines emanate from hot regimes and end on cold regimes, illustrating the path of heat flow. A heatline has two regions that rotate in different directions. The lower one is larger than the other, which means that more heat transfer occurs in this region. As the Hartmann number, increases in the heatlines become weaker because of a reduction in the heat transfer rate caused by applying the magnetic field. The domination of the conduction heat transfer at high Hartmann numbers can be observed from thc heatline patterns since no passive area exists.

It should be mentioned that negative *Nr* values (opposing buoyancy forces) showed more complex and interesting flow patterns, such as multiple cells, which is worth presenting and discussing. In this study positive *Nr* results (aiding buoyancy forces), where the temperature- and species-induced buoyancy forces aide each other, are not considered. For $Nr=0$, the species-induced buoyancy force has no effect on flow; the flow is driven solely by the thermal buoyancy force. However, the effect of species-induced buoyancy increases as *Nr* value increases and reaches a certain value where the effect of thermally induced buoyancy becomes negligible compared with the solutal one. For small *Nr* values, the flow is driven mainly by the thermal buoyancy force. When *Nr* increases a reverse thermal plume appears at $\zeta=90°$. This phenomena is caused by the existence of one counterclockwise eddy at this region. By increasing the Hartmann number, the two main eddies merge into one counterclockwise eddy. The isoconcentrations are more distorted with an increase in the solutal forces. As *Nr* increases the upper region of the heatline is divided into two smaller ones, and this new region disappears as the Hartmann number increases.

The mass flow is given by $\psi_{max}\approx\delta_s v$, where the solutal boundary layer thickness is given by $\delta_s\approx(RaLeNr)^{-1/4}$, and $v\approx(RaLeNr)^{1/2}$ so $\psi_{max}\approx(RaLeNr)^{1/4}$. The solutal boundary layer, δ_s becomes thinner by increasing *Le*. Heatlines are more distorted as *Le* is augmented.

Effects of the Hartmann number, buoyancy ratio, and Lewis number on the local Nusselt number are shown in Fig. 4.43. The local Nusselt number increases as the buoyancy ratio increases, but it decreases as the Hartmann number and Lewis number increase. In general, as ζ increases the local Nusselt number decreases because of the increase in the thermal boundary layer thickness. In the absence of a magnetic field, Nu_{loc} has one local maximum near the bottom wall. The occurrence of maxima for Nu_{loc} is due to dense heatlines based on the conductive heat transport occurring at this portion. This point disappears at high Hartmann numbers. The local Nusselt number profile has a minimum point at $\zeta=90°$ because of the thermal plume that exists in this region. Figure 4.43 also shows that the effect of the Lewis number on Nu_{loc} is negligible at low buoyancy ratios.

The corresponding polynomial representation of such a model of the Nusselt number is:

$$\begin{aligned}
Nu &= a_{13}+a_{23}Y_1+a_{33}Y_2+a_{43}Y_1^2+a_{53}Y_2^2+a_{63}Y_1Y_2\\
Y_1 &= a_{11}+a_{21}Le+a_{31}Nr+a_{41}Le^2+a_{51}Nr^2+a_{61}LeNr\\
Y_2 &= a_{12}+a_{22}Ha^*+a_{32}Nr+a_{42}Ha^{*2}+a_{52}Nr^2+a_{62}Ha^*Nr
\end{aligned} \tag{4.57}$$

where values of a_{ij} can be found in Table 4.5; for example, a_{21} equals −0.16499.

Effects of the Hartmann number, buoyancy ratio, and Lewis number on the average Nusselt number are shown in Figs. 4.44 and 4.45. The presence of a magnetic field causes the thermal plume over the inner wall to disappear and makes the

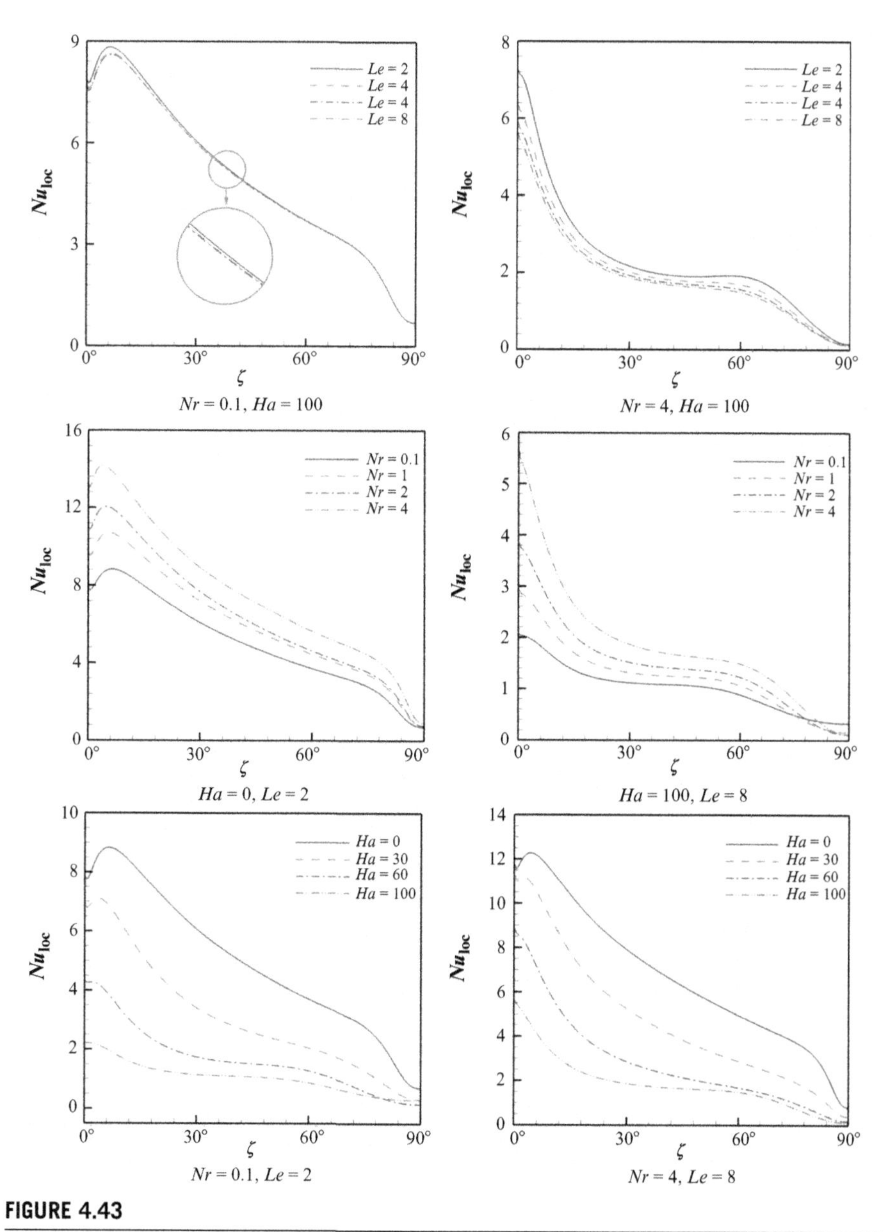

FIGURE 4.43

Effects of the Hartmann number, buoyancy ratio, and Lewis number on the local Nusselt number at $Nt = Nb = 0.5$, $Ra = 10^5$, and $Pr = 10$.

Table 4.5 Constant Coefficients (a_{ij}) for Using Eq. (4.57)

a_{ij}	$i=1$	$i=2$	$i=3$	$i=4$	$i=5$	$i=6$
$j=1$	3.069651	−0.16499	0.578896	0.012949	−0.01115	−0.02536
$j=2$	4.906119	−7.59508	0.585453	3.681993	−0.01115	−0.28078
$j=3$	0.819671	−0.56997	0.947823	0.094258	−0.00422	0.021492

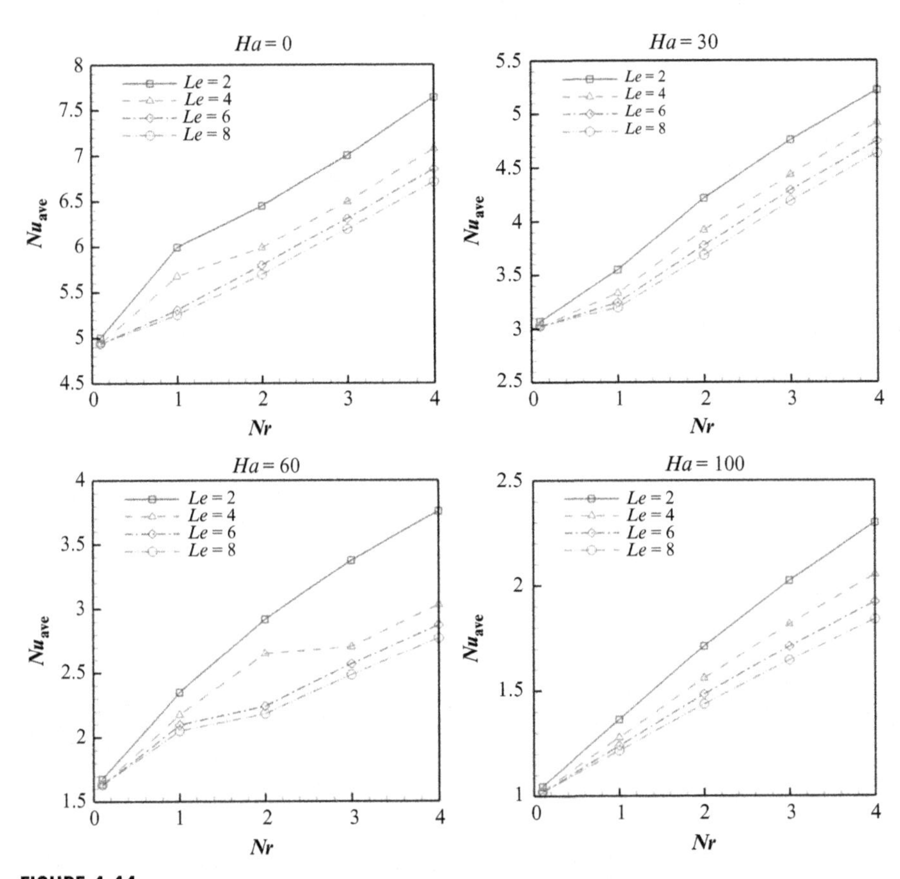

FIGURE 4.44

Effects of the Hartmann number, buoyancy ratio, and Lewis number on the average Nusselt number at $Nt = Nb = 0.5$, $Ra = 10^5$, and $Pr = 10$.

isotherms parallel to each other because of the domination of the conduction mode of heat transfer. Therefore the average Nusselt number decreases as the Hartmann number increases. As the Lewis number increases, the thermal boundary layer thickness increases and, in turn, the Nusselt number decreases. The effect of the buoyancy ratio on Nu_{ave} is in contrast with Ha and Le. The effects of the Hartmann number and Lewis number are more pronounced at higher buoyancy ratios.

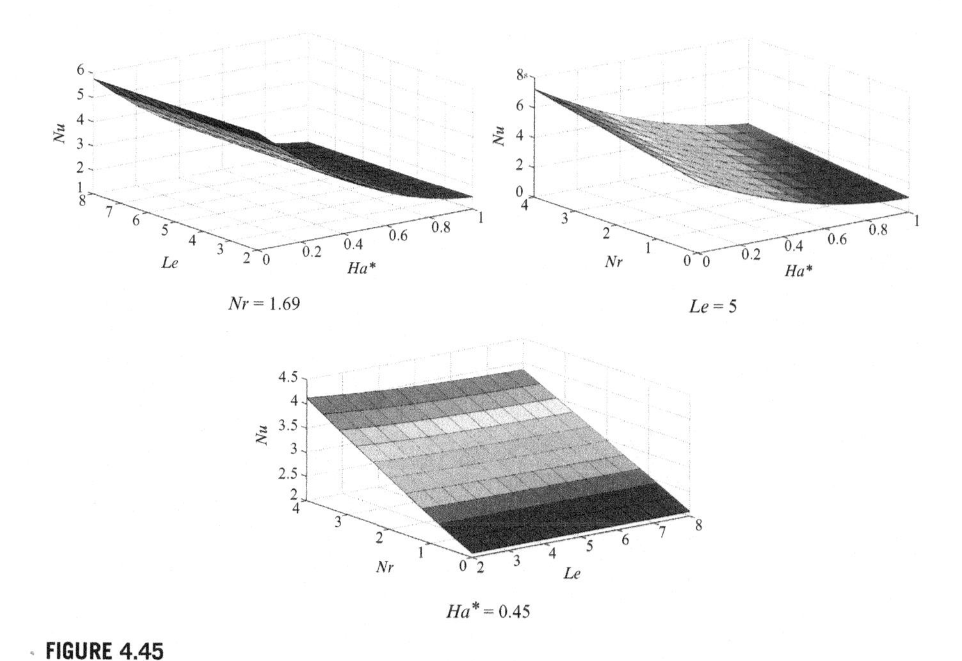

FIGURE 4.45

Variations of Nu_{ave} for various input parameters.

4.3 COMBINED EFFECTS OF FERROHYDRODYNAMICS AND MHD

To express the magnetic field strength, one can consider that the magnetic source represents a magnetic wire placed vertical to the x-y plane at the point $(\overline{a}, \overline{b})$. The components of the magnetic field's intensity $(\overline{H_x}, \overline{H_y})$ and strength $(\overline{H})$ can be considered as [24]:

$$\overline{H_x} = \frac{\gamma}{2\pi} \frac{1}{(x-\overline{a})^2 + (y-\overline{b})^2}(y-\overline{b}) \tag{4.58}$$

$$\overline{H_y} = -\frac{\gamma}{2\pi} \frac{1}{(x-\overline{a})^2 + (y-\overline{b})^2}(x-\overline{a}) \tag{4.59}$$

$$\overline{H} = \sqrt{\overline{H}_x^2 + \overline{H}_y^2} = \frac{\gamma}{2\pi} \frac{1}{\sqrt{(x-\overline{a})^2 + (y-\overline{b})^2}} \tag{4.60}$$

where γ is the magnetic field's strength at the source (of the wire) and $(\overline{a}, \overline{b})$ is the position where the source is located.

4.3.1 MATHEMATICAL MODELING FOR A SINGLE-PHASE MODEL

4.3.1.1 Natural convection

The flow is considered to be steady, two-dimensional, incompressible, and laminar. Using the Boussinesq approximation, the governing equations of heat transfer and fluid flow for a nanofluid can be obtained:

$$\frac{\partial u}{\partial x}+\frac{\partial v}{\partial y}=0 \tag{4.61}$$

$$\rho_{\mathrm{nf}}\left(u\frac{\partial u}{\partial x}+v\frac{\partial u}{\partial y}\right)=-\frac{\partial P}{\partial x}+\mu_{\mathrm{nf}}\left(\frac{\partial^2 u}{\partial x^2}+\frac{\partial^2 u}{\partial y^2}\right)+\mu_0 M\frac{\partial \overline{H}}{\partial x}-\sigma_{\mathrm{nf}}B_y^2 u+\sigma_{\mathrm{nf}}B_x B_y v \tag{4.62}$$

$$\rho_{\mathrm{nf}}\left(u\frac{\partial v}{\partial x}+v\frac{\partial v}{\partial y}\right)=-\frac{\partial P}{\partial y}+\mu_{\mathrm{nf}}\left(\frac{\partial^2 v}{\partial x^2}+\frac{\partial^2 v}{\partial y^2}\right)+\mu_0 M\frac{\partial \overline{H}}{\partial y}-\sigma_{\mathrm{nf}}B_x^2 v+\sigma_{\mathrm{nf}}B_x B_y u+\rho_{\mathrm{nf}}\beta_{\mathrm{nf}}g(T-T_{\mathrm{c}}) \tag{4.63}$$

$$(\rho C_{\mathrm{p}})_{\mathrm{nf}}\left(u\frac{\partial T}{\partial x}+v\frac{\partial T}{\partial y}\right)=k_{\mathrm{nf}}\left(\frac{\partial^2 T}{\partial x^2}+\frac{\partial^2 T}{\partial y^2}\right)+\sigma_{\mathrm{nf}}(uB_y-vB_x)^2 -\mu_0 T\frac{\partial M}{\partial T}\left(u\frac{\partial \overline{H}}{\partial x}+v\frac{\partial \overline{H}}{\partial y}\right)+\mu_{\mathrm{nf}}\left\{2\left(\frac{\partial u}{\partial x}\right)^2+2\left(\frac{\partial v}{\partial x}\right)^2+\left(\frac{\partial u}{\partial x}+\frac{\partial v}{\partial y}\right)^2\right\} \tag{4.64}$$

The terms $\mu_0 M\frac{\partial \overline{H}}{\partial x}$ and $\mu_0 M\frac{\partial \overline{H}}{\partial y}$ in Eqs. (4.62) and (4.63), respectively, represent the components of magnetic force per unit volume and depend on the existence of the magnetic gradient in the corresponding x and y directions. These two terms are well known from ferrohydrodynamic (FHD) and are the so-called Kelvin force. The terms $-\sigma_{\mathrm{nf}}B_y^2 u+\sigma_{\mathrm{nf}}B_x B_y v$ and $-\sigma_{\mathrm{nf}}B_x^2 v+\sigma_{\mathrm{nf}}B_x B_y u$ appearing in Eqs. (4.62) and (4.63), respectively, represent the Lorentz force per unit volume toward the x and y directions and arise due to the electrical conductivity of the fluid. These two terms are known in MHD. The principles of MHD and FHD are combined in the mathematical model presented by Tzirtzilakis et al. [24], and the above-mentioned terms arise together in the governing Eqs. (4.62) and (4.63). The term $\mu_0 T\frac{\partial M}{\partial T}\left(u\frac{\partial \overline{H}}{\partial x}+v\frac{\partial \overline{H}}{\partial y}\right)$ in Eq. (4.64) represents the thermal power per unit volume due to the magnetocaloric effect. Also, the term in Eq. (4.64) represents the Joule heating. For varying the magnetization M with the magnetic field intensity $\overline{H}$ and temperature T, the following relation, derived experimentally [24], is considered:

$$M=K'\overline{H}(T_{\mathrm{c}}'-T) \tag{4.65}$$

where K' is a constant and T_{c}' is the Curie temperature.

In the above equations, μ_0 is the magnetic permeability of a vacuum $(4\pi\times 10^{-7}\ \mathrm{T\ m/A})$, $\overline{H}$ is the magnetic field strength, and $\overline{B}$ is the magnetic induction

$(\overline{B} = \mu_0 \overline{H})$; the bar above the quantities denotes that they are dimensional. The effective density (ρ_{nf}) and heat capacitance $(\rho C_p)_{nf}$ of the nanofluid are defined as [30]:

$$\rho_{nf} = \rho_f(1-\phi) + \rho_s \phi \tag{4.66}$$

$$(\rho C_p)_{nf} = (\rho C_p)_f (1-\phi) + (\rho C_p)_s \phi \tag{4.67}$$

where ϕ is the solid volume fraction of the nanoparticles. Thermal diffusivity of the nanofluid is

$$\alpha_{nf} = \frac{k_{nf}}{(\rho C_p)_{nf}} \tag{4.68}$$

and the thermal expansion coefficient of the nanofluid can be determined as

$$\beta_{nf} = \beta_f(1-\phi) + \beta_s \phi \tag{4.69}$$

The dynamic viscosity of the nanofluids given as

$$\mu_{nf} = \frac{\mu_f}{(1-\phi)^{2.5}} \tag{4.70}$$

The effective thermal conductivity of the nanofluid can be approximated using the Maxwell-Garnetts model as [24]:

$$\frac{k_{nf}}{k_f} = \frac{k_s + 2k_f - 2\phi(k_f - k_s)}{k_s + 2k_f + \phi(k_f - k_s)} \tag{4.71}$$

and the effective electrical conductivity of a nanofluid was presented as

$$\frac{\sigma_{nf}}{\sigma_f} = 1 + \frac{3\left(\frac{\sigma_s}{\sigma_f} - 1\right)\phi}{\left(\frac{\sigma_s}{\sigma_f} + 2\right) - \left(\frac{\sigma_s}{\sigma_f} - 1\right)\phi} \tag{4.72}$$

The stream function and vorticity are defined as:

$$u = \frac{\partial \psi}{\partial y}, \quad v = -\frac{\partial \psi}{\partial x}, \quad \omega = \frac{\partial v}{\partial x} - \frac{\partial u}{\partial y} \tag{4.73}$$

The stream function satisfies the continuity (Eq. 4.61). The vorticity equation is obtained by eliminating the pressure between the two momentum equations, i.e., by taking the y-derivative of Eq. (4.62) and subtracting from it the x-derivative of Eq. (4.63). By introducing the following nondimensional variables:

$$X = \frac{x}{L}, \quad Y = \frac{y}{L}, \quad \Omega = \frac{\omega L^2}{\alpha_f}, \quad \Psi = \frac{\psi}{\alpha_f}, \quad \Theta = \frac{T - T_c}{T_h - T_c},$$

$$U = \frac{uL}{\alpha_f}, \quad V = \frac{vL}{\alpha_f}, \quad H = \frac{\overline{H}}{\overline{H_0}}, \quad H_x = \frac{\overline{H_x}}{\overline{H_0}}, \quad H_y = \frac{\overline{H_y}}{\overline{H_0}} \tag{4.74}$$

where in Eq. (4.74) $\overline{H_0}=\overline{H}(\overline{a},0)=\dfrac{\gamma}{2\pi|b|}$ and $L-r_{out}-r_{in}=r_{in}$. Using the dimensionless parameters, the equations now become:

$$\begin{aligned}\frac{\partial\Psi}{\partial Y}\frac{\partial\Omega}{\partial X}-\frac{\partial\Psi}{\partial X}\frac{\partial\Omega}{\partial Y}&=Pr\left[\frac{\mu_{nf}/\mu_f}{\rho_{nf}/\rho_f}\right]\left(\frac{\partial^2\Omega}{\partial X^2}+\frac{\partial^2\Omega}{\partial Y^2}\right)+RaPr\left[\frac{\beta_{nf}}{\beta_f}\right]\left(\frac{\partial\Theta}{\partial X}\right)\\&+Mn_F\Pr\left(\frac{\rho_f}{\rho_{nf}}\right)\left\{\frac{\partial H}{\partial X}\frac{\partial\Theta}{\partial Y}-\frac{\partial H}{\partial Y}\frac{\partial\Theta}{\partial X}\right\}H-Ha^2Pr\left[\frac{\sigma_{nf}/\sigma_f}{\rho_{nf}/\rho_f}\right]\\&\times\left\{\frac{\partial V}{\partial X}H_x^2+V\left(2H_x\frac{\partial H_x}{\partial X}\right)-\frac{\partial U}{\partial X}H_xH_y-U\frac{\partial H_x}{\partial X}H_y-U\frac{\partial H_y}{\partial X}H_x\right.\\&\left.-\frac{\partial U}{\partial Y}H_y^2-U\left(2H_y\frac{\partial H_y}{\partial Y}\right)+\frac{\partial V}{\partial Y}H_xH_y+V\frac{\partial H_x}{\partial Y}H_y+V\frac{\partial H_y}{\partial Y}H_x\right\}\end{aligned}\tag{4.75}$$

$$\begin{aligned}\frac{\partial\Psi}{\partial Y}\frac{\partial\Theta}{\partial X}-\frac{\partial\Psi}{\partial X}\frac{\partial\Theta}{\partial Y}&=\left(\frac{k_{nf}}{k_f}\right)\bigg/\left(\frac{(\rho C_p)_{nf}}{(\rho C_p)_f}\right)\left(\frac{\partial^2\Theta}{\partial X^2}+\frac{\partial^2\Theta}{\partial Y^2}\right)\\&+Ha^2Ec\left(\frac{\sigma_{nf}}{\sigma_f}\right)\bigg/\left(\frac{(\rho C_p)_{nf}}{(\rho C_p)_f}\right)\{UH_y-VH_x\}^2\\&+Mn_FEc\frac{(\rho C_p)_f}{(\rho C_p)_{nf}}\left\{U\frac{\partial H}{\partial X}+V\frac{\partial H}{\partial Y}\right\}H(\varepsilon_1+\Theta)\\&+Ec\left(\frac{\mu_{nf}}{\mu_f}\right)\bigg/\left(\frac{(\rho C_p)_{nf}}{(\rho C_p)_f}\right)\left\{2\left(\frac{\partial U}{\partial X}\right)^2+2\left(\frac{\partial V}{\partial X}\right)^2+\left(\frac{\partial U}{\partial X}+\frac{\partial V}{\partial Y}\right)^2\right\}\end{aligned}\tag{4.76}$$

$$\frac{\partial^2\Psi}{\partial X^2}+\frac{\partial^2\Psi}{\partial Y^2}=-\Omega\tag{4.77}$$

where $Ra_f=g\beta_fL^3(T_h-T_c)/(\alpha_f\upsilon_f)$, $Pr_f=\upsilon_f/\alpha_f$, $Ha=L\mu_0H_0\sqrt{\sigma_f/\mu_f}$, $\varepsilon_1=T_1/\Delta T$, $Ec=(\mu_f\alpha_f)/\left[(\rho C_p)_f\Delta TL^2\right]$, and $Mn_F=\mu_0H_0^2K'(T_h-T_c)L^2/(\mu_f\alpha_f)$ are the Rayleigh number, Prandtl number, Hartmann number arising from MHD, temperature, Eckert number, and magnetic number arising from FHD for the base fluid, respectively. The thermophysical properties of the nanofluid are given in Table 4.6 [28].

Table 4.6 Thermophysical Properties of Water and Nanoparticles [28]

	ρ (kg/m^3)	C_p (J/kg K)	k (W/m K)	$\beta\times10^3$ (K^{-1})	d_p (nm)	σ (Ω m)$^{-1}$
Pure water	997.1	4179	0.613	21	–	0.05
Fe_3O_4	5200	670	6	1.3	47	25,000

4.3.1.2 Mixed convection

To obtain the governing equations for force convection, the nondimensional variables should be introduced as:

$$X=\frac{x}{L},\ Y=\frac{y}{L},\ \Omega=\frac{\omega L}{u_{\mathrm{r}}},\ \Psi=\frac{\psi}{Lu_{\mathrm{r}}},\ \Theta=\frac{T-T_{\mathrm{c}}}{T_{\mathrm{h}}-T_{\mathrm{c}}},$$
$$U=\frac{u}{u_{\mathrm{r}}},\ V=\frac{vL}{u_{\mathrm{r}}},\ H=\frac{\overline{H}}{\overline{H_0}},\ H_x=\frac{\overline{H_x}}{\overline{H_0}},\ H_y=\frac{\overline{H_y}}{\overline{H_0}} \tag{4.78}$$

where in Eq. (4.78) $\overline{H_0}=\overline{H}(\overline{a},0)=\frac{\gamma}{2\pi|b|}$. Using the dimensionless parameters, the equations now become:

$$\begin{aligned}\frac{\partial\Psi}{\partial Y}\frac{\partial\Omega}{\partial X}-\frac{\partial\Psi}{\partial X}\frac{\partial\Omega}{\partial Y}=&\frac{1}{Re}\left[\frac{\mu_{\mathrm{nf}}/\mu_{\mathrm{f}}}{\rho_{\mathrm{nf}}/\rho_{\mathrm{f}}}\right]\left(\frac{\partial^2\Omega}{\partial X^2}+\frac{\partial^2\Omega}{\partial Y^2}\right)\\&+Mn_{\mathrm{F}}\left(\frac{\rho_{\mathrm{f}}}{\rho_{\mathrm{nf}}}\right)H\left\{\frac{\partial H}{\partial X}\frac{\partial\Theta}{\partial Y}-\frac{\partial H}{\partial Y}\frac{\partial\Theta}{\partial X}\right\}-\frac{Ha^2}{Re}\left[\frac{\sigma_{\mathrm{nf}}/\sigma_{\mathrm{f}}}{\rho_{\mathrm{nf}}/\rho_{\mathrm{f}}}\right]\\&\times\left\{\frac{\partial V}{\partial X}H_x^2+V\left(2H_x\frac{\partial H_x}{\partial X}\right)-\frac{\partial U}{\partial X}H_xH_y-U\frac{\partial H_x}{\partial X}H_y-U\frac{\partial H_y}{\partial X}H_x\right.\\&\left.-\frac{\partial U}{\partial Y}H_y^2-U\left(2H_y\frac{\partial H_y}{\partial Y}\right)+\frac{\partial V}{\partial Y}H_xH_y+V\frac{\partial H_x}{\partial Y}H_y+V\frac{\partial H_y}{\partial Y}H_x\right\}+\frac{Gr}{Re^2}\left[\frac{\beta_{\mathrm{nf}}}{\beta_{\mathrm{f}}}\right]\left(\frac{\partial\Theta}{\partial X}\right)\end{aligned} \tag{4.79}$$

$$\begin{aligned}\frac{\partial\Psi}{\partial Y}\frac{\partial\Theta}{\partial X}-\frac{\partial\Psi}{\partial X}\frac{\partial\Theta}{\partial Y}=&\frac{1}{PrRe}\left(\frac{k_{\mathrm{nf}}}{k_{\mathrm{f}}}\right)/\left(\frac{(\rho C_{\mathrm{p}})_{\mathrm{nf}}}{(\rho C_{\mathrm{p}})_{\mathrm{f}}}\right)\left(\frac{\partial^2\Theta}{\partial X^2}+\frac{\partial^2\Theta}{\partial Y^2}\right)\\&+Ha^2\frac{Ec}{Re}\left(\frac{\sigma_{\mathrm{nf}}}{\sigma_{\mathrm{f}}}\right)/\left(\frac{(\rho C_{\mathrm{P}})_{\mathrm{nf}}}{(\rho C_{\mathrm{p}})_{\mathrm{f}}}\right)\{UH_y-VH_x\}^2\\&+Mn_{\mathrm{F}}Ec\frac{(\rho C_{\mathrm{p}})_{\mathrm{f}}}{(\rho C_{\mathrm{p}})_{\mathrm{nf}}}\left\{U\frac{\partial H}{\partial X}+V\frac{\partial H}{\partial Y}\right\}H(\varepsilon_1+\Theta)\\&+\frac{Ec}{Re}\left(\frac{\mu_{\mathrm{nf}}}{\mu_{\mathrm{f}}}\right)/\left(\frac{(\rho C_{\mathrm{p}})_{\mathrm{nf}}}{(\rho C_{\mathrm{p}})_{\mathrm{f}}}\right)\left\{2\left(\frac{\partial U}{\partial X}\right)^2+2\left(\frac{\partial V}{\partial X}\right)^2+\left(\frac{\partial U}{\partial X}+\frac{\partial V}{\partial Y}\right)^2\right\}\end{aligned} \tag{4.80}$$

$$\frac{\partial^2\Psi}{\partial X^2}+\frac{\partial^2\Psi}{\partial Y^2}=-\Omega \tag{4.81}$$

where $Re=\frac{\rho_{\mathrm{f}}Lu_{\mathrm{r}}}{\mu_{\mathrm{f}}}$, $Ha=LH_0\mu_0\sqrt{\sigma_{\mathrm{f}}/\mu_{\mathrm{f}}}$, $\varepsilon_1=T_1/\Delta T$, $Ec=\left(\rho_{\mathrm{f}}u_{\mathrm{r}}^2\right)/\left[(\rho C_{\mathrm{p}})_{\mathrm{f}}\Delta T\right]$, and $Mn_{\mathrm{F}}=\mu_0H_0^2K'(T-T_{\mathrm{c}})/\left(\rho_{\mathrm{f}}u_{\mathrm{r}}^2\right)$ are the Reynolds number, Grashof number, Prandtl number, Hartmann number arising from MHD, temperature, Eckert number, and magnetic number arising from FHD for the base fluid, respectively.

4.3.2 MATHEMATICAL MODELING FOR A TWO-PHASE MODEL

4.3.2.1 Natural convection

A nanofluid's density ρ is

$$\begin{aligned}\rho &= \phi\rho_{\mathrm{p}} + (1-\phi)\rho_{\mathrm{f}} \\ &\cong \phi\rho_{\mathrm{p}} + (1-\phi)\{\rho_{\mathrm{f_0}}(1-\beta(T-T_{\mathrm{c}}))\}\end{aligned} \tag{4.82}$$

where ρ_{f} is the base fluid's density, T_{c} is a reference temperature, $\rho_{\mathrm{f_0}}$ is the base fluid's density at the reference temperature, and β is the volumetric coefficient of expansion. Taking the density of the base fluid as that of the nanofluid, the density ρ in Eq. (4.82) thus becomes

$$\rho \cong \phi\rho_{\mathrm{p}} + (1-\phi)\{\rho_0(1-\beta(T-T_{\mathrm{c}})\} \tag{4.83}$$

ρ_0 is the nanofluid's density at the reference temperature.

The continuity, momentum under Boussinesq approximation, and energy equations for laminar and steady-state natural convection in a two-dimensional enclosure can be written in dimensional form as follows:

$$\frac{\partial u}{\partial x} + \frac{\partial v}{\partial y} = 0 \tag{4.84}$$

$$\rho_{\mathrm{f}}\left\{u\frac{\partial u}{\partial x} + v\frac{\partial u}{\partial y}\right\} = -\frac{\partial P}{\partial x} + \mu\left(\frac{\partial^2 u}{\partial x^2} + \frac{\partial^2 u}{\partial y^2}\right) + \mu_0 M\frac{\partial \overline{H}}{\partial x} - \sigma B_y^2 u + \sigma B_x B_y v \tag{4.85}$$

$$\begin{aligned}\rho_{\mathrm{f}}\left\{u\frac{\partial v}{\partial x} + v\frac{\partial v}{\partial y}\right\} &= -\frac{\partial P}{\partial y} + \mu\left(\frac{\partial^2 v}{\partial x^2} + \frac{\partial^2 v}{\partial y^2}\right) \\ &\quad - (\phi-\phi_{\mathrm{c}})\left(\rho_{\mathrm{p}} - \rho_{\mathrm{f_0}}\right)g + (1-\phi_{\mathrm{c}})\rho_{\mathrm{f_0}}(T-T_{\mathrm{c}})g + \mu_0 M\frac{\partial \overline{H}}{\partial y} - \sigma B_x^2 v + \sigma B_x B_y u\end{aligned} \tag{4.86}$$

$$\begin{aligned}u\frac{\partial T}{\partial x} + v\frac{\partial T}{\partial y} &= \alpha\left(\frac{\partial^2 T}{\partial x^2} + \frac{\partial^2 T}{\partial y^2}\right) + \frac{(\rho c)_{\mathrm{p}}}{(\rho c)_{\mathrm{f}}}\left[\begin{array}{l} D_{\mathrm{B}}\left\{\dfrac{\partial \phi}{\partial x}\cdot\dfrac{\partial T}{\partial x} + \dfrac{\partial \phi}{\partial y}\cdot\dfrac{\partial T}{\partial y}\right\} \\ + (D_{\mathrm{T}}/T_{\mathrm{c}})\left\{\left(\dfrac{\partial T}{\partial x}\right)^2 + \left(\dfrac{\partial T}{\partial y}\right)^2\right\}\end{array}\right] \\ &\quad + \left(uB_y - vB_x\right)^2 - \mu_0 T\frac{\partial M}{\partial T}\left(u\frac{\partial \overline{H}}{\partial x} + \frac{\partial \overline{H}}{\partial y}\right) + \mu\left\{2\left(\frac{\partial u}{\partial x}\right)^2 + 2\left(\frac{\partial v}{\partial x}\right)^2 + \left(\frac{\partial u}{\partial x} + \frac{\partial v}{\partial y}\right)^2\right\}\end{aligned} \tag{4.87}$$

$$u\frac{\partial \phi}{\partial x} + v\frac{\partial \phi}{\partial y} = D_{\mathrm{B}}\left\{\frac{\partial^2 \phi}{\partial x^2} + \frac{\partial^2 \phi}{\partial y^2}\right\} + \left(\frac{D_{\mathrm{T}}}{T_{\mathrm{c}}}\right)\left\{\frac{\partial^2 T}{\partial x^2} + \frac{\partial^2 T}{\partial y^2}\right\} \tag{4.88}$$

The terms $\mu_0 M\frac{\partial \overline{H}}{\partial x}$ and $\mu_0 M\frac{\partial \overline{H}}{\partial y}$ in Eqs. (4.85) and (4.86), respectively, represent the components of magnetic force per unit volume and depend on the existence of the magnetic gradient in the corresponding x and y directions. These two terms are well known from FHD and are the so-called Kelvin force. The terms $-\sigma_{\mathrm{nf}}B_y^2u+\sigma_{\mathrm{nf}}B_xB_yv$ and $-\sigma_{\mathrm{nf}}B_x^2v+\sigma_{\mathrm{nf}}B_xB_yu$ appearing in Eqs. (4.85) and (4.86), respectively, represent the Lorentz force per unit volume toward the x and y directions and arise because of the electrical conductivity of the fluid. These two terms are known in MHD. The principles of MHD and FHD are combined in the mathematical model presented by Tzirtzilakis et al. [24], and the above-mentioned terms arise together in the governing Eqs. (4.85) and (4.86). The term $\mu_0T\frac{\partial M}{\partial T}\left(u\frac{\partial \overline{H}}{\partial x}+v\frac{\partial \overline{H}}{\partial y}\right)$ in Eq. (4.87) represents the thermal power per unit volume caused by the magnetocaloric effect. Also, the term $\sigma_{\mathrm{nf}}\left(uB_y-vB_x\right)^2$ in Eq. (4.87) represents the Joule heating. For variations in the magnetization M based on the magnetic field intensity $\overline{H}$ and temperature T, the following relation, derived experimentally [24], is considered:

$$M=K'\overline{H}(T_c'-T) \tag{4.89}$$

where K' is a constant and T_c' is the Curie temperature.

In the above equations, μ_0 is the magnetic permeability of a vacuum $(4\pi\times 10^{-7}\ \mathrm{T\ m/A})$, $\overline{H}$ is the magnetic field strength, and $\overline{B}$ is the magnetic induction $(\overline{B}=\mu_0\overline{H})$; the bar above the quantities denotes that they are dimensional.

The stream function and vorticity are defined as follows:

$$u=\frac{\partial\psi}{\partial y},\quad v=-\frac{\partial\psi}{\partial x},\quad \omega=\frac{\partial v}{\partial x}-\frac{\partial u}{\partial y} \tag{4.90}$$

The stream function satisfies the continuity (Eq. 4.84). The vorticity equation is obtained by eliminating the pressure between the two momentum equations, i.e., by taking the y-derivative of Eq. (4.85) and subtracting from it the x-derivative of Eq. (4.86). Also, the following nondimensional variables should be introduced:

$$X=\frac{x}{L},\quad Y=\frac{y}{L},\quad \Omega=\frac{\omega L^2}{\alpha},\quad \Psi=\frac{\psi}{\alpha},\quad \Theta=\frac{T-T_c}{T_h-T_c},\quad \Phi=\frac{\phi-\phi_c}{\phi_h-\phi_c},$$
$$U=\frac{uL}{\alpha},\quad V=\frac{vL}{\alpha},\quad H=\frac{\overline{H}}{\overline{H_0}},\quad H_x\frac{\overline{H_x}}{\overline{H_0}},\quad H_y=\frac{\overline{H_y}}{\overline{H_0}} \tag{4.91}$$

By using these dimensionless parameters the equations become:

$$\left[\frac{\partial\Psi}{\partial Y}\frac{\partial\Omega}{\partial X}-\frac{\partial\Psi}{\partial X}\frac{\partial\Omega}{\partial Y}\right]=Pr\left(\frac{\partial^2\Omega}{\partial X^2}+\frac{\partial^2\Omega}{\partial Y^2}\right)+PrRa\left(\frac{\partial\Theta}{\partial X}-Nr\frac{\partial\Theta}{\partial X}\right)$$
$$+Mn_{\mathrm{F}}\,\mathrm{Pr}\left\{\frac{\partial H}{\partial X}\frac{\partial\Theta}{\partial Y}-\frac{\partial\Psi}{\partial Y}\frac{\partial\Theta}{\partial X}\right\}H$$
$$-Ha^2Pr\times\left\{\frac{\partial V}{\partial X}H_x^2+V\left(2H_x\frac{\partial H_x}{\partial X}\right)-\frac{\partial U}{\partial X}H_xH_y-U\frac{\partial H_x}{\partial X}H_y-U\frac{\partial H_y}{\partial X}H_x\right.$$
$$\left.-\frac{\partial U}{\partial Y}H_y^2-U\left(2H_y\frac{\partial H_y}{\partial Y}\right)+\frac{\partial V}{\partial Y}H_xH_y+V\frac{\partial H_x}{\partial Y}H_y+V\frac{\partial H_y}{\partial Y}H_x\right\} \tag{4.92}$$

$$\frac{\partial \Psi}{\partial Y}\frac{\partial \Theta}{\partial X}-\frac{\partial \Psi}{\partial X}\frac{\partial \Theta}{\partial Y}=\left(\frac{\partial^2 \Theta}{\partial X^2}+\frac{\partial^2 \Theta}{\partial Y^2}\right)+Nb\left(\frac{\partial \Phi}{\partial X}\frac{\partial \Theta}{\partial X}+\frac{\partial \Phi}{\partial Y}\frac{\partial \Theta}{\partial Y}\right)+Nt\left(\left(\frac{\partial \Theta}{\partial X}\right)^2+\left(\frac{\partial \Theta}{\partial Y}\right)^2\right)$$
$$+Ha^2Ec\{UH_y-VH_x\}^2+Mn_FEc\left\{U\frac{\partial H}{\partial X}+V\frac{\partial H}{\partial Y}\right\}H(\varepsilon_1+\Theta)$$
$$+Ec\left\{2\left(\frac{\partial U}{\partial X}\right)^2+2\left(\frac{\partial V}{\partial X}\right)^2+\left(\frac{\partial U}{\partial X}+\frac{\partial V}{\partial Y}\right)^2\right\} \tag{4.93}$$

$$\frac{\partial \Psi}{\partial Y}\frac{\partial \Phi}{\partial X}-\frac{\partial \Psi}{\partial X}\frac{\partial \Phi}{\partial Y}=\frac{1}{Le}\left(\frac{\partial^2 \Phi}{\partial X^2}+\frac{\partial^2 \Phi}{\partial Y^2}\right)+\frac{Nt}{NbLe}\left(\frac{\partial^2 \Theta}{\partial X^2}+\frac{\partial^2 \Theta}{\partial Y^2}\right) \tag{4.94}$$

$$\frac{\partial^2 \Psi}{\partial X^2}+\frac{\partial^2 \Psi}{\partial Y^2}=-\Omega \tag{4.95}$$

where the thermal Rayleigh number, buoyancy ratio, Prandtl number, Brownian motion parameter, thermophoretic parameter, Lewis number, Hartmann number, Eckert number, and magnetic number arising from the FHD of the nanofluid are defined as $Ra=(1-\phi_c)\rho_{f_0}g\beta L^3(T_h-T_c)/(\mu\alpha)$, $Nr=\left(\rho_p-\rho_0\right)(\phi_h-\phi_c)/[(1-\phi_c)\rho_{f_0}\beta L(T_h-T_c)]$, $Pr=\mu/\rho_f\alpha$, $Nb=(\rho c)_pD_B(\phi_h-\phi_c)/((\rho_c)_f\alpha)$, $Nt=(\rho c)_pD_T(T_h-T_c)/[(\rho c)_f\alpha T_c]$, $Le=\alpha/D_B$, $Ha=LH_0\mu_0\sqrt{\sigma/\mu}$, $Ec=(\alpha\mu)/[(\rho C_p)\Delta TL^2]$, and $Mn_F=\mu_0H_0^2K'(T_h-T_c)L^2/(\mu\alpha)$, respectively.

4.3.2.2 Mixed convection

To obtain the governing equations for force convection, the nondimensional variables should be introduced as:

$$X=\frac{x}{L},\quad Y=\frac{y}{L},\quad \Omega=\frac{\omega L}{u_r},\quad \Psi=\frac{\psi}{u_rL},\quad \Theta=\frac{T-T_c}{T_h-T_c},\quad \Phi=\frac{\phi-\phi_c}{\phi_h-\phi_c},$$
$$U=\frac{u}{u_r},\quad V=\frac{v}{u_r},\quad H=\frac{\overline{H}}{\overline{H_0}},\quad H_x\frac{\overline{H_x}}{\overline{H_0}},\quad H_y=\frac{\overline{H_y}}{\overline{H_0}} \tag{4.96}$$

By using these dimensionless parameters the equations become:

$$\left[\frac{\partial \Psi}{\partial Y}\frac{\partial \Omega}{\partial X}-\frac{\partial \Psi}{\partial X}\frac{\partial \Omega}{\partial Y}\right]=\frac{1}{Re}\left(\frac{\partial^2 \Omega}{\partial X^2}+\frac{\partial^2 \Omega}{\partial Y^2}\right)+\frac{Gr}{Re^2}\left(\frac{\partial \Theta}{\partial X}-Nr\frac{\partial \Theta}{\partial X}\right)+Mn_F\left\{\frac{\partial H}{\partial X}\frac{\partial \Theta}{\partial Y}-\frac{\partial H}{\partial Y}\frac{\partial \Theta}{\partial X}\right\}H$$
$$-\frac{Ha^2}{Re}\times\left\{\frac{\partial V}{\partial X}H_x^2+V\left(2H_x\frac{\partial H_x}{\partial X}\right)-\frac{\partial U}{\partial X}H_xH_y-U\frac{\partial H_x}{\partial X}H_y-U\frac{\partial H_y}{\partial X}H_x\right.$$
$$\left.-\frac{\partial U}{\partial Y}H_y^2-U\left(2H_y\frac{\partial H_y}{\partial Y}\right)+\frac{\partial V}{\partial Y}H_xH_y+V\frac{\partial H_x}{\partial Y}H_y+V\frac{\partial H_y}{\partial Y}H_x\right\} \tag{4.97}$$

$$\frac{\partial \Psi}{\partial Y}\frac{\partial \Theta}{\partial X}-\frac{\partial \Psi}{\partial X}\frac{\partial \Theta}{\partial Y}=\frac{1}{RePr}\left(\frac{\partial^2 \Theta}{\partial X^2}+\frac{\partial^2 \Theta}{\partial Y^2}\right)+\frac{Nb}{Re}\left(\frac{\partial \Phi}{\partial X}\frac{\partial \Theta}{\partial Y}+\frac{\partial \Phi}{\partial Y}\frac{\partial \Theta}{\partial Y}\right)$$
$$+\frac{Nt}{Re}\left(\left(\frac{\partial \Theta}{\partial X}\right)^2+\left(\frac{\partial \Theta}{\partial Y}\right)^2\right)+\frac{Ha^2 Ec}{Re}\{UH_y-VH_x\}^2$$
$$+Mn_F Ec\left\{U\frac{\partial H}{\partial X}+V\frac{\partial H}{\partial Y}\right\}H(\varepsilon_1+\Theta) \tag{4.98}$$
$$+\frac{Ec}{Re}\left\{2\left(\frac{\partial U}{\partial X}\right)^2+2\left(\frac{\partial V}{\partial X}\right)^2+\left(\frac{\partial U}{\partial X}+\frac{\partial V}{\partial Y}\right)^2\right\}$$

$$\frac{\partial \Psi}{\partial Y}\frac{\partial \Phi}{\partial X}-\frac{\partial \Psi}{\partial X}\frac{\partial \Phi}{\partial Y}=\frac{1}{ReLe}\left(\frac{\partial^2 \Phi}{\partial X^2}+\frac{\partial^2 \Phi}{\partial Y^2}\right)+\frac{Nt}{ReNbLe}\left(\frac{\partial^2 \Theta}{\partial X^2}+\frac{\partial^2 \Theta}{\partial Y^2}\right) \tag{4.99}$$

$$\frac{\partial^2 \Psi}{\partial X^2}+\frac{\partial^2 \Psi}{\partial Y^2}=-\Omega \tag{4.100}$$

where the thermal Grashof number, buoyancy ratio, Prandtl number, Brownian motion parameter, thermophoretic parameter, Lewis number, Reynolds number, Hartmann number, Eckert number, and magnetic number arising from FHD of the nanofluid are defined as $Gr=(1-\phi_c)\rho_{f_0} g\beta L^3\rho_f(T_h-T_c)/(\mu^2)$, $Nr=\left(\rho_p-\rho_0\right)(\phi_h-\phi_c)/\left[(1-\phi_c)\rho_{f_0}\beta L(T_h-T_c)\right]$, $Pr=\mu/\rho_f\alpha$, $Nb=(\rho c)_p D_B(\phi_h-\phi_c)/((\rho c)_f \upsilon)$, $Nt=(\rho c)_p D_T(T_h-T_c)/\left[(\rho c)_f \upsilon T_c\right]$, $Re=\frac{\rho_f L u_r}{\mu}$, $Le=\upsilon/D_B$, $Ha=LH_0\mu_0\sqrt{\sigma/\mu}$, $Ec=\left(u_r^2\right)/\left[(C_p)\Delta T\right]$, and $Mn_F=\mu_0 H_0^2 K'(T_h-T_c)L^2/(\mu\alpha)$, respectively.

4.3.3 APPLICATION OF CVFEM FOR THE COMBINED EFFECTS OF FHD AND MHD

4.3.3.1 Combined effects of FHD and MHD in a semiannulus enclosure with a sinusoidal hot wall

4.3.3.1.1 Problem definition

The physical model, along with important geometrical parameters and the mesh of the enclosure used in the present CVFEM program, are shown in Fig 4.46 [35]. The inner and outer walls are maintained at constant temperatures T_h and T_c, respectively. The shape of the inner cylinder's profile is assumed to mimic the following pattern:

$$r=r_{in}+A\cos(N(\zeta)) \tag{4.101}$$

in which r_{in} is the radius of the base circle, r_{out} is the radius of the outer cylinder, A and N are amplitude and number of undulations, respectively; ζ is the rotation angle. In this study A and N equal 0.2 and 4, respectively. For the expression of the magnetic field's strength, that the magnetic source represents a magnetic wire placed vertically to the x-y plane at the point $(\bar{a},\bar{b})$. The components of the

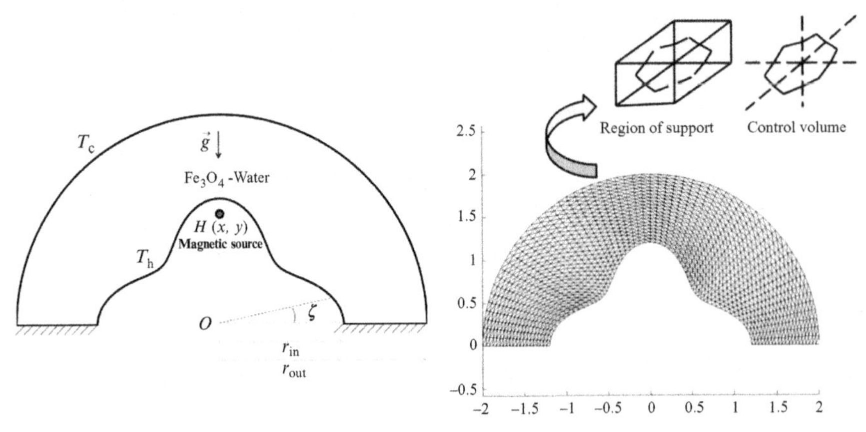

FIGURE 4.46

Geometry and the boundary conditions (a) and the mesh of the enclosure (b) considered in this work.

magnetic field intensity $(\overline{H}_x, \overline{H}_y)$ and the magnetic field strength $(\overline{H})$ can be considered as [35]:

$$\overline{H}_x = \frac{\gamma}{2\pi}\frac{1}{(x-\overline{a})^2 + (y-\overline{b})^2}(y-\overline{b}) \tag{4.102}$$

$$\overline{H}_y = -\frac{\gamma}{2\pi}\frac{1}{(x-\overline{a})^2 + (y-\overline{b})^2}(x-\overline{a}) \tag{4.103}$$

$$\overline{H} = \sqrt{\overline{H}_x^2 + \overline{H}_y^2} = \frac{\gamma}{2\pi}\frac{1}{\sqrt{(x-\overline{a})^2 + (y-\overline{b})^2}} \tag{4.104}$$

where γ is the magnetic field's strength at the source (of the wire) and $(\overline{a}, \overline{b})$ is the position where the source is located. The contours of the magnetic field's strength are shown in Fig. 4.47. In this study the magnetic source is located at $(-0.05\text{cols}, 0.5\text{rows})$.

4.3.3.1.2 Effects of active parameters

Effects of an external magnetic field on ferrofluid flow and heat transfer in a semi-annulus enclosure with a sinusoidal hot wall is studied. The mathematical model used to formulate the problem is consistent with the principles of FHD and MHD. The thermophysical properties of Fe_3O_4 nanoparticles and the base fluid (water) are shown in Table 4.6 [28]. Various values of volume fraction of the nanoparticles ($\phi = 0\%$ and 4%), Rayleigh numbers ($Ra = 10^3$, 10^4, and 10^5), magnetic numbers

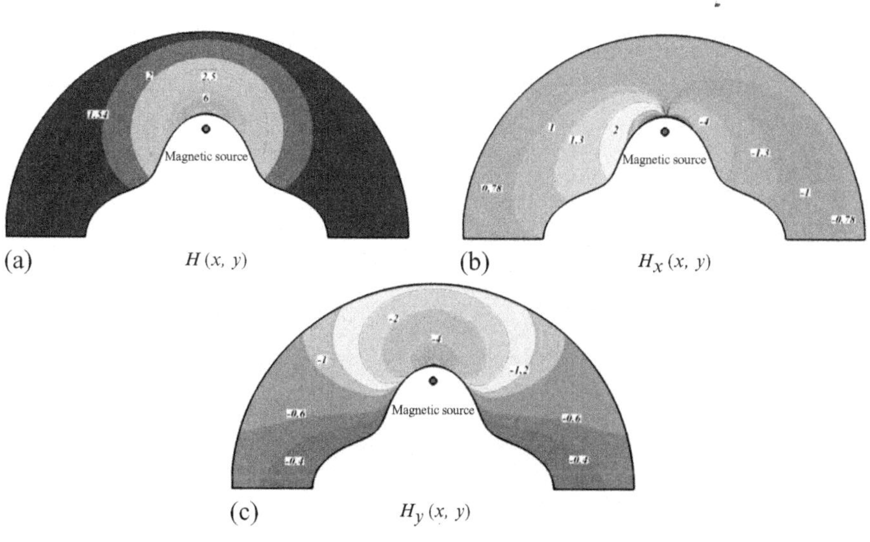

FIGURE 4.47

Contours of the magnetic field strength H (a), the magnetic field intensity component in the x direction H_x (b), and the magnetic field intensity component in the y direction H_y (c).

arising from FHD ($Mn_F = 0$, 50, and 500), and Hartmann numbers arising from MHD ($Ha = 0$, 1, 3, and 5). In all calculations, the Prandtl number (Pr), temperature (ε_1), and Eckert number (Ec) are set to 6.8, 0.0, and 1×10^{-5}, respectively.

The isotherms and streamlines of the nanofluid and the pure fluid are compared in Fig.4.48. The velocity components of a nanofluid are increased because the energy transport in the fluid is augmented. Thus the absolute values of stream functions increase with an increase in the volume fraction of the nanofluid. Also, there is no sensible change in the isotherm.

Effects of the Rayleigh number, Hartmann number, and magnetic number on isotherms and streamlines are shown in Figs. 4.49–4.51. At low Rayleigh numbers, the conduction heat transfer mechanism is dominant in the absence of a magnetic field. As the Hartmann number increases the absolute value of the maximum stream function decreases. As the magnetic number increases three thermal plumes appear over the hot wall. Also, the primary vortex on each side turns into two smaller vortices. As the Rayleigh number increases, the role of convection in heat transfer becomes more significant, and consequently the thermal boundary layer on the surface of the inner wall becomes thinner. In addition, a plume starts to appear on the top of the inner circular wall at $\zeta = 90°$. As Lorentz forces increase the thermal plume converts to three plumes. At high Hartmann numbers two secondary eddies exist near the vertical center line. At $Ra = 10^5$, the effects of Kelvin forces are not sensible, while increasing Lorentz forces causes two equals eddies to appear near the vertical center line.

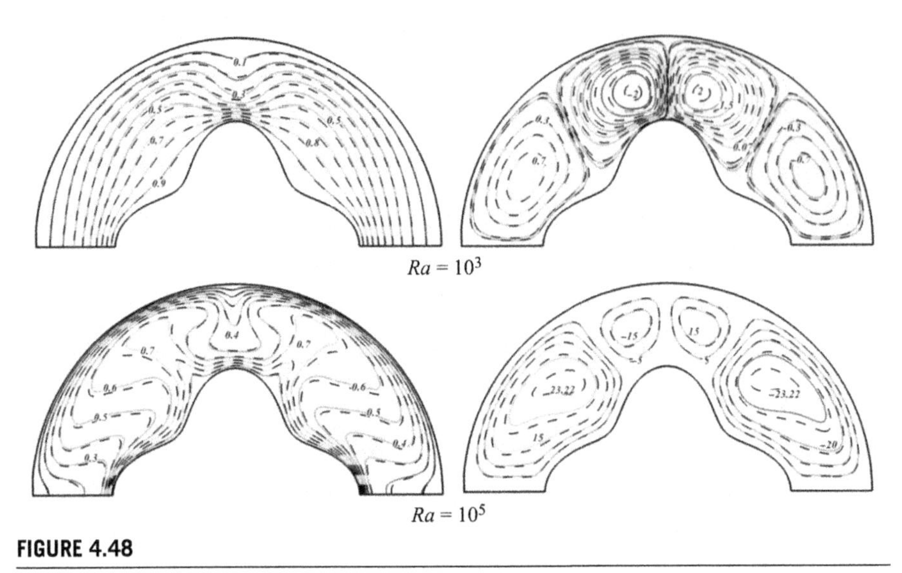

FIGURE 4.48

Comparison of the isotherms and streamlines between a nanofluid ($\phi = 0.04$) (---) and a pure fluid ($\phi = 0$) (—) when $Mn_F = 500$, $Ha = 5$, and $Pr = 6.8$.

Effects of the magnetic number, Hartmann number, and Rayleigh number on local and average Nusselt numbers are depicted in Fig. 4.52 and Table 4.7. The Nusselt number increases as the Rayleigh number increases, whereas it decreases as the Hartmann number increases. Increasing the magnetic number causes the Nusselt number to increase, expect for $Ra = 10^5$. Table 4.8 illustrates the effects of the magnetic number, Hartmann number, and Rayleigh number on heat transfer enhancement. For low Rayleigh numbers, enhancement increases as the Hartmann number increases, whereas it decreases as the magnetic number increases. The opposite trend is observed for the highest Rayleigh number.

4.3.3.2 Combined effects of FHD and MHD when considering thermal radiation

4.3.3.2.1 Problem definition

The schematic diagram and the mesh of the semiannulus enclosure used in the present CVFEM program are shown in Fig. 4.53. The inner wall is under constant heat flux (q'') and the outer wall is maintained at a constant temperature (T_c), while the two other walls are thermally insulated. In this study γ equals to 45°. For the expression of the magnetic field's strength, one can consider that the magnetic source represents a magnetic wire placed vertically to the x-y plane at the point $(\overline{a}, \overline{b})$. The contours of the magnetic field's strength are shown in Fig. 4.54. In this study the magnetic source is located at $(-0.05\text{cols}, 0.5\text{rows})$.

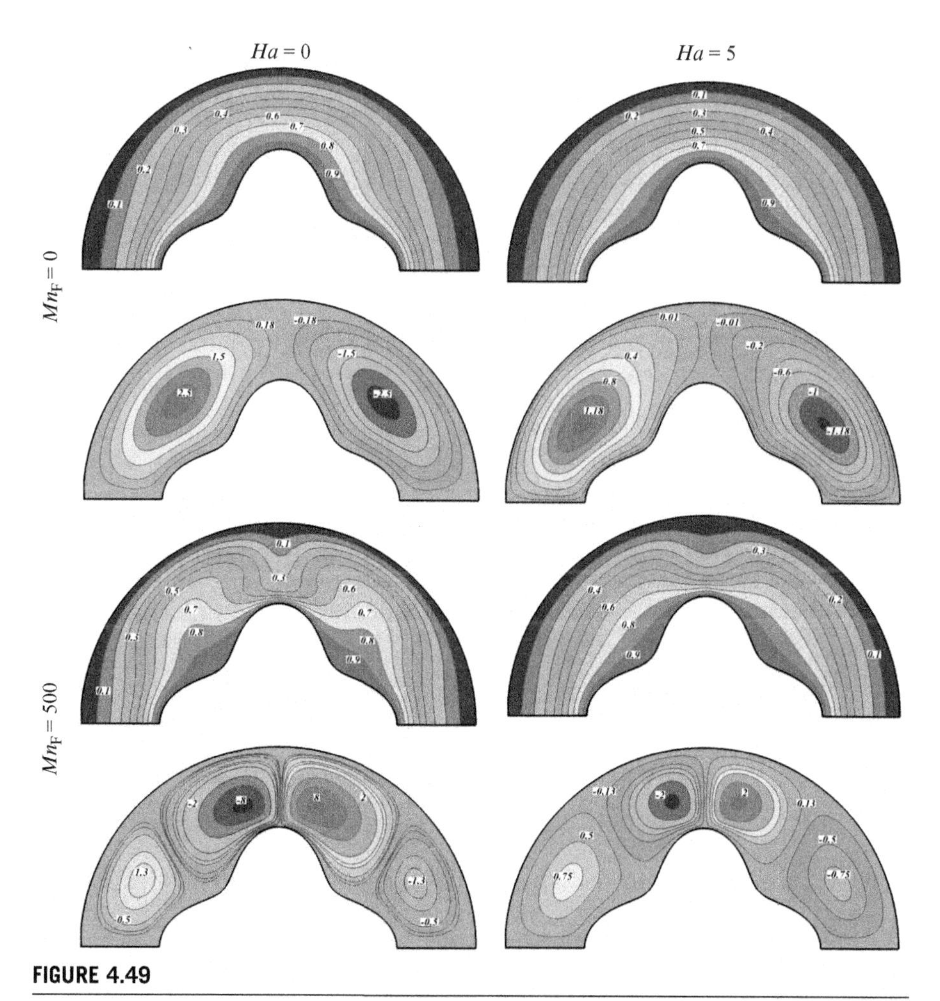

FIGURE 4.49

Isotherm (up) and streamline (down) contours for different Hartmann numbers and magnetic numbers when $Ra = 10^3$, $\phi = 0.04$, and $Pr = 6.8$.

The governing equations are:

$$\begin{aligned}
\frac{\partial \Psi}{\partial Y}\frac{\partial \Omega}{\partial X} - \frac{\partial \Psi}{\partial X}\frac{\partial \Omega}{\partial Y} &= Pr\left[\frac{\mu_{\mathrm{nf}}/\mu_{\mathrm{f}}}{\rho_{\mathrm{nf}}/\rho_{\mathrm{f}}}\right]\left(\frac{\partial^2 \Omega}{\partial X^2} + \frac{\partial^2 \Omega}{\partial Y^2}\right) \\
&+ Ra\,Pr\left[\frac{\beta_{\mathrm{nf}}}{\beta_{\mathrm{f}}}\right]\left(\frac{\partial \Theta}{\partial X}\right) + Mn_{\mathrm{F}}Pr\left(\frac{\rho_{\mathrm{f}}}{\rho_{\mathrm{nf}}}\right)\left\{\frac{\partial H}{\partial X}\frac{\partial \Theta}{\partial Y} - \frac{\partial H}{\partial Y}\frac{\partial \Theta}{\partial X}\right\}H - Ha^2 Pr\left[\frac{\sigma_{\mathrm{nf}}/\sigma_{\mathrm{f}}}{\sigma_{\mathrm{nf}}/\rho_{\mathrm{f}}}\right] \\
&\times\left\{\frac{\partial V}{\partial X}H_x^2 + V\left(2H_x\frac{\partial H_x}{\partial X}\right) - \frac{\partial U}{\partial X}H_xH_y - U\frac{\partial H_x}{\partial X}H_y - U\frac{\partial H_y}{\partial X}H_x\right. \\
&\left. - \frac{\partial U}{\partial Y}H_y^2 - U\left(2H_y\frac{\partial H_y}{\partial Y}\right) + \frac{\partial V}{\partial Y}H_xH_y + V\frac{\partial H_x}{\partial Y}H_y + V\frac{\partial H_y}{\partial Y}H_x\right\}
\end{aligned} \tag{4.105}$$

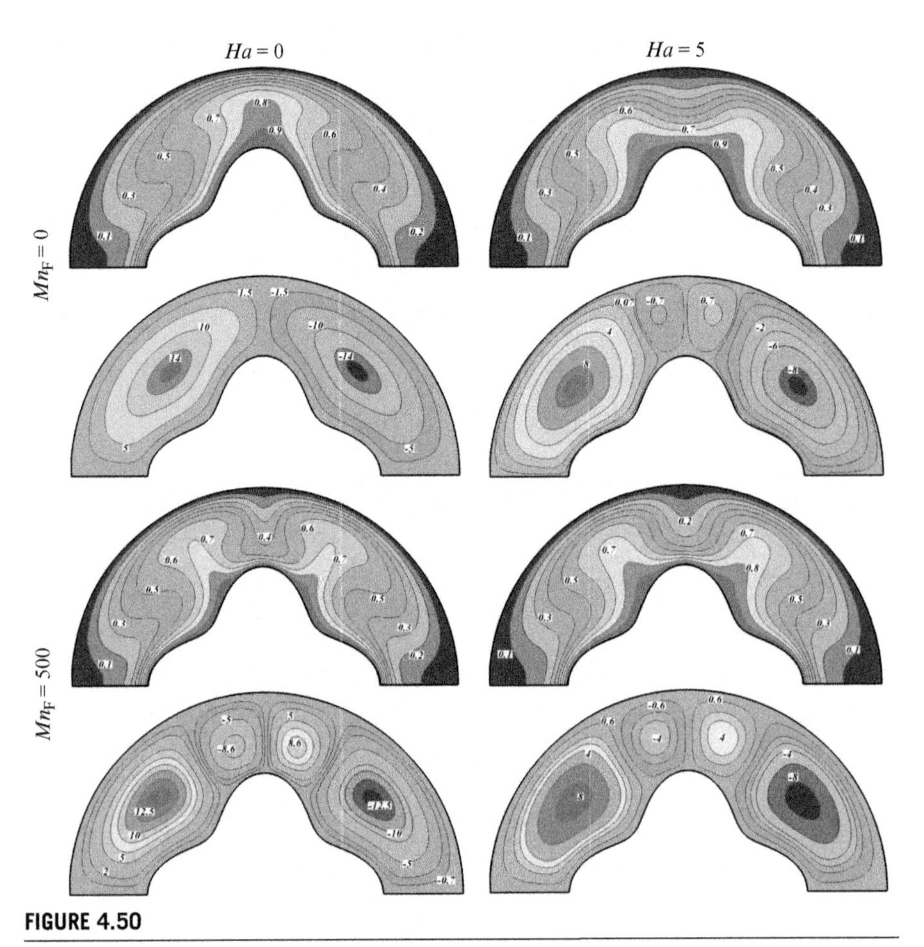

FIGURE 4.50

Isotherm (up) and streamline (down) contours for different Hartmann numbers and magnetic numbers when $Ra = 10^4$, $\phi = 0.04$, and $Pr = 6.8$.

$$
\begin{aligned}
\frac{\partial \Psi}{\partial Y}\frac{\partial \Theta}{\partial X} - \frac{\partial \Psi}{\partial X}\frac{\partial \Theta}{\partial Y} = & \left[\frac{\frac{k_{nf}}{k_f}}{\frac{(\rho C_p)_{nf}}{(\rho C_p)_f}}\right]\left(\frac{\partial^2 \Theta}{\partial X^2} + \frac{\partial^2 \Theta}{\partial Y^2}\right) \\
& + Ha^2 EC \left[\frac{\frac{\sigma_{nf}}{\sigma_f}}{\frac{(\rho C_p)_{nf}}{(\rho C_p)_f}}\right]\left\{U H_y - V H_x\right\}^2 \\
& + Mn_F\, Ec \frac{\left(\rho C_p\right)_f}{\left(\rho C_p\right)_{nf}}\left\{U\frac{\partial H}{\partial X} + V\frac{\partial H}{\partial Y}\right\} H(\varepsilon_1 + \Theta) \\
& + Ec \left[\frac{\frac{\mu_{nf}}{\mu_f}}{\frac{(\rho C_p)_{nf}}{(\rho C_p)_f}}\right]\left\{2\left(\frac{\partial U}{\partial X}\right)^2 + 2\left(\frac{\partial V}{\partial X}\right)^2 + \left(\frac{\partial U}{\partial X} + \frac{\partial V}{\partial Y}\right)^2\right\} \\
& + \left(\frac{k_{nf}}{k_f}\frac{\left(\rho C_p\right)_f}{\left(\rho C_p\right)_{nf}} + \frac{4}{3}Rd\left(\frac{k_{nf}}{k_f}\right)^{-1}\right)\frac{\partial^2 \Theta}{\partial Y^2}
\end{aligned}
\tag{4.106}
$$

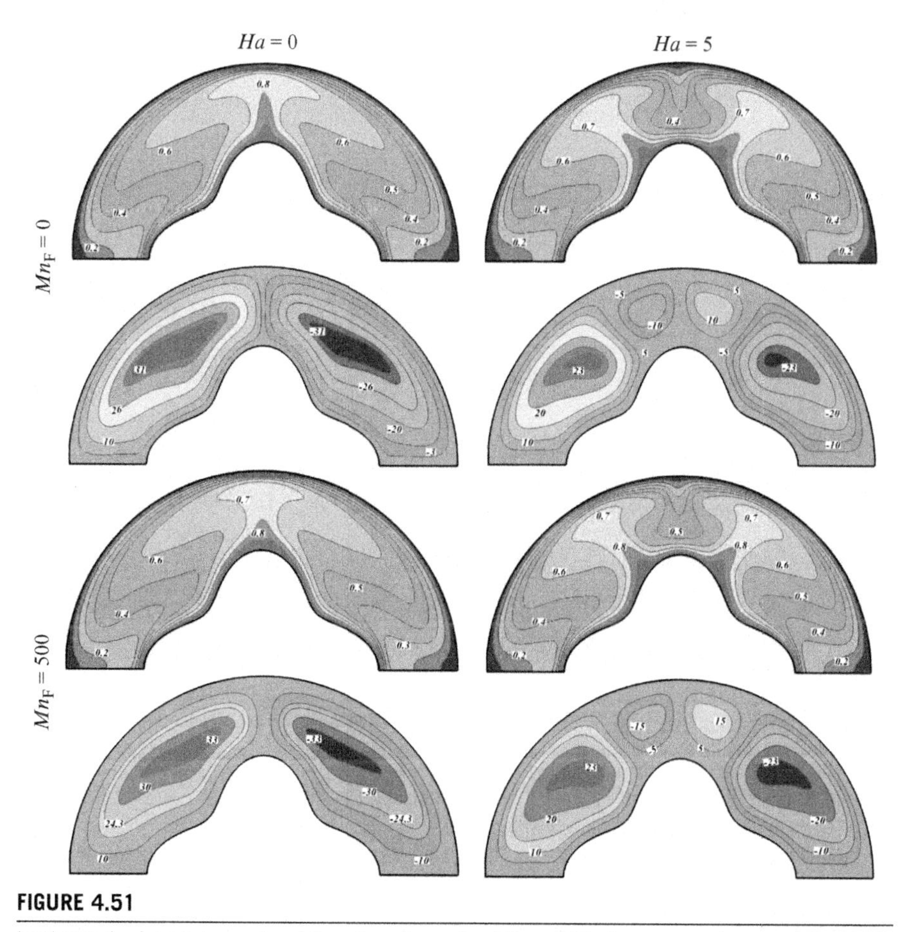

FIGURE 4.51

Isotherm (up) and streamline (down) contours for different Hartmann numbers and magnetic numbers when $Ra = 10^5$, $\phi = 0.04$, and $Pr = 6.8$.

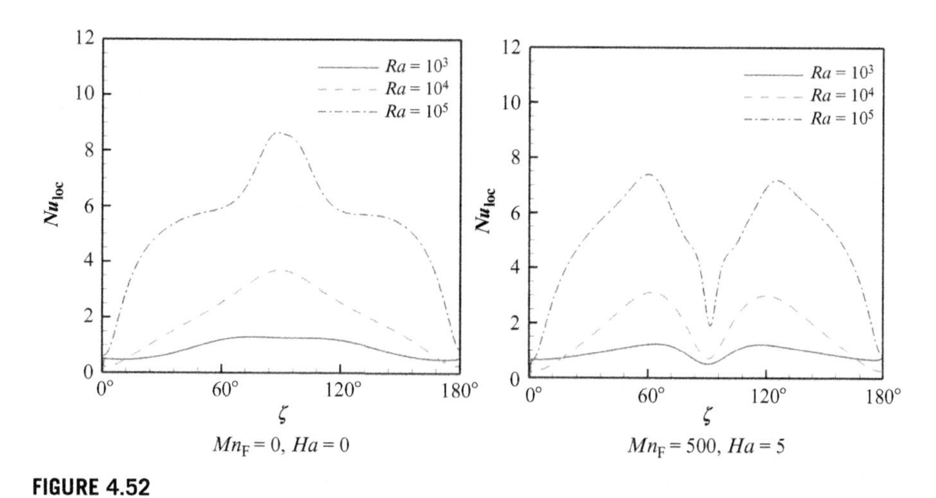

FIGURE 4.52

Effects of the magnetic number, Hartmann number, and Rayleigh number on the local Nusselt number Nu_{loc} along a hot wall when $\phi = 0.04$.

Table 4.7 Effects of the Rayleigh Number, Hartmann Number, and Magnetic Number on the Average Nusselt Number When $\phi = 0.04$

Ra	*Ra*	*Ha*	Nu_{ave}
10^3	0	0	0.984445
10^3	0	5	0.899518
10^3	500	0	1.478063
10^4	500	5	0.999565
10^4	0	0	2.135067
10^4	0	5	1.702675
10^4	500	0	2.365715
10^5	500	5	1.926001
10^5	0	0	5.926064
10^5	0	5	5.490067
10^5	500	0	5.843355
10^5	500	5	5.495555

Table 4.8 Effects of the Rayleigh Number, Hartmann Number, and Magnetic Number on Heat Transfer Enhancement

Ra	Mn_F	*Ha*	*En*
10^3	0	0	7.461432
10^3	0	5	8.552658
10^3	500	0	4.123763
10^3	500	5	2.677329
10^4	0	0	7.135186
10^4	0	5	5.86332
10^4	500	0	6.471199
10^4	500	5	4.675113
10^5	0	0	11.38515
10^5	0	5	11.32785
10^5	500	0	11.60219
10^5	500	5	10.95681

$$\frac{\partial^2 \Psi}{\partial X^2} + \frac{\partial^2 \Psi}{\partial Y^2} = -\Omega \tag{4.107}$$

where $Ra = g\beta_f L^4 q''/(k_f \alpha_f \nu_f)$, $Pr = \upsilon_f/\alpha_f$, $Ha = LB_x\sqrt{\sigma_f/\mu_f}$, $\varepsilon_1 = T_1/(q''L/k_f)$, $Ec = (\mu_f \alpha_f)/\left[(\rho C_p)_f (q''L/k_f) L^2\right]$, $Mn_F = \mu_0 H_0^2 K'(q''L/k_f)L^2/(\mu_f \alpha_f)$, and $Rd = 4\sigma_e T_c^3/(\beta_R k_f)$ are the Rayleigh number, Prandtl number, Hartmann number arising from MHD, temperature, Eckert number, magnetic number, and radiation parameter arising from FHD for the base fluid, respectively.

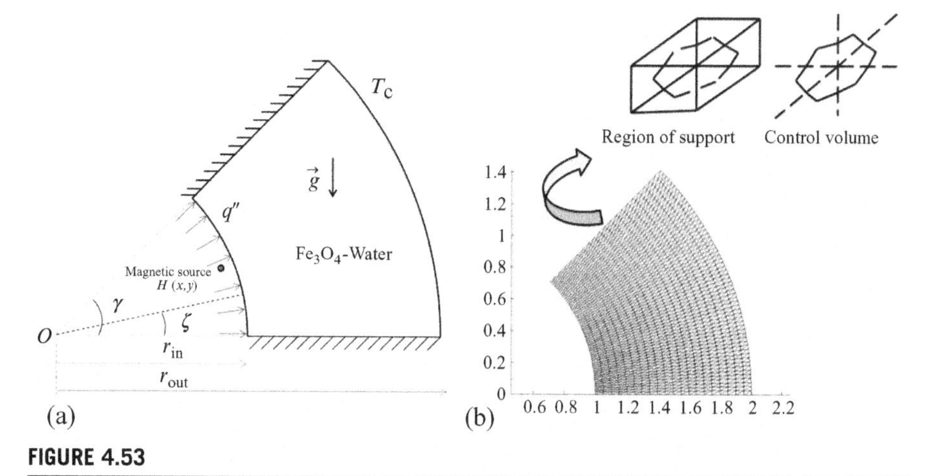

FIGURE 4.53

Geometry and the boundary conditions (a) and the mesh of the enclosure (b) considered in this work.

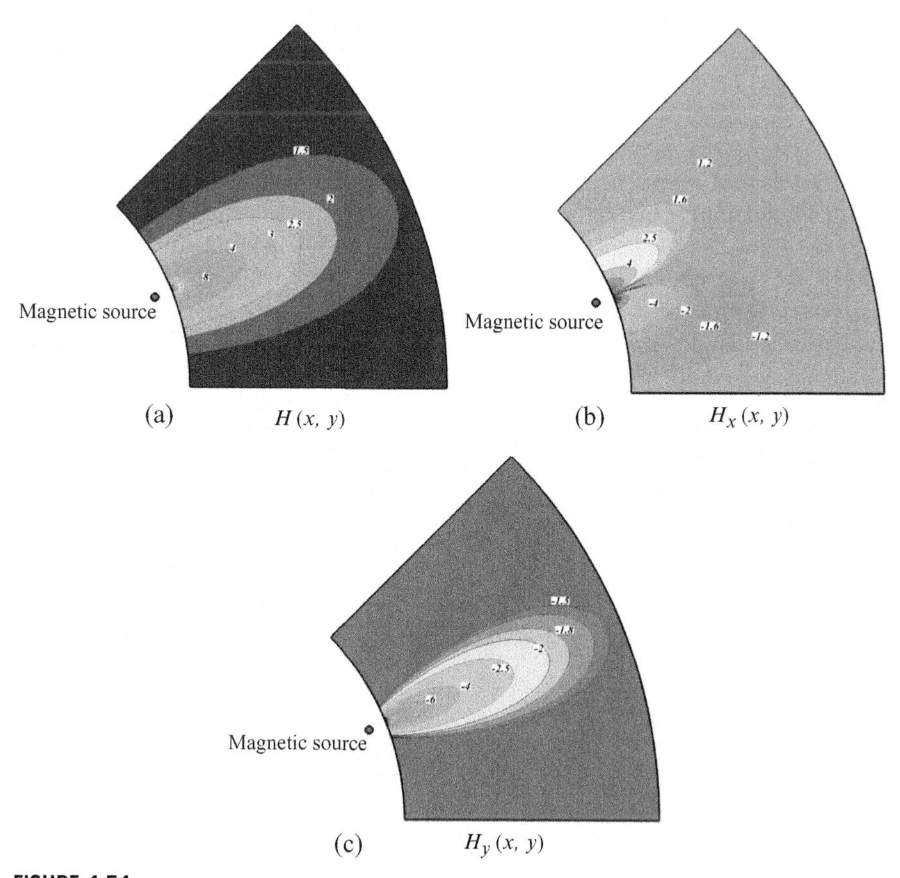

FIGURE 4.54

Contours of the magnetic field strength H (a), the magnetic field intensity component in the x direction H_x (b), and the magnetic field intensity component in the y direction H_y (c).

4.3.3.2.2 Effects of active parameters

Ferrofluid (Fe_3O_4-water) flow and heat transfer in a semiannulus enclosure in the presence of an external magnetic field is investigated. The CVFEM is used to solve the governing equations. The mathematical model used to formulate the problem is consistent with the principles of FHD and MHD. Various values of the volume fraction of nanoparticles ($\phi = 0\%$ and 4%), Rayleigh numbers ($Ra = 10^3,\ 10^4$, and 10^5), magnetic number arising from FHD ($Mn_F = 0,\ 20,\ 60$, and 100), radiation parameter ($Rd = 0.0$ and 0.02), and Hartmann number arising from MHD ($Ha = 0,\ 2,\ 6$, and 10) are calculated. In all calculations, the Prandtl number (Pr), temperature (ε_1), and Eckert number (Ec) are set to 6.8, 0.0, and 1×10^{-5}, respectively.

Figure 4.55 compares the isotherms and streamlines between a nanofluid and a pure fluid. The velocity components of the nanofluid are increased because of an increase in the energy transport in the fluid as the volume fraction increases. Thus the absolute values of stream functions indicate that the strength of the flow increases as the volume fraction of the nanofluid increases. Also, the thermal boundary layer thickness increases with an increase in the volume fraction of the nanofluid. Figure 4.56 shows the effect of the Hartmann number on isotherms and streamlines. As the Hartmann number increases, the absolute value of the maximum stream function decreases. This figure also shows that streamlines become more disturbed near the magnetic source. Isotherm and streamline contours for different Rayleigh numbers, Hartmann numbers, radiation parameters, and magnetic numbers are shown in Figs. 4.57 and 4.58.

When $Ra = 10^3$, the heat transfer in the enclosure is mainly dominated by conduction. At $Mn_F = 0$, $Ha = 0$, the streamlines show one rotating eddy. In the presence of a magnetic field, the center of the vortex moves upward. As the Rayleigh number increases, the role of convection in heat transfer becomes more significant

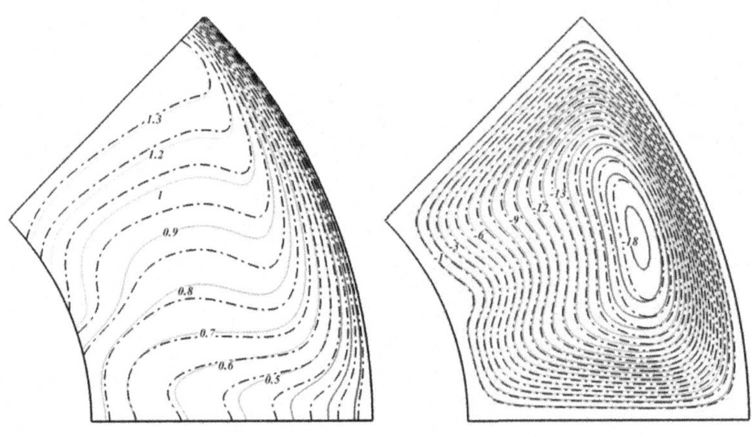

FIGURE 4.55

Comparison of the isotherms and streamlines between a nanofluid ($\phi = 0.04$) (—) and a pure fluid ($\phi = 0$) (-·-·-) when $Ra = 10^5$, $Mn_F = 100$, $Ha = 10$, and $Pr = 6.8$.

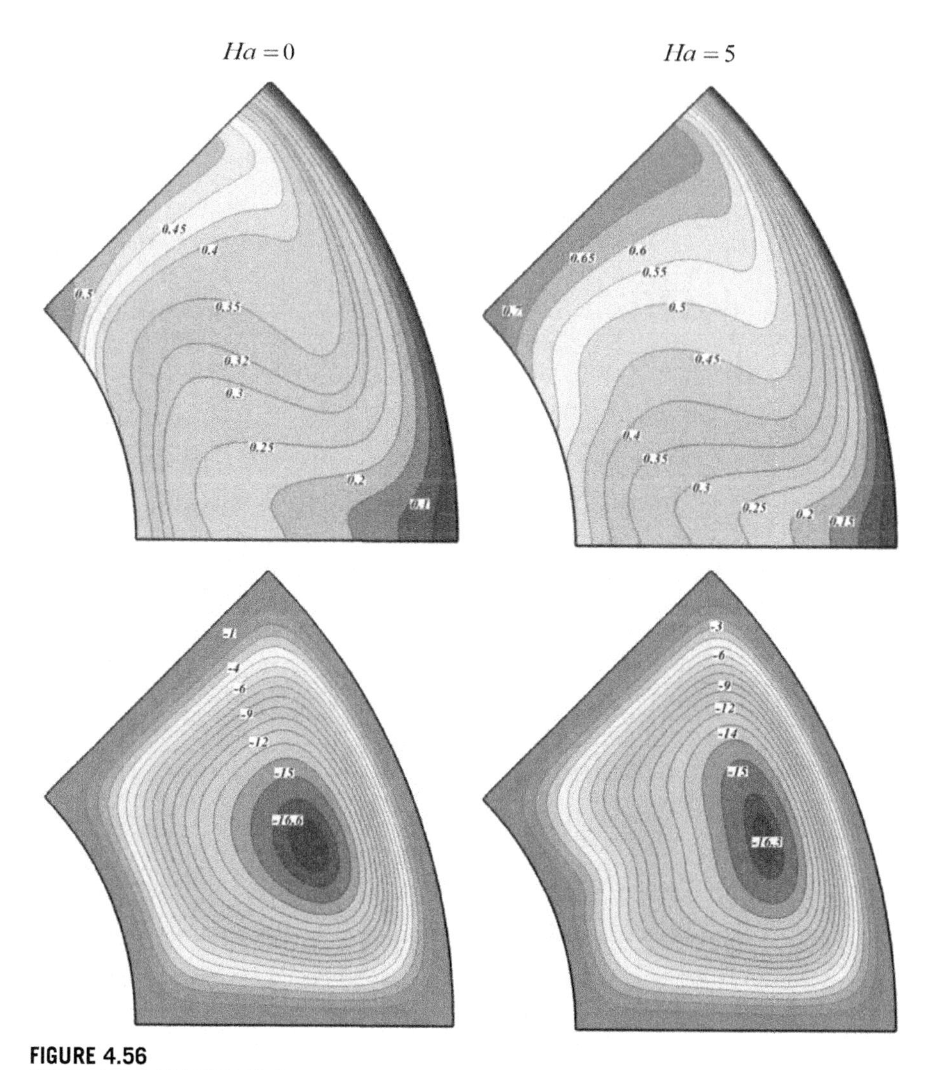

FIGURE 4.56

Effect of the Hartmann number on isotherms and streamlines when $Ra = 10^5$, $Mn = 100$, $\phi = 0.04$, and $Rd = 0.02$.

and, consequently, the thermal boundary layer on the surface of the inner wall becomes thinner. In addition, a plume starts to appear on the top of the inner circular wall at $\zeta = 45°$. The isotherms and streamlines become denser near the magnetic source ($\zeta = 22.5°$). Figure 4.58 shows that the absolute value of the maximum stream function increases with augmentation of the radiation parameter. The center of the vortex moves down and left when the radiation parameter is increased.

Figure 4.59 depicts the effects of the magnetic number, Hartmann number, radiation parameter, and Rayleigh number on the local Nusselt number. The local

Nusselt number increases as the Rayleigh number increases, whereas it decreases as the Hartmann number and radiation parameter increase. The local minimum point at $\zeta = 22.5°$ in these profiles is related to the existence of a magnetic source in this region. Effects of the volume fraction of the nanofluid, magnetic number,

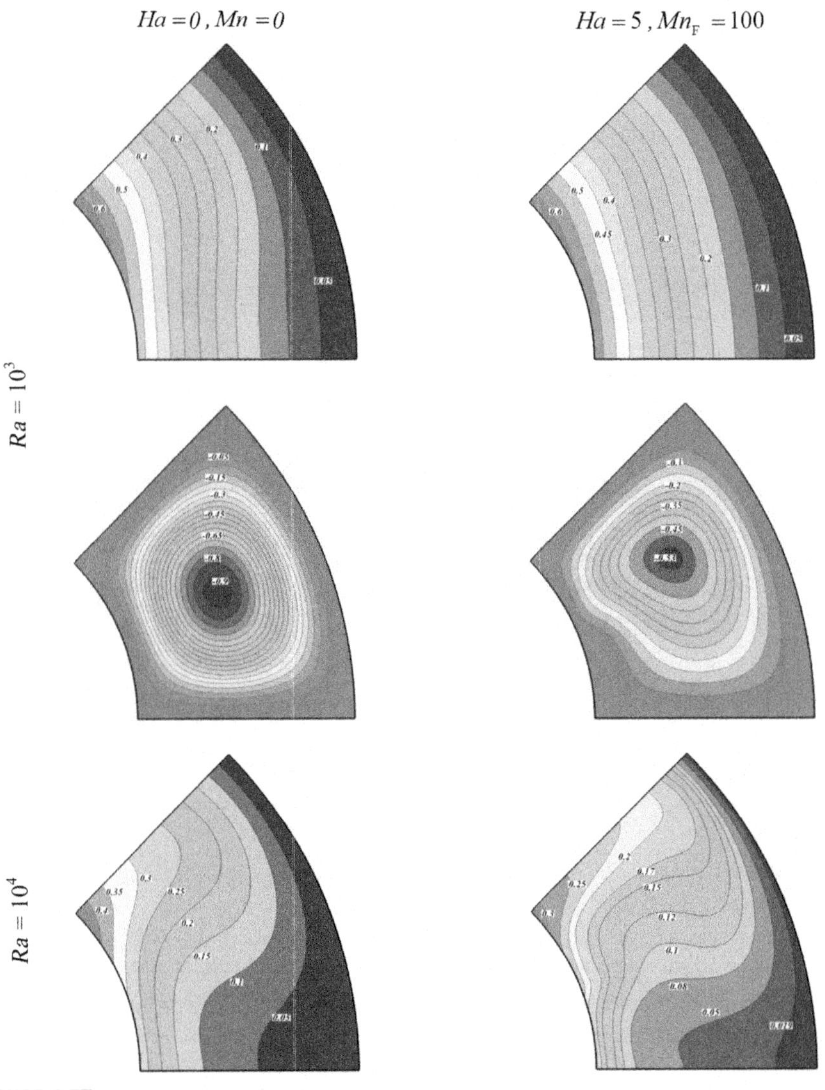

FIGURE 4.57

Isotherm (up) and streamline (down) contours for different Rayleigh numbers, Hartmann numbers, and magnetic numbers when $Rd = 0.0$, $\phi = 0.04$, and $Pr = 6.8$.

Continued

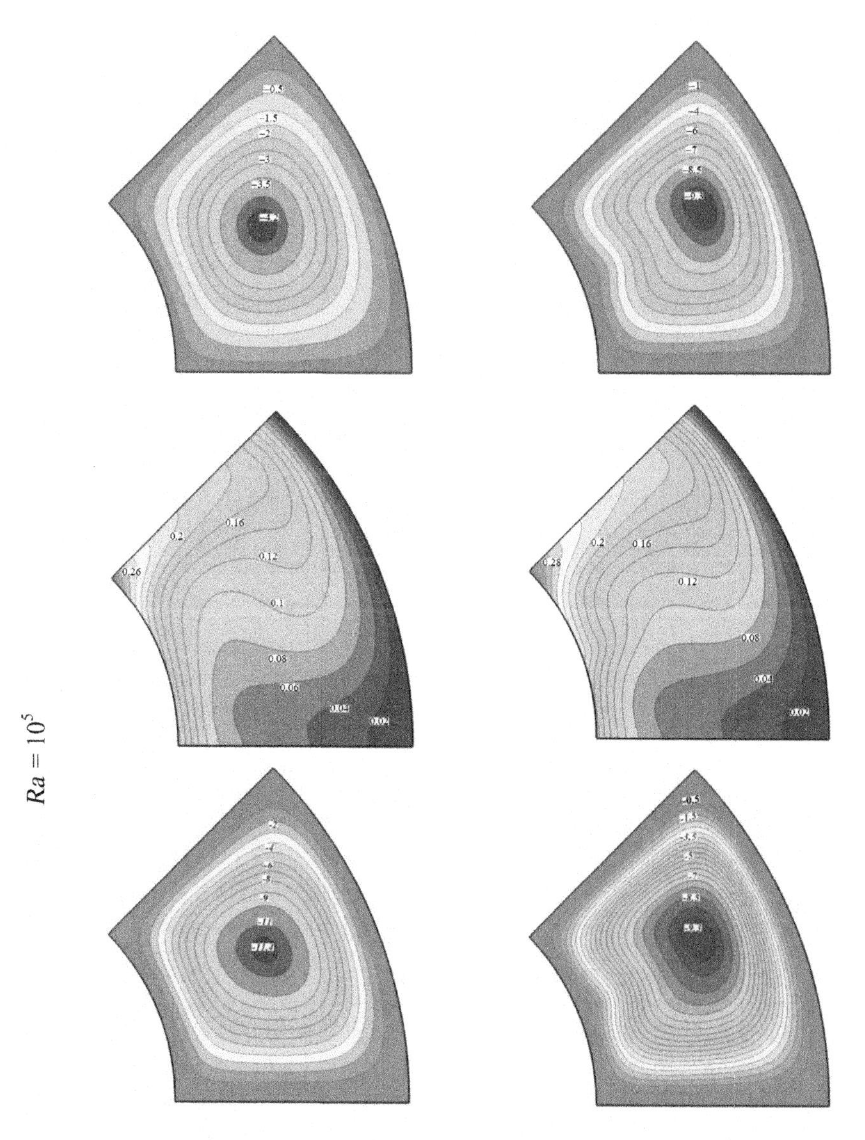

FIGURE 4.57, cont'd

Hartmann number, radiation parameter, and Rayleigh number on the average Nusselt number are shown in Fig. 4.60 and Table 4.9. The average Nusselt number increases with an increase in the volume fraction of the nanofluid, the magnetic number, and the Rayleigh number; it decreases as the Hartmann number and radiation parameter increase.

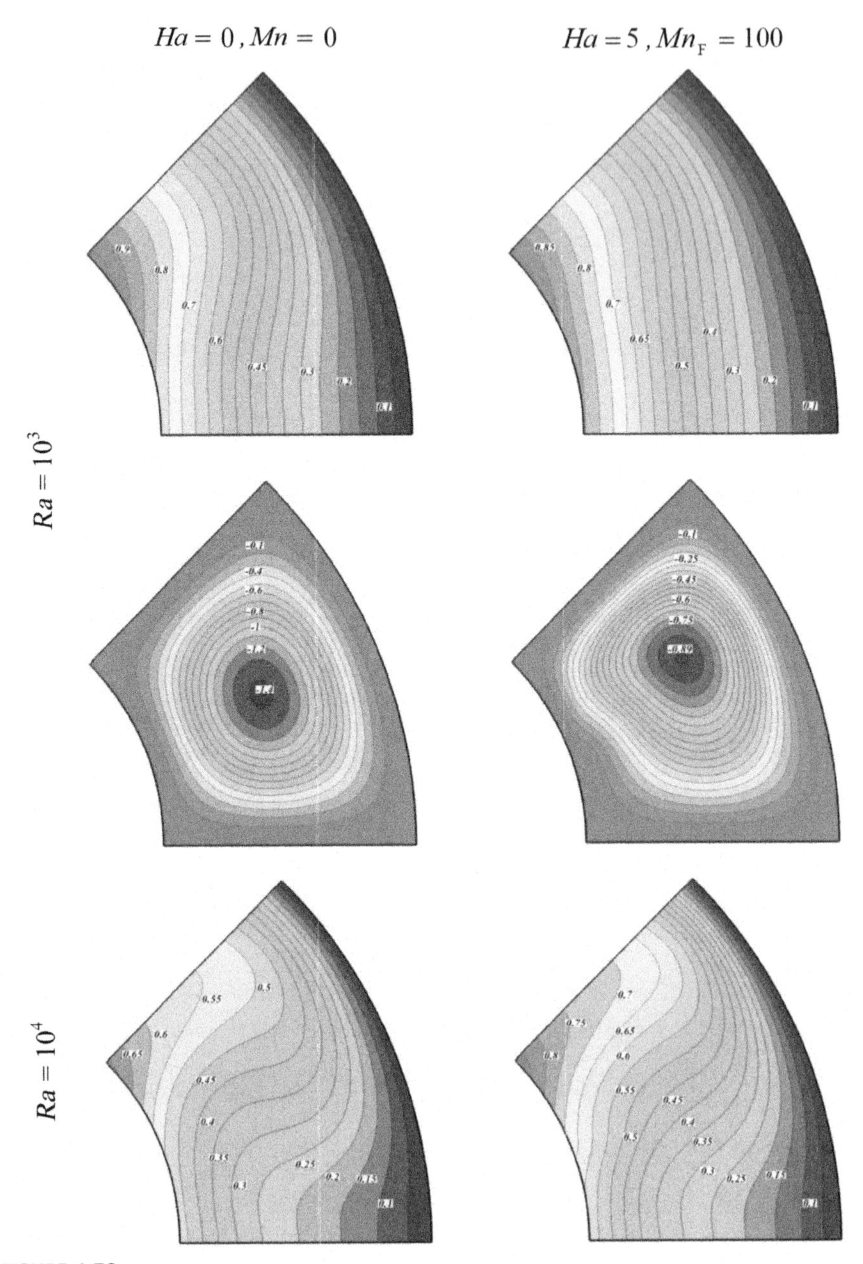

FIGURE 4.58

Isotherm (up) and streamline (down) contours for different Rayleigh numbers, Hartmann numbers, and magnetic numbers when $Rd = 0.2$, $\phi = 0.04$, and $Pr = 6.8$.

Continued

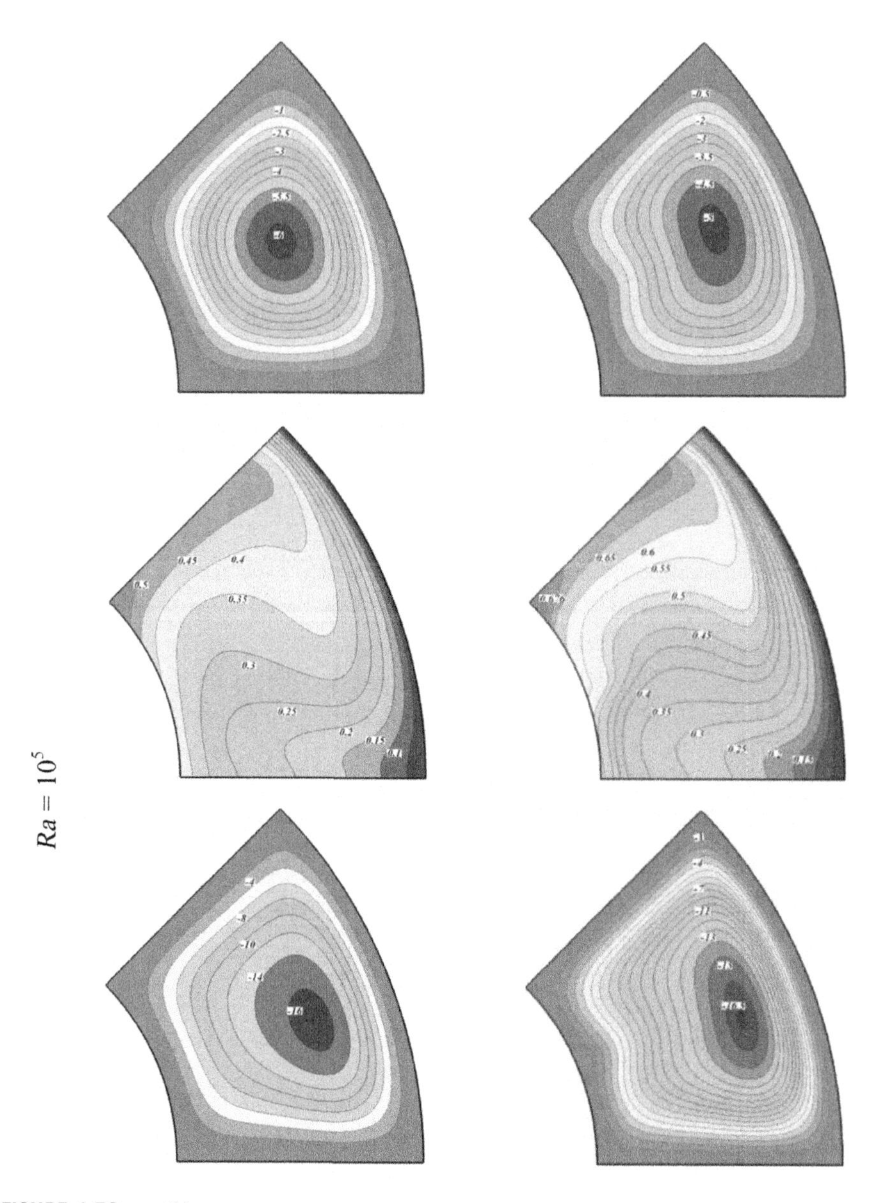

FIGURE 4.58, cont'd

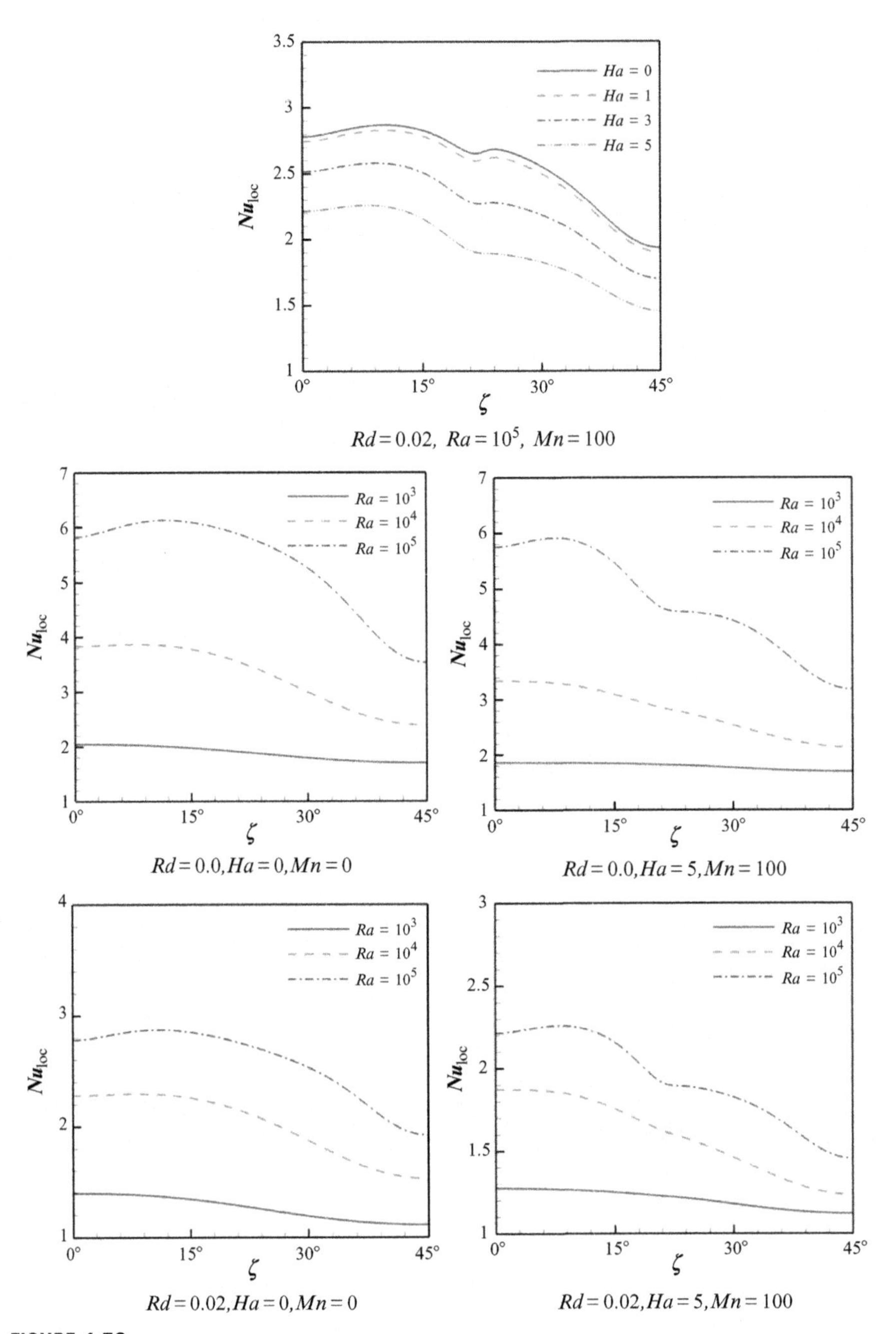

FIGURE 4.59

Effects of the magnetic number, Hartmann number, radiation parameter, and Rayleigh number on the local Nusselt number Nu_{loc} along a hot wall when $\phi = 0.04$.

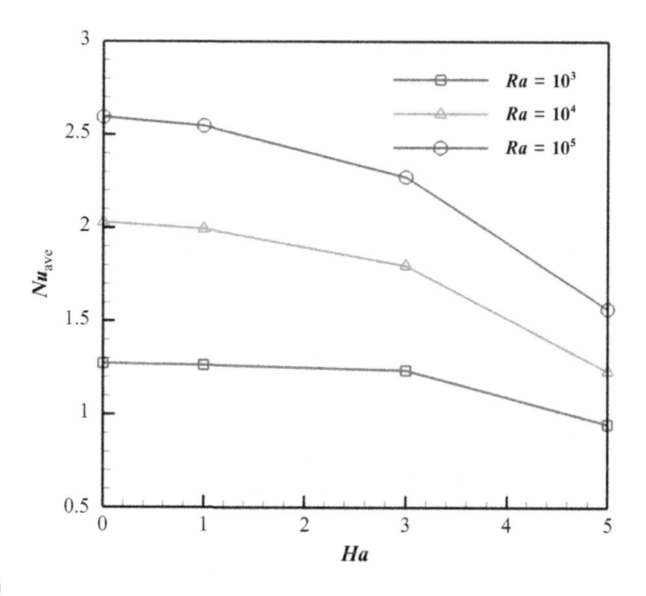

FIGURE 4.60

Effects of the Hartmann number and Rayleigh number on the average Nusselt number Nu_{ave} along a hot wall when $\phi = 0.04$.

4.3.3.3 Effect of a nonuniform magnetic field on ferrofluid flow and convective heat transfer in a semiannulus enclosure

4.3.3.3.1 Problem definition

The schematic diagram and the mesh of the semiannulus enclosure used in the present CVFEM program are shown in Fig. 4.61. The inner and outer walls are maintained at constant temperatures T_h and T_c, respectively, while the two other walls are thermally insulated. To express the magnetic field's strength one can consider that the magnetic source represents a magnetic wire placed vertical to the x-y plane at the point $(\bar{a}, \bar{b})$. The contours of the magnetic field are shown in Fig. 4.62. In this study the magnetic source is located at $(1.05\,\text{cols}, 0.5\,\text{rows})$.

4.3.3.3.2 Effects of active parameters

Effects of applying an external magnetic field on a ferrofluid (Fe_3O_4-water) flow and heat transfer in a semiannulus enclosure is studied numerically using the CVFEM. The mathematical model used to formulate the problem is consistent with the principles of FHD and MHD. Various nanoparticle volume fractions ($\phi = 0\%$ and 4%), Rayleigh numbers ($Ra = 10^3$, 10^4, and 10^5), magnetic numbers arising from FHD ($Mn_F = 0$, 20, 60, and 100), and Hartmann numbers arising from MHD ($Ha = 0$, 2, 6, and 10) are calculated. In all calculations the Prandtl number (Pr), temperature (ε_1), and Eckert number (Ec) are set to 6.8, 0.0, and 1×10^{-5}, respectively.

Table 4.9 Effects of the Hartmann Number, Radiation Parameter, Nanoparticle Volume Fraction, and Magnetic Number on the Average Nusselt Number

		$Mn_F=0$, $Ha=0$		$Mn_F=100$, $Ha=5$		$Mn_F=100$, $Ha=0$		$Mn_F=0$, $Ha=0$	
		Rd		*Rd*		*Rd*		*Rd*	
Ra	ϕ	**0**	**0.02**	**0**	**0.02**	**0**	**0.02**	**0**	**0.02**
10^3	0	1.611611	0.995565	1.519592	0.928174	1.711999	1.010865	1.510564	0.953924
10^3	0.04	1.890241	1.270623	1.801949	1.215039	1.972767	1.285426	1.796937	1.234266
10^4	0	2.906977	1.609061	2.477273	1.244099	2.934773	1.636442	2.459938	1.226934
10^4	0.04	3.312842	2.022216	2.802561	1.596926	3.34569	2.048861	2.784897	1.582475
10^5	0	4.799285	2.103907	4.316693	1.570677	4.775208	2.106432	4.331907	1.564888
10^5	0.04	5.372904	2.595023	4.789146	1.942158	5.345175	2.588536	4.807063	1.940414

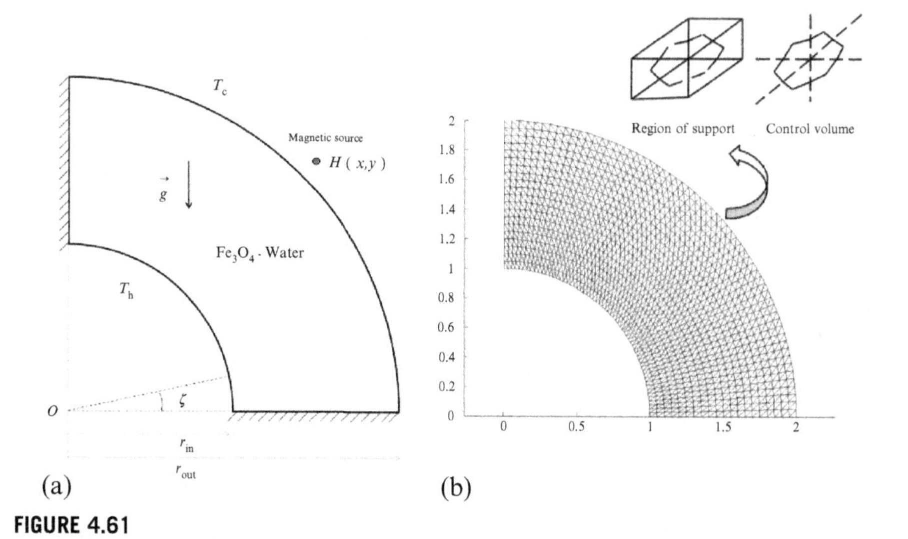

FIGURE 4.61

Geometry and the boundary conditions (a) and the mesh of the enclosure (b) considered in this work.

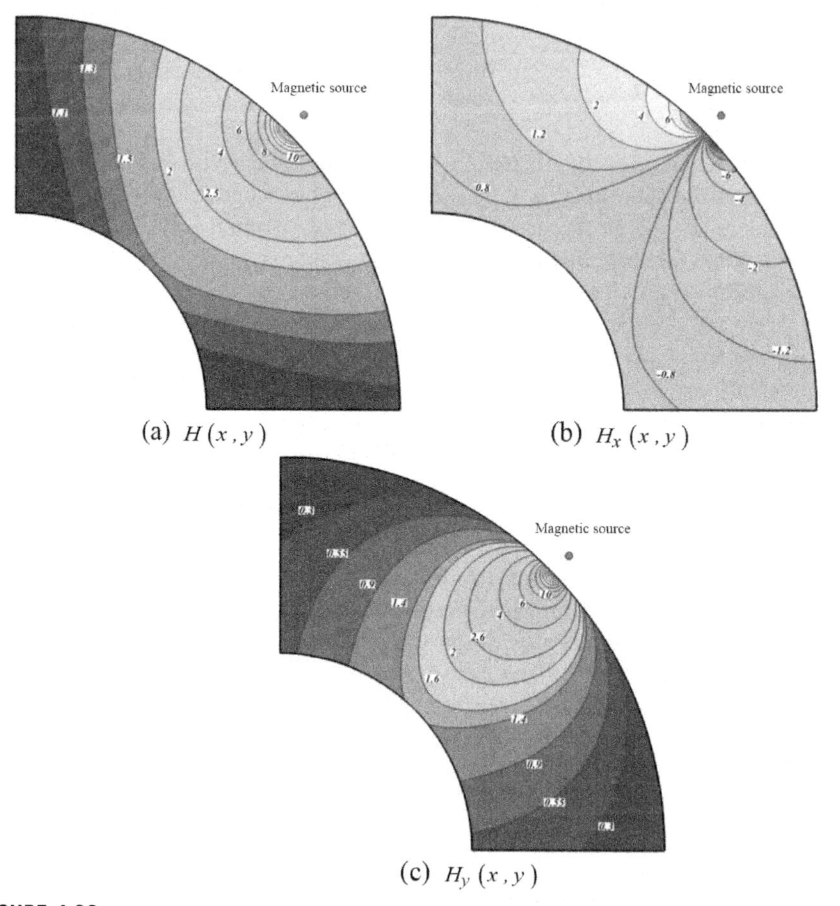

FIGURE 4.62

Contours of the magnetic field strength H (a), the magnetic field intensity component in the x direction H_x (b), and the magnetic field intensity component in the y direction H_y (c).

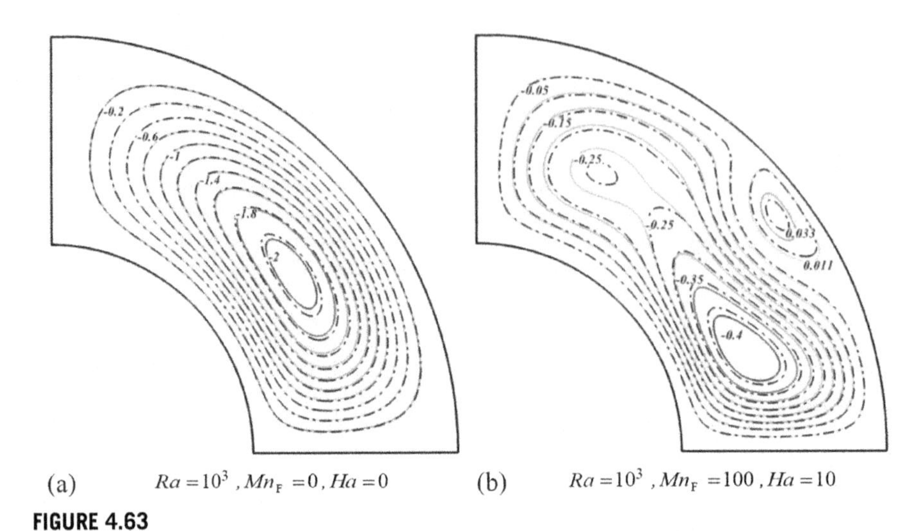

FIGURE 4.63

Comparison of the streamlines between nanofluid ($\phi = 0.04$) (–·–·–) and pure fluid ($\phi = 0$) (—) when (a) $Ra = 10^3$, $Mn_F = 0$, $Ha = 0$; (b) $Ra = 10^3$, $Mn_F = 100$, $Ha = 10$ and $Pr = 6.8$.

Figure 4.63 compares the streamlines of a nanofluid and a pure fluid. The velocity components of the nanofluid are increased because of an increase in the energy transport in the fluid with an increasing volume fraction. Thus the absolute values of stream functions indicate that the strength of flow increases as the volume fraction of the nanofluid increases.

Isotherm and streamline contours for different Rayleigh, Hartmann, and magnetic numbers are shown in Figs. 4.64–4.66. When $Ra = 10^3$, the heat transfer in the enclosure is mainly dominated by conduction. At $Mn_F = 0$, $Ha = 0$, the streamlines show one rotating eddy. When the Hartmann number increases the primary vortex turns into three smaller eddies. One of them, located near the magnetic source, rotates counterclockwise. When the magnetic number increases the primary vortex turns into two eddies; the clockwise eddy is larger than the other. As the Rayleigh number increases to 10^4, the role of convection in heat transfer becomes more significant and, consequently, the thermal boundary layer of the inner wall becomes thinner. In addition, a plume starts to appear on the top of the inner circular wall at $\zeta = 90°$. As the Hartmann number increases the thermal plume turns to two plumes, which are located at $\zeta = 45°$ and $\zeta = 90°$. For high Rayleigh numbers $\left(Ra = 10^5\right)$, Lorentz forces have no significant effect on flow and heat transfer in the absence of Kelvin forces. When $Mn_F = 100$, thermal plume slants to $\zeta = 45°$ as the Hartmann number increases. Another reverse plume is also observed at $\zeta = 90°$. This occurs because of the two existing eddies, which rotate in different directions.

Figure 4.67 depicts that the effects of the magnetic number, Hartmann number, and Rayleigh number on the local Nusselt number. The local Nusselt number increases with an increase in the magnetic number and Rayleigh number, whereas it decreases with an increase in the Hartmann number. The local maximum and

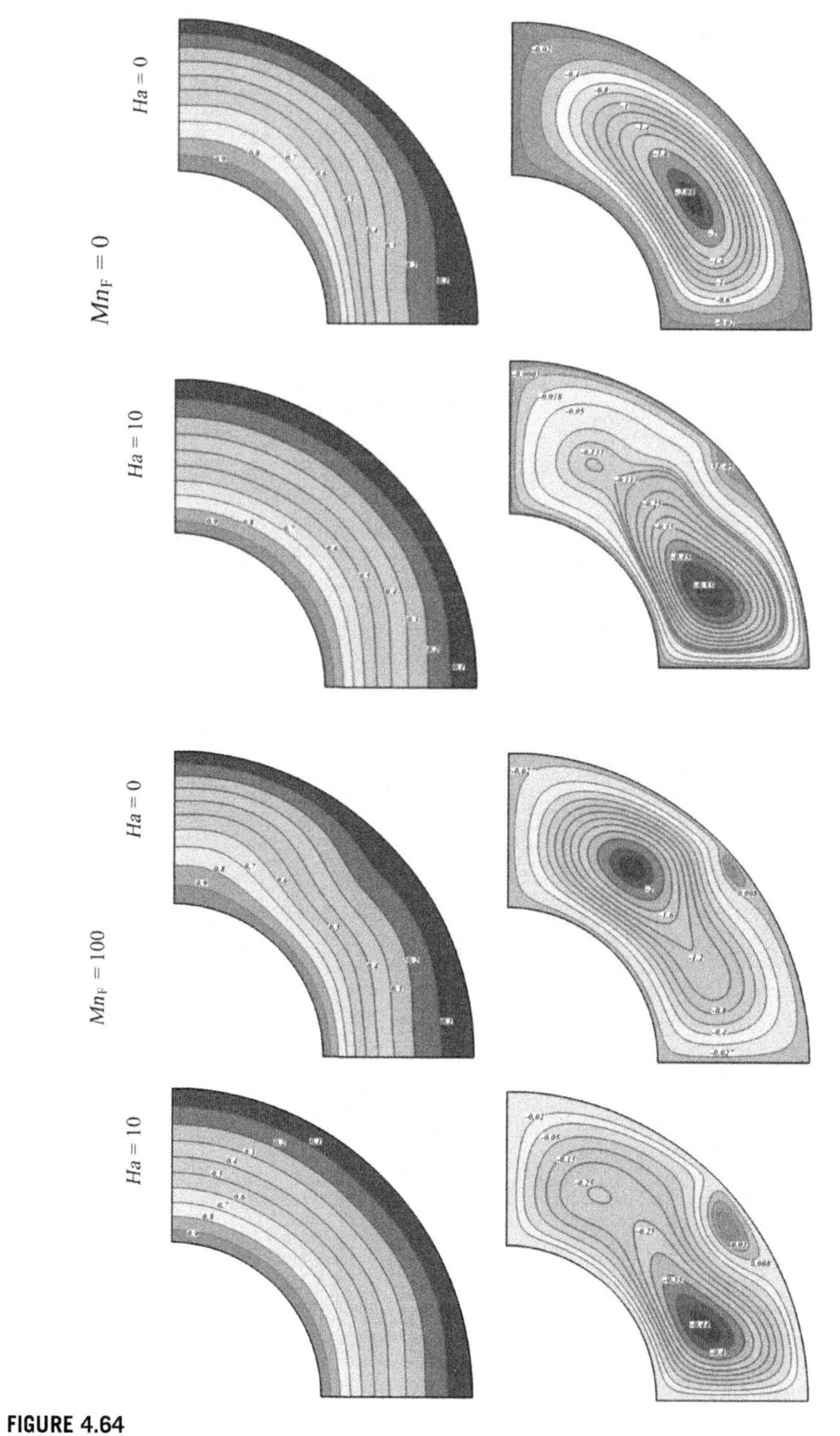

FIGURE 4.64

Isotherm (left) and streamline (right) contours for different Hartmann numbers and magnetic numbers when $Ra = 10^3$.

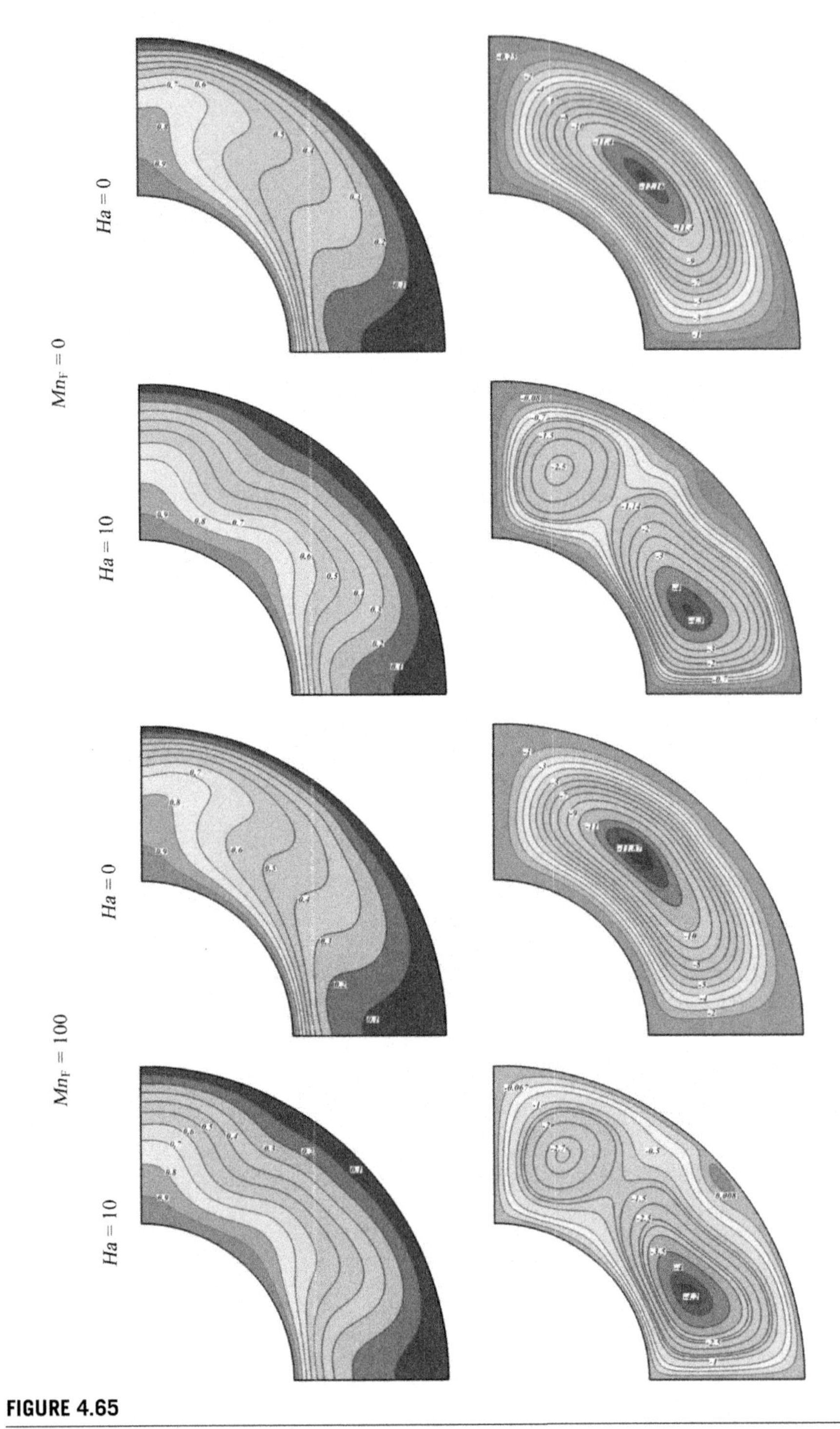

FIGURE 4.65

Isotherm (left) and streamline (right) contours for different Hartmann numbers and magnetic numbers when $Ra = 10^4$.

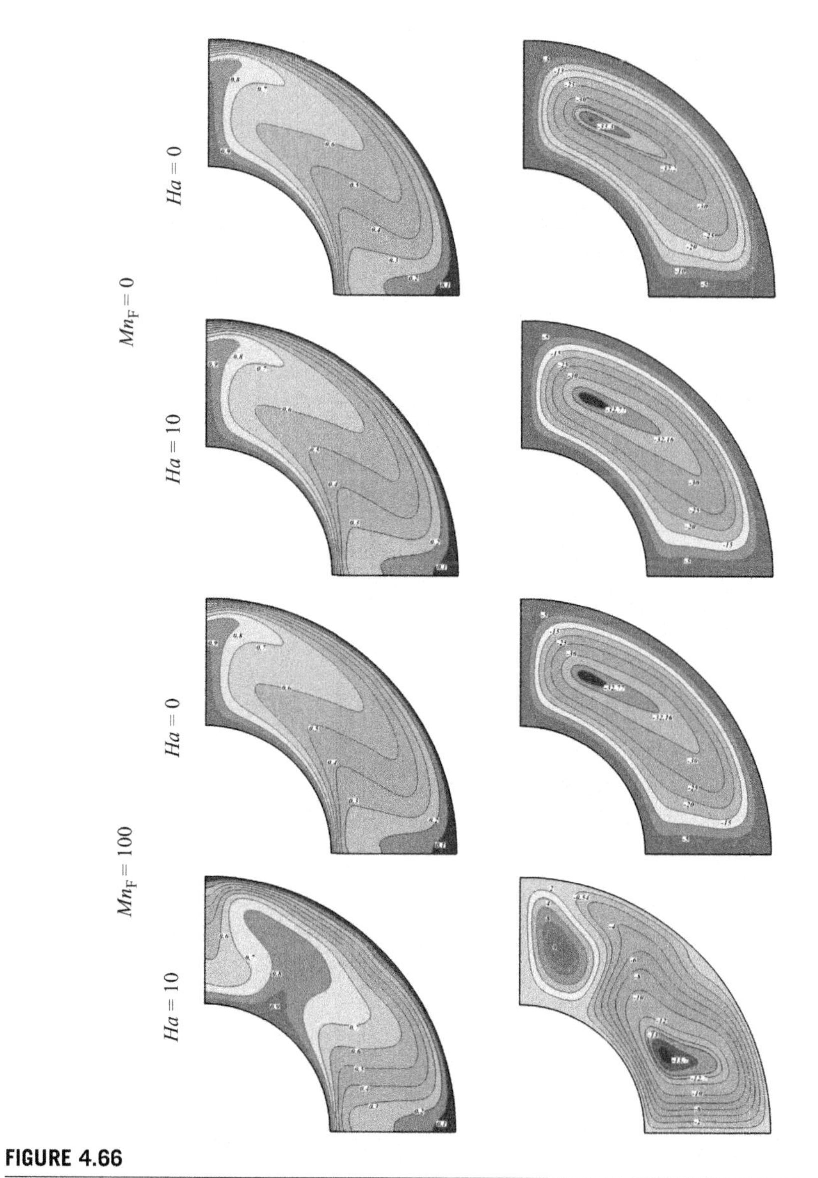

FIGURE 4.66

Isotherm (left) and streamline (right) contours for different Hartmann numbers and magnetic numbers when $Ra = 10^5$.

minimum points in these profiles are related to the thermal plume that forms over the inner cylinder. Figure 4.68 and Table 4.10 show the effects of the Hartmann number, magnetic number, and Rayleigh number on the average Nusselt number. The Nusselt number is an increasing function of Ra, Mn_F and ϕ, whereas it is a decreasing function of the Hartmann number. Effects of the magnetic number, Rayleigh number, and Hartmann number on heat transfer enhancement are shown in Fig. 4.69. The effect of

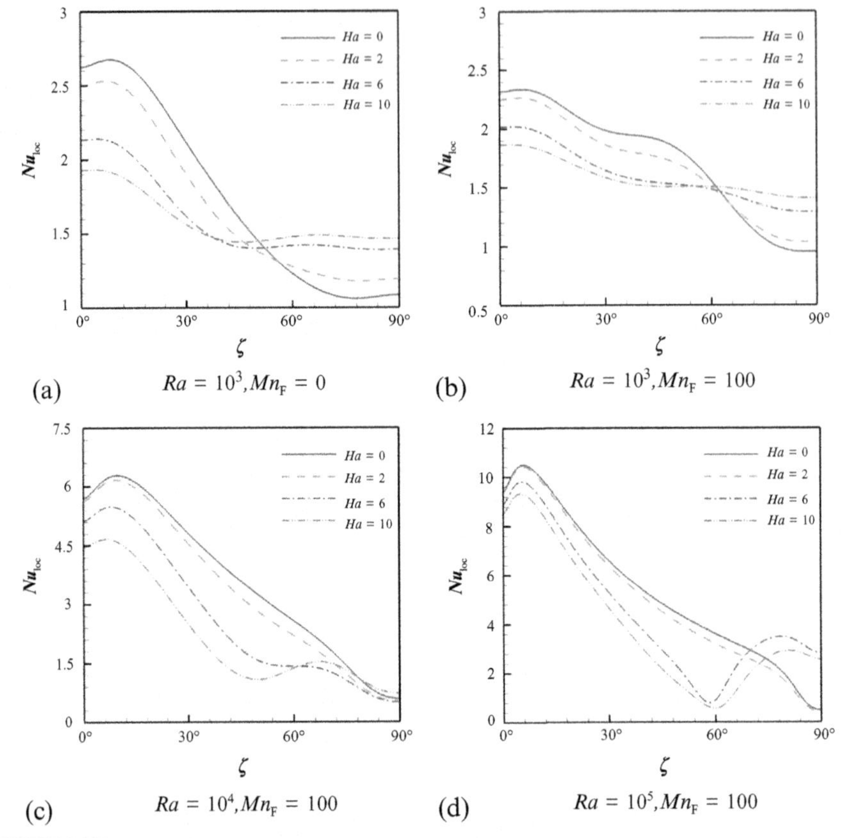

FIGURE 4.67

Effects of Magnetic number, Hartmann number and Rayleigh number on local Nusselt number Nu_{ave} along hot wall when (a) $Ra = 10^3$, $Mn_F = 0$; (b) $Ra = 10^3$, $Mn_F = 100$; (c) $Ra = 10^4$, $Mn_F = 100$ and (d) $Ra = 10^5$, $Mn_F = 100$.

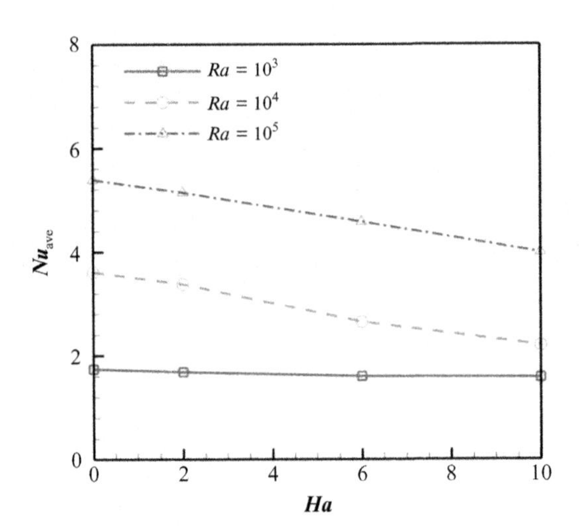

FIGURE 4.68

Effects of the Hartmann number and Rayleigh number on the average Nusselt number Nu_{ave} along a hot wall when $Mn_F = 100$.

Table 4.10 Effects of the Hartmann Number and Magnetic Number on the Average Nusselt Number Nu_{ave} Along a Hot Wall When $Ra = 10^5$

	Ha		
Mn_F	0	2	10
0	5.380431	5.136342	3.997727
20	5.382286	5.138101	3.997764
60	5.385799	5.141522	3.99787
100	5.389023	5.144803	3.99802

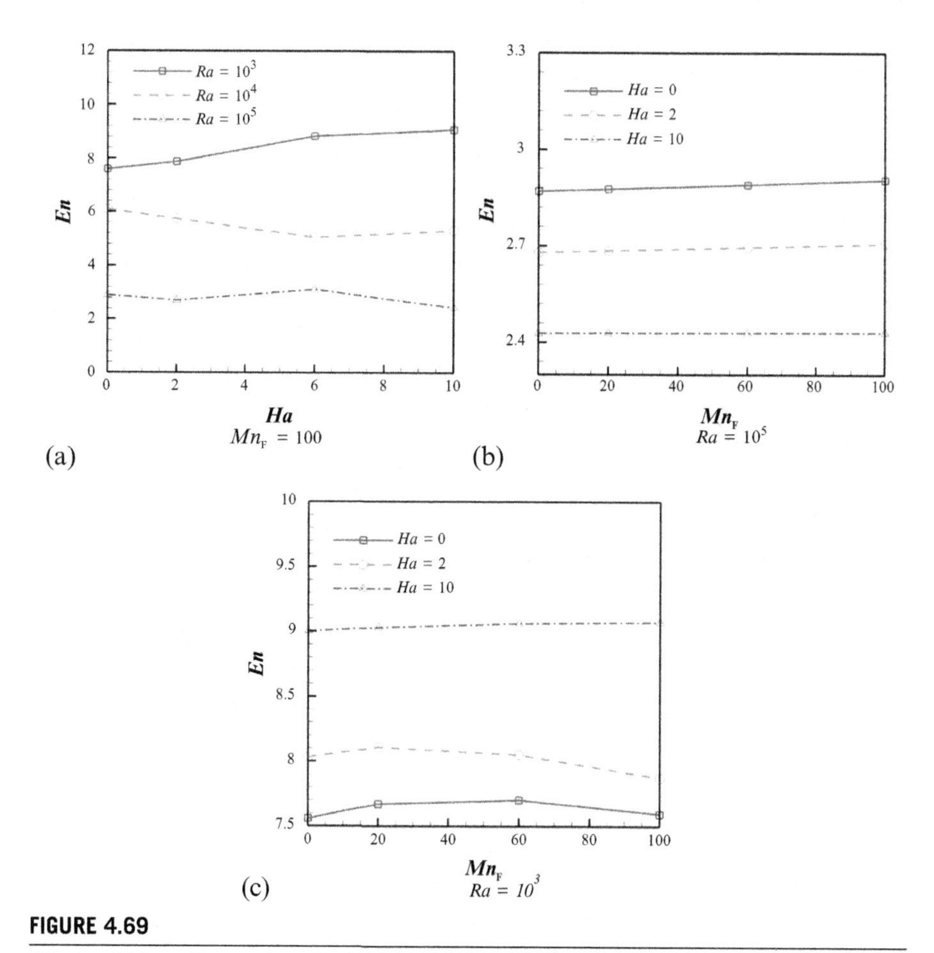

FIGURE 4.69

Effects of Magnetic number, Rayleigh number and Hartmann number on heat transfer enhancement when (a) $Mn_F = 100$; (b) $Ra = 10^5$ and (c) $Ra = 10^3$.

nanoparticles is more pronounced at low Rayleigh numbers than at high Rayleigh numbers because of the greater enhancement rate. This observation can be explained by noting that at low Rayleigh numbers, heat transfer is dominated by conduction. Therefore the addition of nanoparticles with high thermal conductivity increases the conduction and makes the enhancement more effective. At high Rayleigh numbers, heat transfer enhancement increases with an increase in the magnetic number, whereas it decreases with an increase in the Hartmann number. At low Rayleigh numbers, En has a direct relationship with the Hartmann number, whereas the magnetic number behaves differently. It means that in the absence of Lorentz forces, at first En increases then decreases with an increase in Mn_F.

REFERENCES

[1] H. Ozoe, K. Okada, The effect of the direction of the external magnetic field on the three-dimensional natural convection in a cubical enclosures, Int. J. Heat Mass Transfer 32 (10) (1989) 1939–1954.

[2] N. Rudraiah, R.M. Barron, M. Venkatachalappa, C.K. Subbaraya, Effect of a magnetic field on free convection in a rectangular enclosure, Int. J. Eng. Sci 33 (8) (1995) 1075–1084.

[3] N.M. Al-Najem, K. Khanafer, M.M. El-Refaee, Numerical study of laminar natural convection in tilted cavity with transverse magnetic field, Int. J. Numer. Methods Heat Fluid Flow 8 (1998) 651–672.

[4] M.C. Ece, E. Buyuk, Natural-convection flow under a magnetic field in an inclined rectangular enclosure heated and cooled on adjacent walls, Fluid Dyn. Res. 38 (8) (2006) 564–590.

[5] M. Sheikholeslami, I. Hashim, S. Soleimani, Numerical investigation of the effect of magnetic field on natural convection in a curved-shape enclosure. Math. Probl. Eng. 2013 (2013) 78, http://dx.doi.org/10.1155/2013/831725, 11 pp. (article ID 831725).

[6] M. Sheikholeslami, H.R. Ashorynejad, D. Domairry, I. Hashim, Investigation of the laminar viscous flow in a semi-porous channel in the presence of uniform magnetic field using optimal homotopy asymptotic method, Sains Malays. 41 (10) (2012) 1177–1229.

[7] M. Sheikholeslami, D.D. Ganji, H.B. Rokni, Nanofluid flow in a semi-porous channel in the presence of uniform magnetic field, IJE Trans. C Asp. 26 (6) (2013) 653–662.

[8] H.R. Ashorynejad, M. Sheikholeslami, I. Pop, D.D. Ganji, Nanofluid flow and heat transfer due to a stretching cylinder in the presence of magnetic field, Heat Mass Transfer 49 (2013) 427–436.

[9] M. Sheikholeslami, M. Gorji-Bandpy, D.D. Ganji, S. Soleimani, Effect of a magnetic field on natural convection in an inclined half-annulus enclosure filled with Cu-water nanofluid using CVFEM, Adv. Powder Technol. 24 (2013) 980–991.

[10] M. Sheikholeslami, M. Gorji-Bandpy, D.D. Ganji, S. Soleimani, MHD natural convection in a nanofluid filled inclined enclosure with sinusoidal wall using CVFEM, Neural Comput. Appl. 24 (2014) 873–882.

[11] M. Sheikholeslami, M. Gorji-Bandpy, D.D. Ganji, S. Soleimani, S.M. Seyyedi, Natural convection of nanofluids in an enclosure between a circular and a sinusoidal cylinder in the presence of magnetic field, Int. Commun. Heat Mass Transfer 39 (2012) 1435–1443.

[12] M. Sheikholeslami, M. Gorji-Bandpay, D.D. Ganji, Magnetic field effects on natural convection around a horizontal circular cylinder inside a square enclosure filled with nanofluid, Int. Commun. Heat Mass Transfer 39 (2012) 978–986.

[13] M. Sheikholeslami, M. Gorji-Bandpy, D.D. Ganji, MHD free convection in an eccentric semi-annulus filled with nanofluid. J. Taiwan Inst. Chem. Eng. 45 (2014) 1204–1216, http://dx.doi.org/10.1016/j.jtice.2014.03.010.

[14] M. Sheikholeslami, M. Gorji-Bandpy, D.D. Ganji, Lattice Boltzmann method for MHD natural convection heat transfer using nanofluid, Powder Technol. 254 (2014) 82–93.

[15] M. Sheikholeslami, M. Gorji-Bandpy, D.D. Ganji, Numerical investigation of MHD effects on Al_2O_3-water nanofluid flow and heat transfer in a semi-annulus enclosure using LBM, Energy 60 (2013) 501–510.

[16] M. Sheikholeslami, M. Gorji-Bandpy, D.D. Ganji, MHD free convection in an eccentric semi-annulus filled with nanofluid, J. Taiwan Inst. Chem. Eng. 45 (2014) 1204–1216.

[17] M. Sheikholeslami, D.D. Ganji, H.R. Ashorynejad, H.B. Rokni, Analytical investigation of Jeffery-Hamel flow with high magnetic field and nano particle by Adomian decomposition method, Appl. Math. Mech. Engl. Ed. 33 (1) (2012) 1553–1564.

[18] R.E. Rosensweig, Ferrohydrodynamics, Cambridge University Press, London, 1985.

[19] R. Hiegeister, W. Andra, N. Buske, R. Hergt, I. Hilger, U. Richter, W. Kaiser, Application of magnetite ferrofluids for hyperthermia, J. Magn. Magn. Mater. 201 (1999) 420–422.

[20] K. Nakatsuka, B. Jeyadevan, S. Neveu, H. Koganezawa, The magnetic fluid for heat transfer applications, J. Magn. Magn. Mater. 252 (2002) 360–362.

[21] S. Shuchi, K. Sakatani, H. Yamaguchi, An application of a binary mixture of magnetic fluid for heat transport devices, J. Magn. Magn. Mater. 289 (2005) 257–259.

[22] F. Selimefendigil, H.F. Oztop, Effect of a rotating cylinder in forced convection of ferrofluid over a backward facing step, Int. J. Heat Mass Transfer 71 (2014) 142–148.

[23] W. Wrobel, E. Fornalik-Wajs, J.S. Szmyd, Experimental and numerical analysis of thermo-magnetic convection a vertical annular enclosure, Int. J. Heat Fluid Flow 31 (2010) 1019–1031.

[24] E.E. Tzirtzilakis, M. Xenos, V.C. Loukopoulos, N.G. Kafoussias, Turbulent biomagnetic fluid flow in a rectangular channel under the action of a localized magnetic field, Int. J. Eng. Sci. 44 (2006) 1205–1224.

[25] E.E. Tzirtzilakis, V.D. Sakalis, N.G. Kafoussias, P.M. Hatzikonstantinou, Biomagnetic fluid flow in a 3D rectangular duct, Int. J. Numer. Methods Fluids 44 (2004) 1279–1298.

[26] H. Aminfar, M. Mohammadpourfard, Y. Narrimani Kahnamouei, A 3D numerical simulation of mixed convection of a magnetic nanofluid in the presence of non-uniform magnetic field in a vertical tube using two phase mixture model, J. Magn. Magn. Mater. 323 (2011) 1963–1972.

[27] M. Sheikholeslami, M. Gorji-Bandpy, Free convection of ferrofluid in a cavity heated from below in the presence of an external magnetic field, Powder Technol. 256 (2014) 490–498.

[28] H. Aminfar, M. Mohammadpourfard, F. Mohseni, Two-phase mixture model simulation of the hydro-thermal behavior of an electrical conductive ferrofluid in the presence of magnetic fields, J. Magn. Magn. Mater. 324 (2012) 830–842.

[29] M. Lajvardi, J. Moghimi-Rad, I. Hadi, A. Gavili, T. Dallal, F. Zabihi, J. Sabbaghzadeh, Experimental investigation for enhanced ferrofluid heat transfer under magnetic field effect, J. Magn. Magn. Mater. 322 (2010) 3508–3513.

[30] M. Sheikholeslami, M. Gorji Bandpy, R. Ellahi, M. Hassan, S. Soleimani, Effects of MHD on Cu-water nanofluid flow and heat transfer by means of CVFEM, J. Magn. Magn. Mater. 349 (2014) 188–200.

[31] M. Sheikholeslami, M. Gorji-Bandpy, D.D. Ganji, S. Soleimani, Natural convection heat transfer in a cavity with sinusoidal wall filled with CuO-water nanofluid in presence of magnetic field, J. Taiwan Inst. Chem. Eng. 45 (2014) 40–49.
[32] M. Sheikholeslami, D.D. Ganji, M. Gorji-Bandpy, S. Soleimani, Magnetic field effect on nanofluid flow and heat transfer using KKL model, J. Taiwan Inst. Chem. Eng. 45 (2014) 795–807.
[33] M. Sheikholeslami, M. Gorji-Bandpy, D.D. Ganji, S. Soleimani, Heat flux boundary condition for nanofluid filled enclosure in presence of magnetic field, J. Mol. Liq. 193 (2014) 174–184.
[34] M. Sheikholeslami, M. Gorji-Bandpy, D.D. Ganji, P. Rana, S. Soleimani, Magnetohydrodynamic free convection of Al_2O_3-water nanofluid considering thermophoresis and Brownian motion effects, Comput. Fluids 94 (2014) 147–160.
[35] M. Sheikholeslami, D.D. Ganji, Ferrohydrodynamic and magnetohydrodynamic effects on ferrofluid flow and convective heat transfer, Energy 75 (2014) 400–410.

CHAPTER

Flow and heat transfer in porous media

5

5.1 INTRODUCTION

Natural convective flow in differentially heated enclosures filled with Darcian or non-Darcian, fluid-saturated porous media has received considerable attention in the literature. This topic is of practical interest in building and geothermal sciences, engineering, and agriculture. Nithiarasu et al. [1] examined the effects of variable porosity on convective flow patterns inside a porous cavity. In their study the flow was triggered by sustaining a temperature gradient between isothermal lateral walls. They found that the nature of porosity variation in a medium with variable porosity significantly affects heat transfer and flow results. Pakdee and Rattanadecho [2] performed numerical investigations of transient natural convection flow through a fluid-saturated porous medium in a rectangular cavity with a convection surface condition. They found that the heat transfer coefficient, Rayleigh number, and Darcy number considerably influenced characteristics of the flow and heat transfer mechanisms. Furthermore, the flow pattern had a local effect on the heat convection rate. Baytas and Pop [3] studied natural convection on a trapezoidal porous enclosure in situations such as the top enclosure being cooled, the bottom surface being heated, and the remaining two nonparallel plane sidewalls of the enclosure being adiabatic. Although their study dealt with heat transfer analysis of various applications in trapezoidal porous spaces, a comprehensive analysis of heat transfer and flow circulation for applications related to the extraction of molten metals, salt water, and olive oil confined within a porous bed is yet to appear in the literature about various tilt angles.

A geometric pattern can be useful in improving heat transfer. Natural convection heat transfer inside a wavy enclosure is one of several devices used to enhance heat and mass transfer efficiency. Flow and heat transfer from irregular surfaces are often encountered in many engineering applications to enhance heat transfer, such as microelectronic devices, flat-plate solar collectors and flat-plate condensers in refrigerators, geophysical applications, electric machinery, cooling systems in microelectronic devices, and so on. Saidi et al. [4] presented numerical and experimental results of flow over and heat transfer from a sinusoidal cavity. They reported that the total heat exchange between the wavy wall of the cavity and the flowing fluid was reduced by the presence of a vortex. Das and Mahmud [5] conducted a numerical investigation of natural convection in an enclosure consisting of two isothermal, horizontal, wavy walls and two adiabatic, vertical, straight walls. They reported that the amplitude-to-wavelength ratio affected the local heat transfer rate, but it had no significant influence on the average heat transfer rate. Adjlout et al. [6] conducted a numerical study of natural convection in an inclined cavity with a hot wavy wall and a cold flat wall. One of their interesting findings was the decrease in the average heat transfer with a wavy surface compared with a flat wall. Ziabakhsh et al. [7] studied heat transfer over an unsteady, stretching, permeable surface with a prescribed wall temperature.

Natural convection in an enclosure with vertical wavy walls was studied by Mahmud et al. [8] using a numerical investigation. They simulated the behavior of fluid flow and heat transfer performance for different values of aspect ratio, Grashof number, and amplitude-to-wavelength ratio. They showed that the average heat transfer decreases with an increase in surface waviness, and a higher heat transfer rate is observed at lower aspect ratios for a certain Grashof number. Yao [9] theoretically studied natural convection along a vertical wavy surface. He found the heat transfer rate for a wavy surface was constantly smaller than that of a corresponding flat plate. The influence of the geometric parameters on the mean Nusselt number is clearly shown from his results. Dalal and Das [10] numerically analyzed natural convection in a cavity with a wavy wall heated from below. Their results showed that the presence of undulation in the right wall affected the local heat transfer rate and flow field as well as the thermal field. Rostami [11] investigated the unsteady heat transfer and fluid flow characteristics in an enclosure with two vertical wavy and two horizontal straight walls. The conditions were simulated at different Grashof numbers, Prandtl numbers, wave ratios, and aspect ratios. Ziabakhsh et al. [12] presented the analytic solution for the nonlinear Brinkman equation for stagnation-point flow in a porous medium.

Using the finite element method, Rathish Kumar [13] numerically analyzed free convection induced by a vertical wavy surface with uniform heat flux in a porous enclosure. The results revealed that small sinusoidal drifts from the smoothness of a vertical wall, with a phase angle of 60° and high frequency, enhanced free convection from a vertical wall with uniform heat flux. Chen et al. [14] numerically analyzed steady-state free convection inside a cavity made of two horizontal straight walls and two vertical bent-wavy walls and filled with a fluid-saturated porous medium. Their results showed that the dependence of the local Nusselt number on the Darcy number and porosity was not small at large a Darcy-Rayleigh number. The problem of laminar, isothermal, incompressible, and viscous flow in a rectangular domain bounded by two moving porous walls was investigated analytically by Ganji et al. [15]. Ziabakhsh et al. [16] studied the problem of flow and diffusion of chemically reactive species over a nonlinearly stretching sheet immersed in a porous medium.

5.2 GOVERNING EQUATIONS FOR FLOW AND HEAT TRANSFER IN POROUS MEDIA

Porous medium is assumed to be homogeneous, thermally isotropic, and saturated with a fluid that is in local thermodynamic equilibrium with a solid matrix. Fluid flow is laminar and incompressible. The pressure work and viscous dissipation are all assumed to be negligible. The thermophysical properties of a porous medium are considered constant; however, the Boussinesq approximation takes into account the effect of density variation on the buoyancy force. Furthermore, a solid matrix is made of spherical particles, whereas the porosity and permeability of the medium are assumed to be uniform throughout the enclosure. Using standard symbols, the governing equations describing the heat transfer phenomenon are given by

$$\frac{\partial u}{\partial x}+\frac{\partial v}{\partial y}=0 \tag{5.1}$$

$$\frac{1}{\varepsilon}\frac{\partial u}{\partial t}+\frac{u}{\varepsilon^2}\frac{\partial u}{\partial x}+\frac{v}{\varepsilon^2}\frac{\partial u}{\partial y}=-\frac{1}{\rho_f}\frac{\partial P}{\partial x}+\frac{\upsilon}{\varepsilon}\left(\frac{\partial^2 u}{\partial x^2}+\frac{\partial^2 u}{\partial y^2}\right)-\frac{\mu u}{\rho_f k}, \tag{5.2}$$

$$\frac{1}{\varepsilon}\frac{\partial v}{\partial t}+\frac{u}{\varepsilon^2}\frac{\partial v}{\partial x}+\frac{v}{\varepsilon^2}\frac{\partial v}{\partial y}=-\frac{1}{\rho_f}\frac{\partial P}{\partial y}+\frac{\upsilon}{\varepsilon}\left(\frac{\partial^2 v}{\partial x^2}+\frac{\partial^2 v}{\partial y^2}\right)+g\beta(T-T_\infty)-\frac{\mu v}{\rho_f k}, \tag{5.3}$$

$$\sigma\frac{\partial T}{\partial t}+u\frac{\partial T}{\partial x}+v\frac{\partial T}{\partial y}=\alpha\left(\frac{\partial^2 T}{\partial x^2}+\frac{\partial^2 T}{\partial y^2}\right), \tag{5.4}$$

$$\sigma=\frac{\left[\varepsilon(\rho C_p)_f+(1-\varepsilon)(\rho C_p)_s\right]}{(\rho C_p)_f}, \tag{5.5}$$

where k is medium permeability, β is thermal expansion coefficient, α is effective thermal diffusivity of the porous medium, μ and υ are viscosity and kinematic viscosity of the fluid, respectively. In this study the heat capacity ratio σ is set at unity because the thermal properties of the solid matrix and the fluid are assumed to be identical. The momentum equation consists of the Brinkmann term, which accounts for the viscous effects of the presence of a solid body [17]. This form of the momentum equation is known as the Brinkmann-extended Darcy model. Lauriat and Prasad [18] used the Brinkmann-extended Darcy formulation to investigate the effects of buoyancy on natural convection in a vertical enclosure. Although the viscous boundary layer in the porous medium is very thin for most engineering applications, inclusion of this term is essential for heat transfer calculations [19]. The inertial effect was neglected, however, because the flow was relatively low. The variables are transformed into the dimensionless quantities defined as

$$X=\frac{x}{L},\quad Y=\frac{y}{L},\quad \tau=\frac{t\alpha}{H^2},\quad U=\frac{uH}{\alpha},$$
$$V=\frac{uH}{\alpha},\quad \Omega=\frac{\omega H^2}{\alpha},\quad \Psi=\frac{\psi}{\alpha},\quad \Theta=\frac{T-T_c}{T_h-T_c}, \tag{5.6}$$

where ω and ψ represent dimensional vorticity and stream function, respectively. Symbol α denotes thermal diffusivity.

$$\frac{\partial^2\Psi}{\partial X^2}+\frac{\partial^2\Psi}{\partial Y^2}=-\Omega, \tag{5.7}$$

$$\varepsilon\frac{\partial\Omega}{\partial t}+\frac{\partial\Psi}{\partial Y}\frac{\partial\Omega}{\partial X}-\frac{\partial\Psi}{\partial X}\frac{\partial\Omega}{\partial Y}=\varepsilon\, Pr\left(\frac{\partial^2\Omega}{\partial X^2}+\frac{\partial^2\Omega}{\partial Y^2}\right)+Ra\, Pr\varepsilon^2\left(\frac{\partial\Theta}{\partial X}\right)-\frac{Pr\varepsilon^2}{Da}\Omega, \quad (5.8)$$

$$\sigma\frac{\partial\Theta}{\partial t}+\frac{\partial\Psi}{\partial Y}\frac{\partial\Theta}{\partial X}-\frac{\partial\Psi}{\partial X}\frac{\partial\Theta}{\partial Y}=\alpha\left(\frac{\partial^2\Theta}{\partial X^2}+\frac{\partial^2\Theta}{\partial Y^2}\right), \quad (5.9)$$

where the Darcy number Da is defined as k/H^2 and $Pr=\upsilon/\alpha$ is a Prandtl number, where $\alpha=k_e/(\rho C_p)_f$ is the thermal diffusivity. The Rayleigh number Ra is defined as $Ra=g\beta L^3(T_h-T_c)/(\alpha\upsilon)$.

5.3 APPLICATION OF THE CVFEM FOR MAGNETOHYDRODYNAMIC NANOFLUID FLOW AND HEAT TRANSFER

5.3.1 MODELING FREE CONVECTION BETWEEN THE INCLINED HOT ROOF OF A BASEMENT AND A COLD ENVIRONMENT

5.3.1.1 Problem definition

The physical model along with important geometrical parameters and the mesh of the enclosure used in the present control volume finite element method (CVFEM) program are shown in Fig. 5.1. The enclosure has a-to-height aspect ratio of 2. The two sidewalls with length H are thermally insulated, whereas the lower flat and upper sinusoidal walls are maintained at constant temperatures T_h and T_c, respectively. Under all circumstances $T_h > T_c$ condition is maintained. The shape of the upper sinusoidal wall profile is assumed to mimic the following pattern:

$$Y=H-\{a(H+\sin(\pi x-\pi/2))\} \quad (5.10)$$

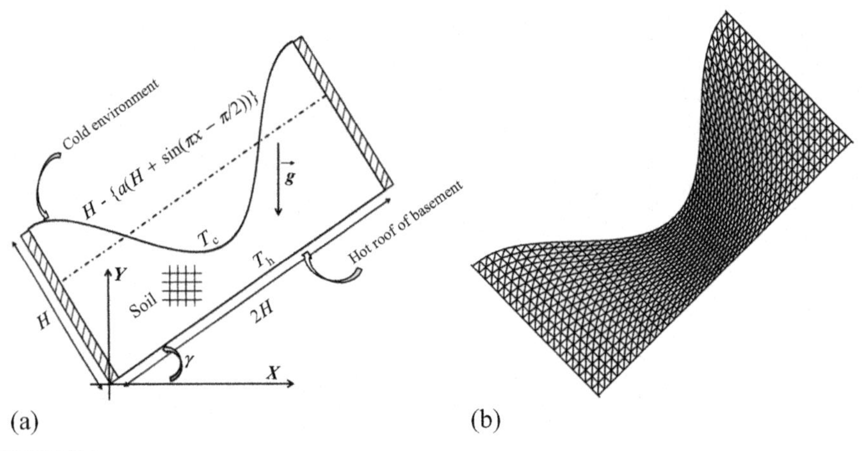

FIGURE 5.1

Geometry and the boundary conditions (a) with the mesh of the enclosure (b) considered in this work.

where a is the dimensionless amplitude of the sinusoidal wall. This geometry could be considered as heat transfer between the roof of a basement and a cold environment.

5.3.1.2 Effects of active parameters

Natural convection flow through a fluid-saturated porous medium in an enclosure with one sinusoidal wall convection was numerically simulated. Effects of the dimensionless amplitude of the sinusoidal wall, Rayleigh number, porosity, Darcy number, and inclination angle of the enclosure on the heat transfer and fluid flow are shown in terms of streamlines and isotherms, whereas the heat transfer characteristic from the straight hot wall is investigated using the average and local Nusselt numbers. Table 5.1 compares the average Nusselt numbers calculated by Nithiarasu et al. [1] and those calculated in this study. These comparisons illustrate an excellent agreement between the present calculations and those of previous works.

Figure 5.2 shows the effect of dimensionless amplitude (a) of the sinusoidal wall along with the inclination angle (γ) on the isotherms and streamlines at $Da = 10^{-3}$, $\varepsilon = 0.5$, and $Ra = 10^3$. For all inclination angles, the temperature distribution and temperature contours are nearly smooth curves, and isothermal lines are nearly parallel to each other and follow the geometry of the sinusoidal surfaces, which is a characteristic of the conduction-dominant mechanism of heat transfer at low Rayleigh numbers. At $\gamma = 0°$ two counter-rotating vortex cores are observed. This bicellular flow pattern divides the enclosure into two symmetric parts with respect to the vertical center line of the enclosure. An increase in a for this inclination angle results in stronger circulation in the enclosure because of the decreasing distance between the hot and cold walls. As seen in Fig. 5.2, the effect of the enclosure wall amplitude on the streamline is more pronounced at $\gamma = 45°$ and $90°$ while this effect is negligible on the temperature distribution contours as same as $\gamma = 0°$. At $\gamma = 45°$ a single vortex is formed within the enclosure for $a = 0.1$ and 0.2. Based on the streamlines, it can be determined that an increase a from 0.1 to 0.2 causes the center cell of the vortex to move upward. Afterward, as a enhances further up to 0.3, this single cell divides into two cells with different strengths: a stronger cell in the upper region and a weaker cell in the lower region of the enclosure. However, at $\gamma = 90°$, the division of the main cell

Table 5.1 Comparison of the average Nusselt numbers calculated by Nithiarasu et al. [1] and those calculated in this study.

Ra	$\varepsilon = 0.6$		$\varepsilon = 0.6$	
	Present Study	Nithiarasu et al. [1]	Present Study	Nithiarasu et al. [1]
10^3	1.0199	1.01	1.0249	1.015
10^4	1.4222	1.418	1.5571	1.530
10^5	3.1674	3.083	3.6497	3.595

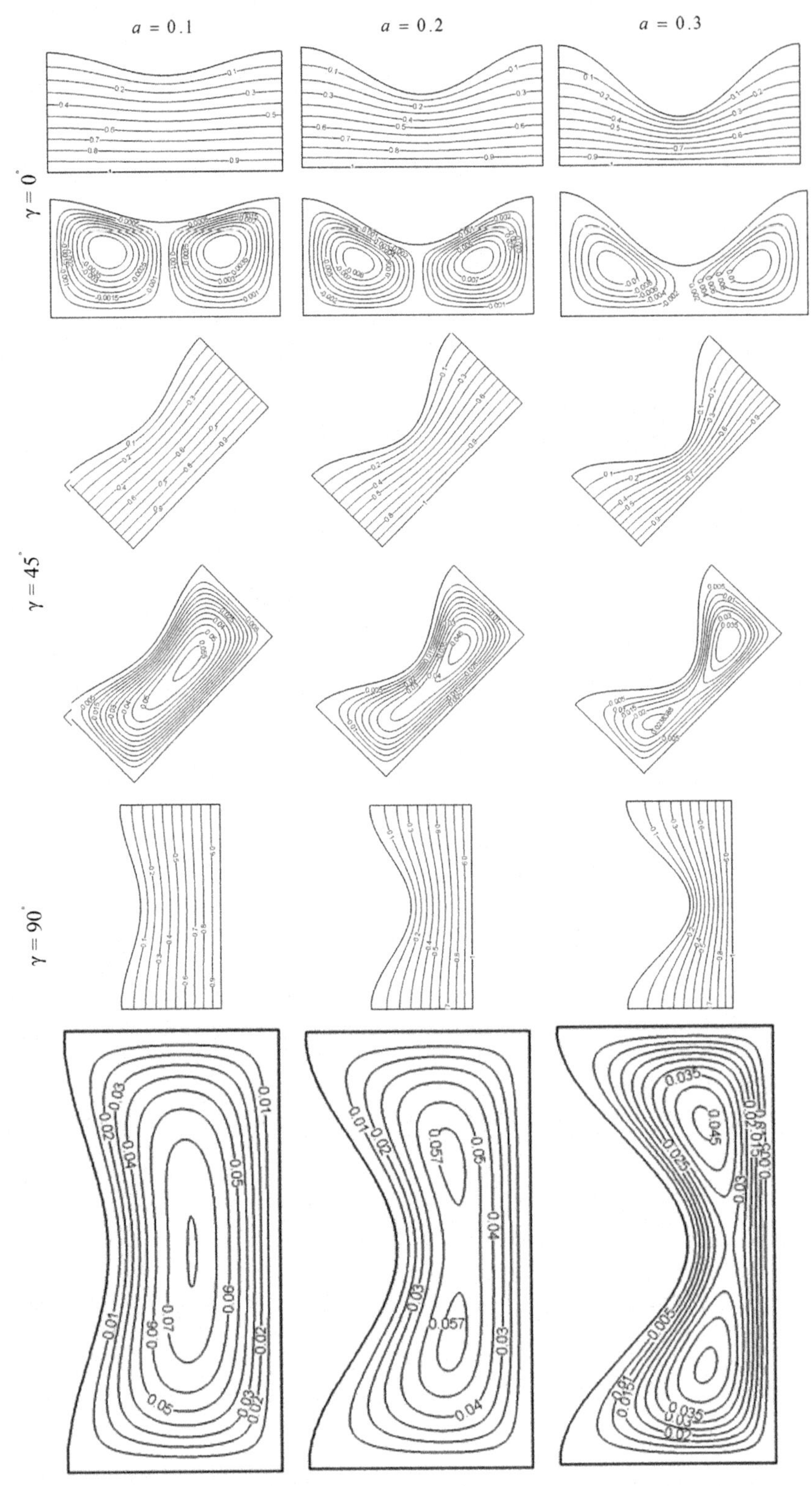

FIGURE 5.2

Isotherms (up) and streamlines (down) contours for different values of dimensionless amplitude (a) and inclination angle (γ) at $Da = 0.001$, $\varepsilon = 0.5$, and $Ra = 10^3$.

into two vortices of equal size and strength occurs at $a = 0.2$. In contrast with $\gamma = 0°$, increment of a at $\gamma = 45°$ and $90°$ decreases the strength of vortexes inside the enclosure because the sinusoidal wall suppresses flow circulation at these inclination angles.

Figure 5.3 investigates the effects of a and γ are at high Rayleigh numbers, that is, $Ra = 10^5$. In general, as the Rayleigh number increases up to 10^5, the buoyancy-driven circulation inside the enclosure become stronger, as seen from the greater magnitudes of stream function. In addition, more distortion occurs in the isothermal lines. Figure 5.3 also shows that at $\gamma = 0°$ an increase in a decreases the intensity of the two counterclockwise vortices. This finding is in contrast with the previous result for $Ra = 10^3$, which may be due to the domination of the conduction mechanism at this Rayleigh number as the hot and cold walls approach each other. The streamlines indicate that for $\gamma = 45°$ and $90°$ the effect of the dimensionless amplitude of the sinusoidal wall on the vortices inside the enclosure is in contrast with that of $\gamma = 45°$ and $90°$,

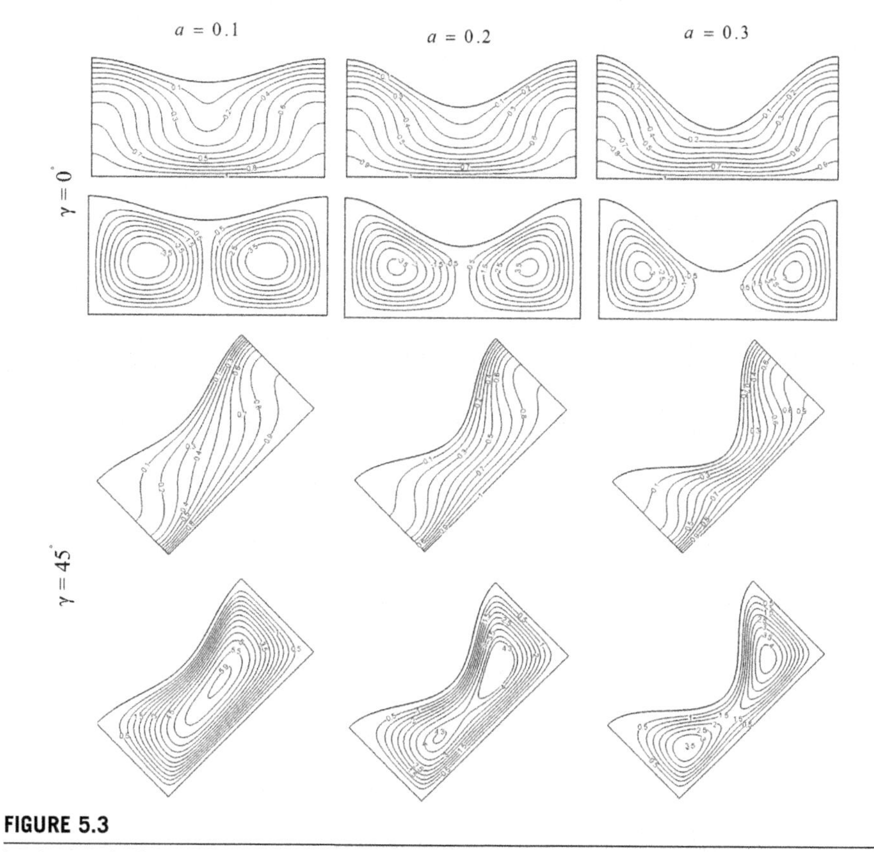

FIGURE 5.3

Isotherms (up) and streamlines (down) contours for different values of dimensionless amplitude (a) and inclination angle (γ) at $Da = 0.001$, $\varepsilon = 0.5$, and $Ra = 10^5$.

Continued

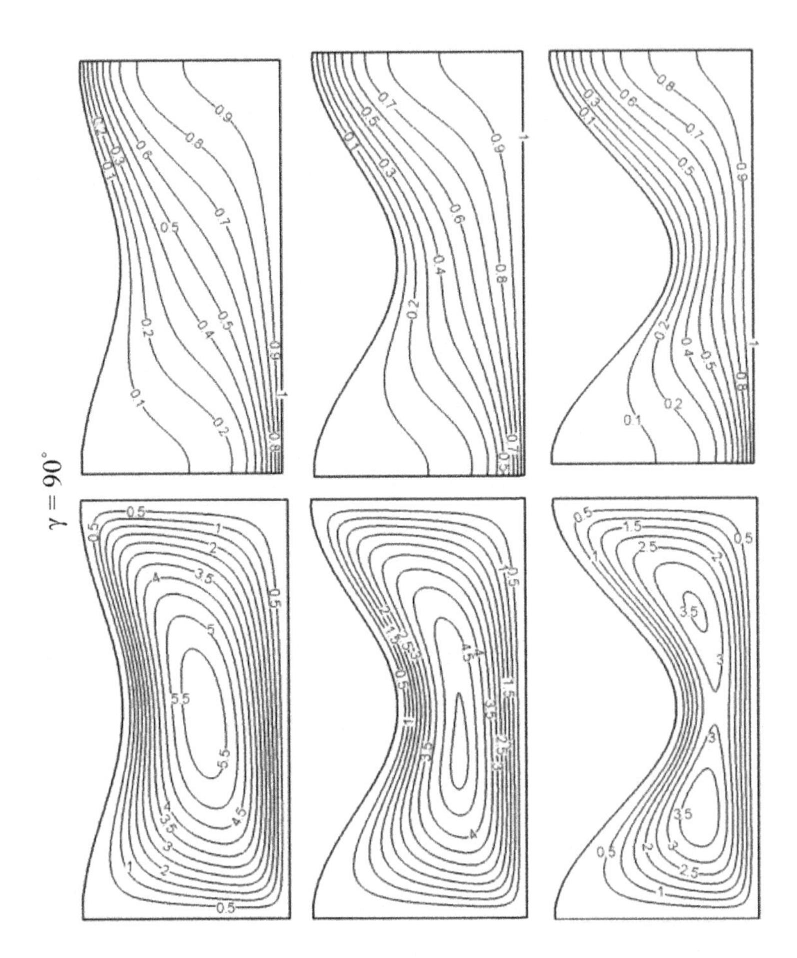

FIGURE 5.3, cont'd

which means that the division of the main vortex into two smaller vortices occurs at smaller value of $a\,(a=0.2)$, whereas this division occurs at greater value of $a\,(a=0.3)$. The magnitude of the stream function also illustrates that for $\gamma=45°$ the upper cell at $a=0.2$ and 0.3 is stronger than the lower one, whereas at $\gamma=90°$, the opposite trend is observed. The isotherms for $\gamma=45°$ and 90° show that the thermal boundary layer in the proximity of the hot wall in the upper region of the enclosure is thinner at $\gamma=45°$ in comparison with $\gamma=90°$, whereas it is thicker at $\gamma=90°$ in the lower region of the enclosure. It should be noted that the thermal boundary layer thickness in these regions is related to the intensity of the two existing vortices; as stated before, at $\gamma=45°$ the upper vortex is stronger than that at $\gamma=90°$. A similar justification could be adopted for the boundary layer at the lower part of the enclosure, where the thermal boundary layer is thinner than that of $\gamma=45°$.

Effects of the Darcy number on fluid flow and temperature inside the enclosure are depicted in Fig. 5.4. In this case, the effect of the Darcy number is investigated at

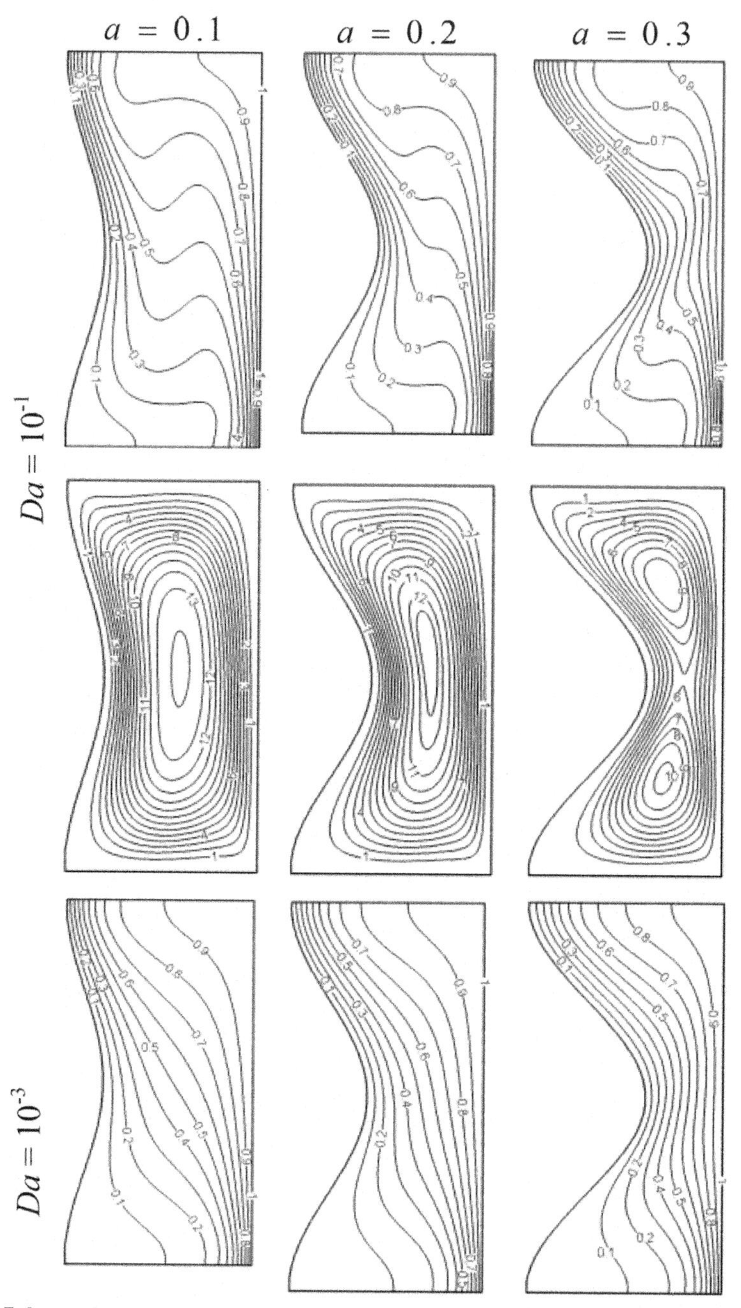

FIGURE 5.4

Isotherms (up) and streamlines (down) contours for different values of dimensionless amplitude (a) and Darcy number (Da) at $\gamma = 90$, $\varepsilon = 0.5$, and $Ra = 10^5$.

Continued

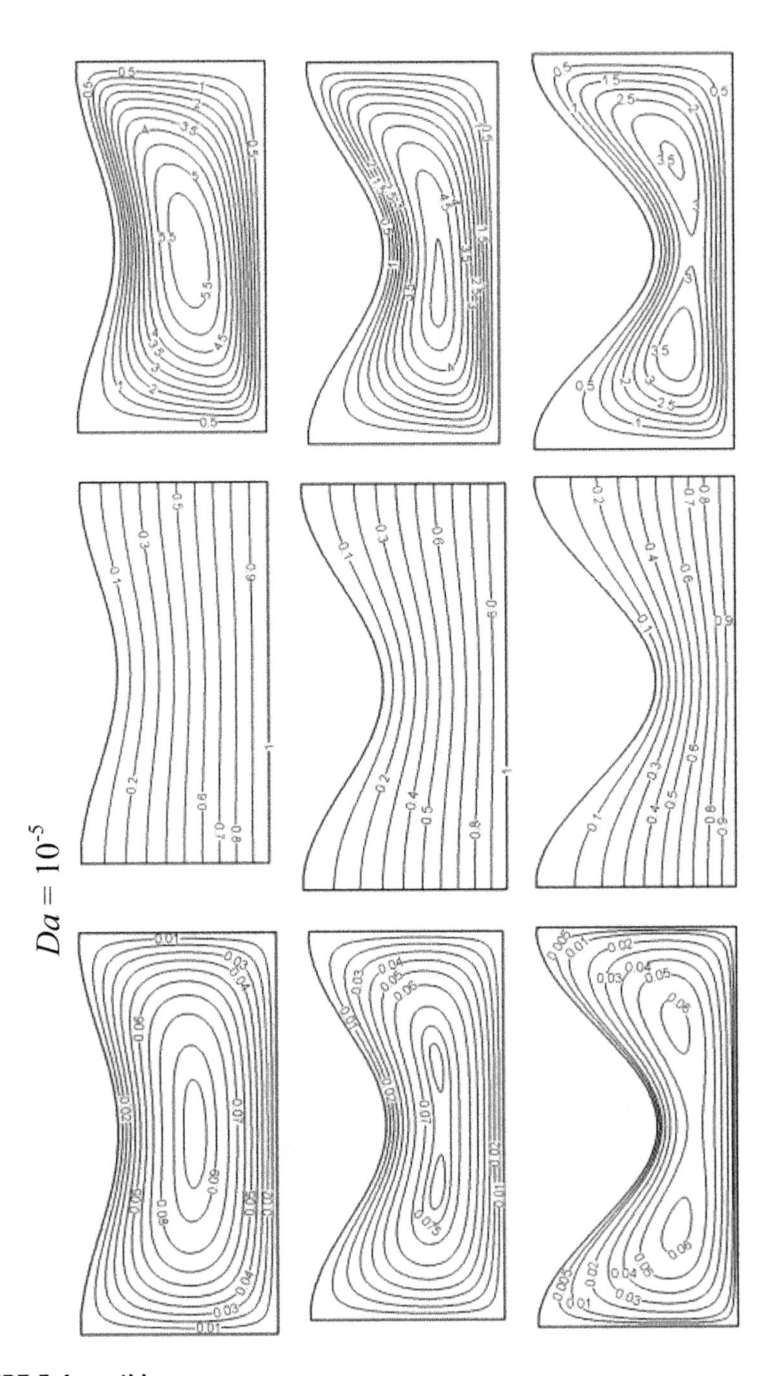

FIGURE 5.4, cont'd

different dimensionless amplitudes (a) and Darcy numbers (Da) and fixed inclination angles of enclosure ($\gamma = 90°$), porosity $\varepsilon = 0.5$, and $Ra = 10^5$. The streamlines and isotherms are depicted for Darcy numbers $Da = 10^{-1}$, 10^{-3}, and 10^{-5} when the dimensionless amplitude varies from 0.1 to 0.3. Based on the definition of a Darcy number, the direct proportion of the Darcy number to permeability is clear. The permeability of a porous medium works as a measure the flow strength and can be considered as conductivity for fluid flow. Hence high permeability causes strong flow circulation in the enclosure, whereas low permeability inhibits flow circulation inside the enclosure and leads to a weak flow. At low Darcy number, conduction heat transfer is the main heat transfer mechanism. Increasing Darcy number leads to increase buoyancy forces. As a result, the isothermal contours are more distorted at high Darcy numbers, whereas they follow the form of the enclosure at low Darcy numbers. Figure 5.4 also shows that the dimensionless amplitude of the sinusoidal wall of the enclosure has a significant effect on the flow characteristics inside the enclosure; this influence is completely obvious even at low Darcy numbers, at which the main heat transfer mechanism is conduction. The streamline for different Darcy numbers indicates that the division of the main cell to two vortices occurs when $a = 0.2$ at $Da = 10^{-5}$, whereas this division corresponds to $a = 0.3$ at higher Darcy numbers, i.e., $Da = 10^{-3}$ and 10^{-1}.

The influence of porosity (ε) along with the effect of dimensionless amplitude is reported in Fig. 5.5 for $\gamma = 90°$, $Da = 10^{-3}$, and $Ra = 10^5$. As a general observation, the effect of various porosity values on streamlines and isotherms is less significant when compared with the effect of other governing parameters such as Darcy or Rayleigh numbers. From this figure it is clear that at a higher porosity the drag terms are less significant, leading to higher dimensionless flow velocities. Hence increasing ε results in higher flow circulation inside the enclosure and a greater magnitude of stream function. In addition, the effect of ε on the isotherm is as same as increasing Ra.

The distribution of a local Nusselt number on the hot wall is depicted in Figs. 5.6–5.10. Figure 5.6 shows the effect of the inclination angle and Rayleigh number on the local Nusselt number at $Da = 10^{-3}$ $\varepsilon = 0.5$, and $a = 0.3$. The figure shows that the inclination of the enclosure has no significant effect on the local Nusselt number, and the local Nusselt number is symmetric with respect to the center line of the enclosure because of the dominant conduction heat transfer mechanism in which the buoyancy forces and therefore the effect of the inclination angle are negligible. For $Ra = 10^5$ at $\gamma = 90°$, the Nu_{loc} is again symmetric because of the symmetry of the geometry and boundary conditions with respect to the enclosure's center line. In this Rayleigh number for $\gamma = 45°$ and $90°$, the local Nusselt number is greater at the lower part of the enclosure, where the isotherms become denser by the returning flow accompanying the cold fluid. The local Nusselt number in the lower part of the enclosure is greater for $\gamma = 90°$ in comparison with $\gamma = 45°$, while an opposite trend is seen for the upper part of the enclosure. It may correspond to the strength of the upper and lower vortices for these cases, as illustrated in Fig. 5.3.

Figure 5.7 shows the effects of dimensionless amplitude on the local Nusselt number at $Da=0.001$, $\varepsilon=0.5$, $Ra=10^5$, and $\gamma=90°$. When a approaches zero, the local Nusselt number profile becomes monotonous, whereas the profile experiences a peak near the center of the enclosure for higher values of a. Effects of porosity on the local Nusselt number at $Da=0.001$, $a=0.3$, $Ra=10^5$, and $\gamma=90°$ are shown in Fig. 5.8. The general trend of the local Nusselt number is the same for various values of ε, but increasing the porosity of the porous medium leads to the maximum value of the local Nusselt number. Figure 5.9 illustrates the effect of the Darcy number on the Nu_{loc} at $\varepsilon=0.5$, $a=0.3$, $Ra=10^5$, and $\gamma=90°$. As stated previously,

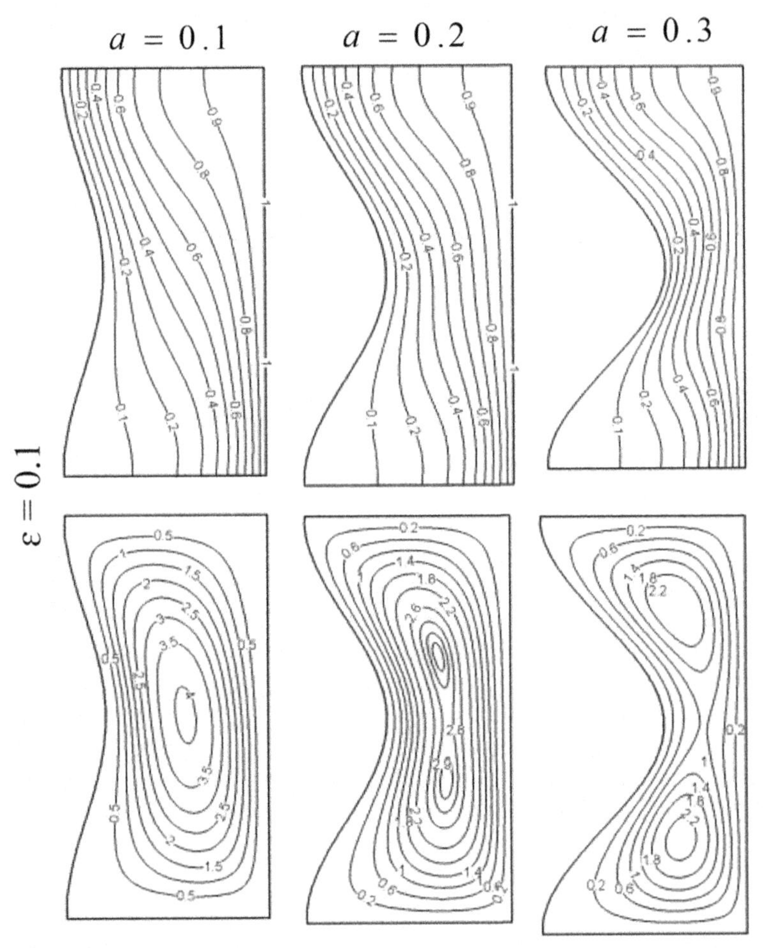

FIGURE 5.5

Isotherms (up) and streamlines (down) contours for different values of dimensionless amplitude (a) and porosity (ε) at $\gamma=90°$, $Da=0.001$, and $Ra=10^5$.

Continued

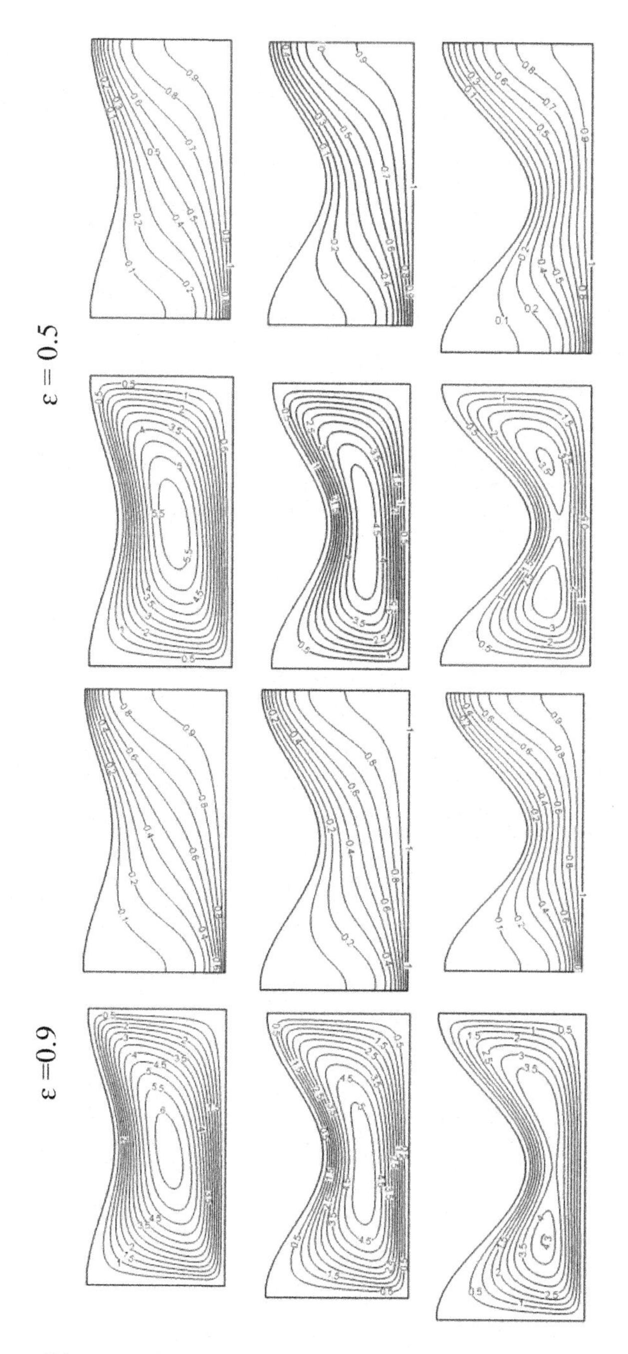

FIGURE 5.5, cont'd

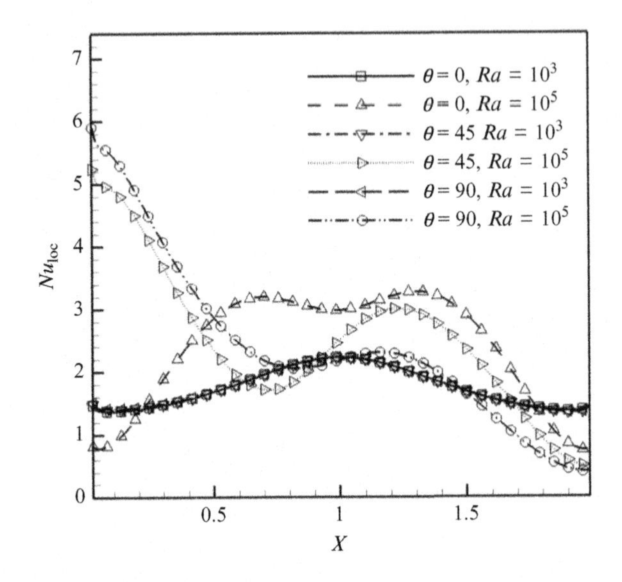

FIGURE 5.6

Effects of the Rayleigh number and inclination angle on the local Nusselt number at $Da = 0.001$, $\varepsilon = 0.5$, and $a = 0.3$.

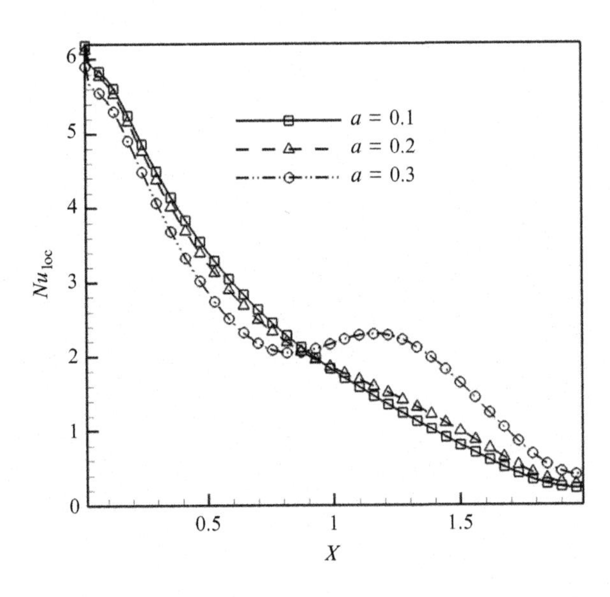

FIGURE 5.7

Effects of the dimensionless amplitude on the local Nusselt number at $Da = 0.001$, $\varepsilon = 0.5$, $Ra = 10^5$, and $\gamma = 90°$.

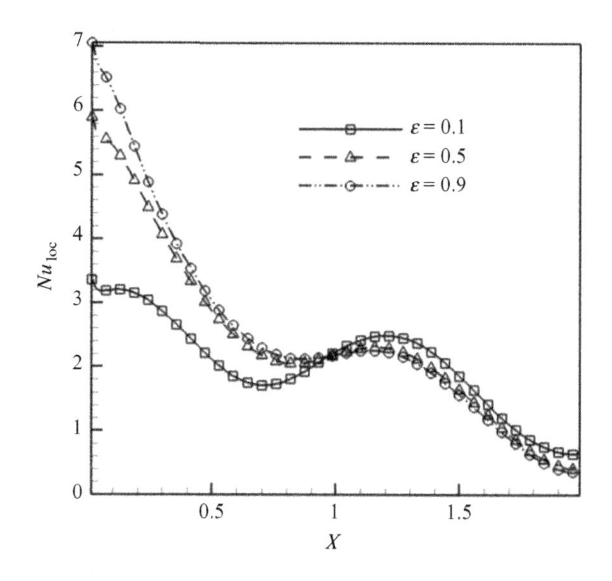

FIGURE 5.8

Effects of the porosity on the local Nusselt number at $Da = 0.001$, $a = 0.3$, $Ra = 10^5$, and $\gamma = 90°$.

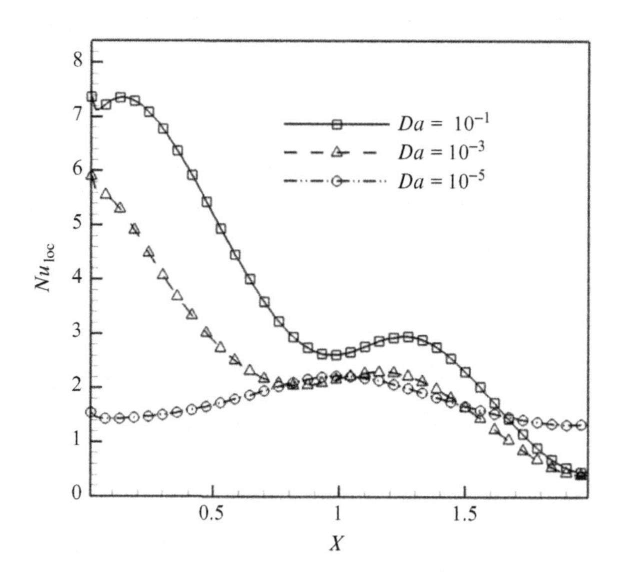

FIGURE 5.9

Effects of the Darcy number on the local Nusselt number at $\varepsilon = 0.5$, $a = 0.3$, $Ra = 10^5$, and $\gamma = 90°$.

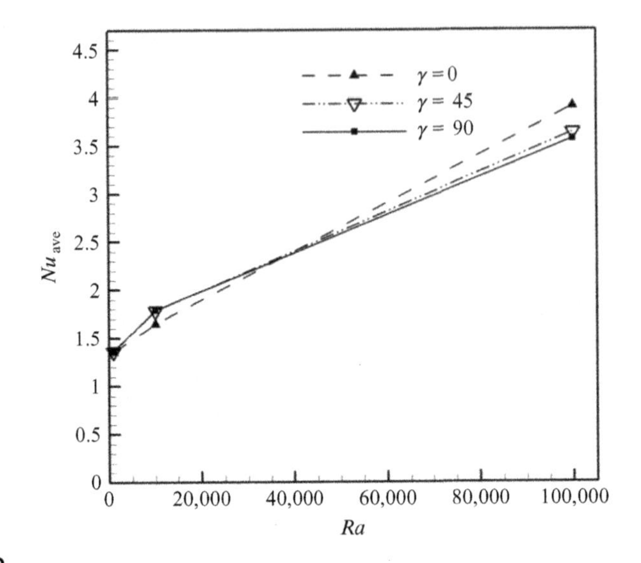

FIGURE 5.10

Effects of the Ra and γ on the average Nusselt number at $\varepsilon = 0.5$, $a = 0.2$, and $Da = 0.1$.

an increase in the Darcy number makes the convection heat transfer mechanism more pronounced; hence the values of the local Nusselt number of the straight heated wall increases.

Figures 5.10–5.12 depict the effect of governing parameters on the average Nusselt number of the heated wall of the enclosure. Figure 5.10 indicates that at a low Rayleigh number ($Ra = 10^3$) the inclination angle of the enclosure has no significant effect on the average Nusselt number. The figure also indicates that for this case the effect of (γ) on the Nu_{ave} at $Ra = 10^4$ is in contrast with that of $Ra = 10^5$; at $Ra = 10^5$, the average Nusselt number increases with a decrease in the inclination angle, whereas it is the opposite for $Ra = 10^4$. Investigating the effect of both porosity and Darcy numbers of the porous enclosure on the average Nusselt number is depicted against the porosity for different Darcy numbers at $\gamma = 90°$, $a = 0.2$, and $Ra = 10^5$ in Fig. 5.11. As can be seen, increasing the porosity enhances the Nu_{ave}, especially at greater Darcy numbers.

Figure 5.12 compares the effect of different values of a on the Nusselt number. As a increases, the magnitude of Nu_{ave} decreases because the peak of the wall profile inhibits flow circulation inside the enclosure. In addition, the effect of various values of a on the Nu_{ave} is more pronounced at lower Rayleigh numbers.

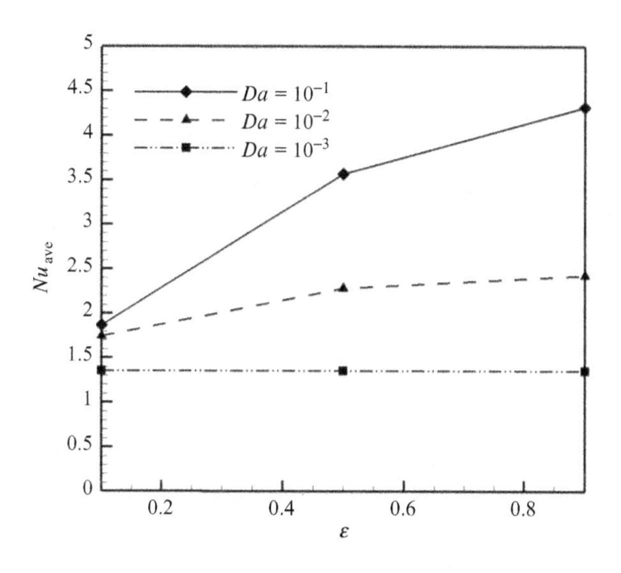

FIGURE 5.11

Effects of the ε and Da on the average Nusselt number at $\gamma = 90°$, $a = 0.2$, and $Ra = 10^5$.

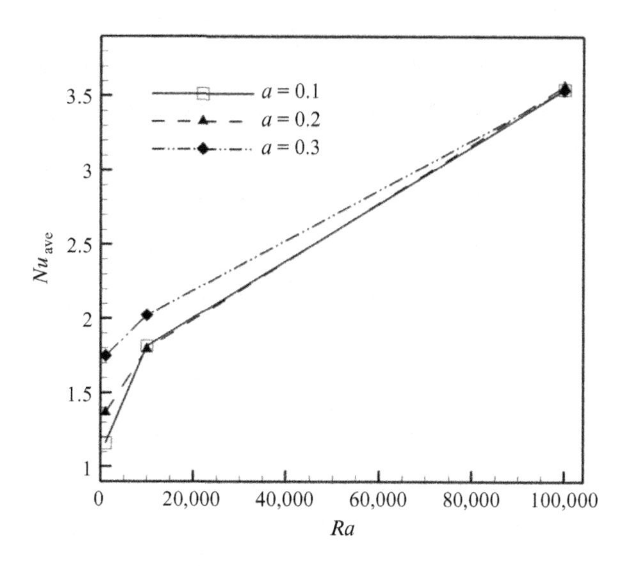

FIGURE 5.12

Effects of a and Ra on the average Nusselt number at $\gamma = 90°$ $Da = 0.1$ and $\varepsilon = 0.5$.

5.3.2 MODELING FLUID FLOW DUE TO CONVECTIVE HEAT TRANSFER FROM A HOT PIPE BURIED IN SOIL

5.3.2.1 Problem definition

The physical model along with important geometrical parameters and the mesh of the enclosure used in the present CVFEM program are shown in Fig. 5.13. The enclosure has a width-to-height aspect ratio of 1. The top wall of the enclosure is maintained at a constant cold temperature (T_c), whereas the other walls are thermally insulated. The inner pipe is maintained at a constant hot temperature (T_h). Under all circumstances $T_h > T_c$ condition is maintained. To assess the shape of the inner circular and outer rectangular boundaries, which comprise the right and top walls, a superelliptic function can be used:

$$\left(\frac{X}{a}\right)^{2n} + \left(\frac{Y}{b}\right)^{2n} = 1 \tag{5.11}$$

When $a = b$ and $n = 1$, the geometry becomes a circle. As n increases from 1, the geometry approaches a rectangle for $a \neq b$ and a square for $a = b$.

5.3.2.2 Effects of active parameters

In all computations, the porosity (ε) and Prandtl number (Pr) of the porous media are set at 0.5 and 1, respectively. Figure 5.14 shows the isotherms and streamlines at various Darcy numbers (Da) and the radius of the hot pipe at a Rayleigh number ($Ra = 10^3$). As seen in the figure, at this Rayleigh number the isotherms are smooth and the temperature gradient does not change considerably throughout the computational domain. Moreover, the isotherms are parallel to each other and take the shape of the inner and outer cylinders, which is a characteristic of the conduction-dominant mechanism of heat transfer at low Rayleigh numbers. At $r_{in} = 0.1$, two counter-rotating vortices cores are observed. This bicellular flow pattern divides the

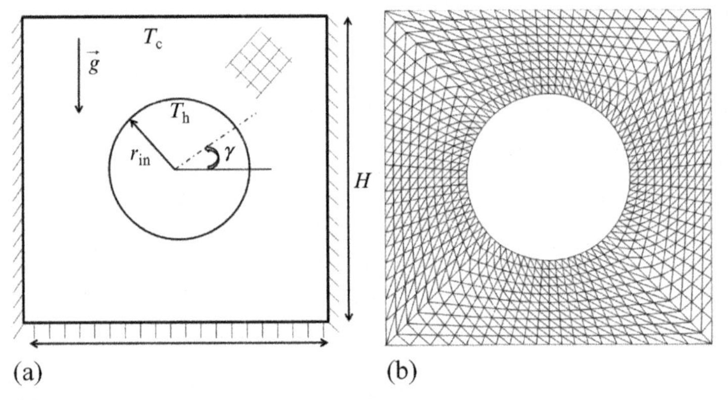

FIGURE 5.13

The geometry and boundary conditions (a) with the mesh of the geometry (b) considered in this work.

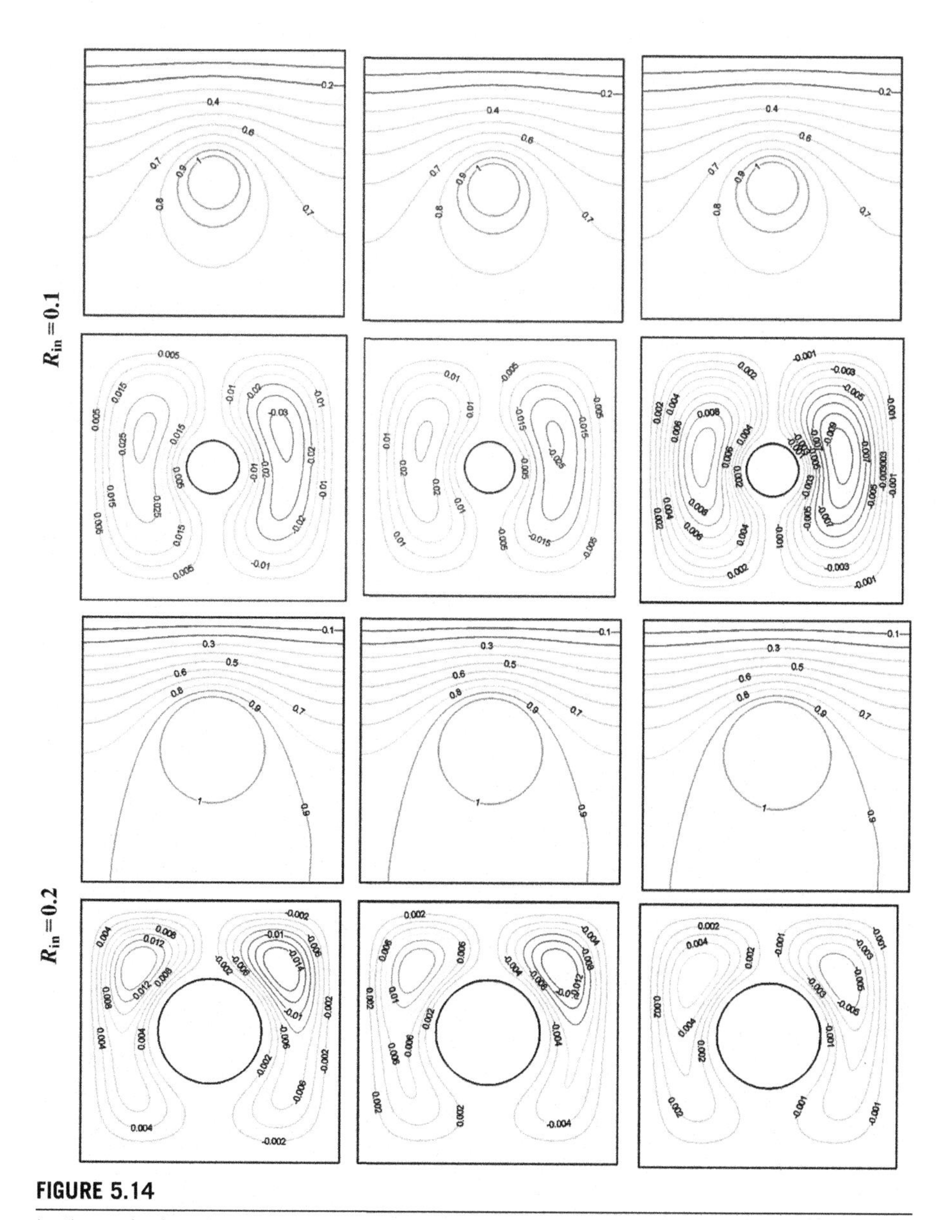

FIGURE 5.14

Isotherm (up) and streamline (down) contours for different values of Da and r_{in} at $\varepsilon = 0.5$ and $Ra = 10^3$.

Continued

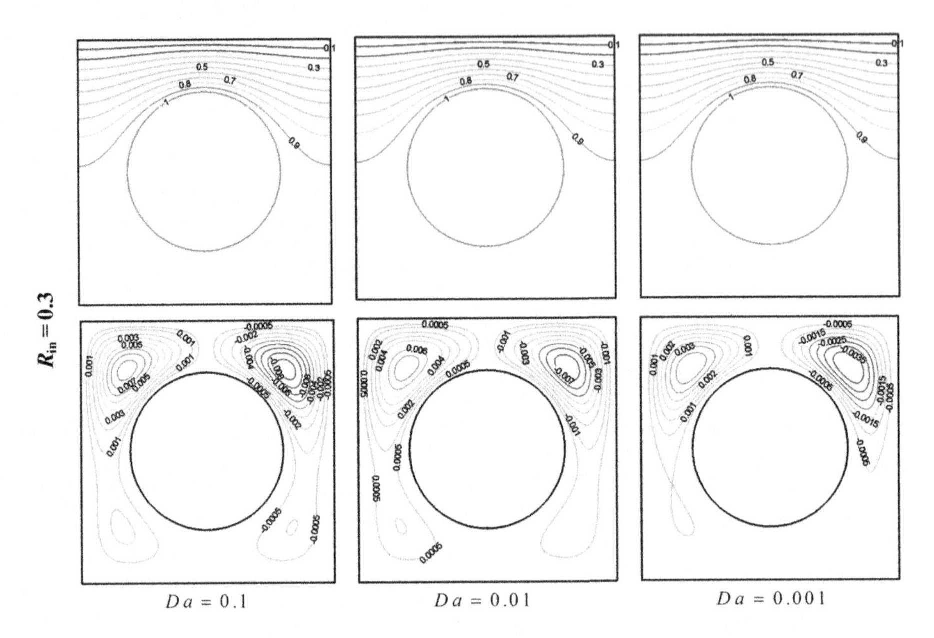

FIGURE 5.14, cont'd

enclosure into two symmetric parts with respect to the vertical center line of the enclosure. Increasing the value of the inner radius at this Rayleigh number results in the formation of a pair of vortices at the bottom of the enclosure. Figure 5.14 also indicates that the effect of the Darcy number on the isotherms is relatively small, but it is more pronounced on the streamlines at low Rayleigh numbers. Based on the definition of Darcy number, the direct proportionality of the Darcy number to the permeability is clear. The permeability of a porous medium works to measure the flow strength and can be considered as conductivity for a fluid flow. Hence high permeability causes strong flow circulation in the enclosure, whereas low permeability inhibits flow circulation inside the enclosure and leads to weak flow. Based on the stream function value, it is obvious that a decrease in Da from 0.1 to 0.001 weakens the flow circulation inside the domain and diminishes the absolute values of the stream function throughout the physical domain.

The effect of Da and r_{in} on the isotherms and streamlines at $Ra = 10^4$ is shown in Fig. 5.15. As can be seen, the general pattern of the fluid flow and temperature contours are similar to those of $Ra = 10^3$. The figure shows that increasing the Ra growth the values of the stream function relatively, which is the result of the effect of convective heat transfer at higher Rayleigh numbers; the effect of the Rayleigh number increase on isotherms, however, is negligible.

Figure 5.16 illustrates isotherms and streamlines at various values of Da and r_{in} at $Ra = 10^5$. At this Rayleigh number, a thermal plume forms over the inner hot pipe, which indicates the domination of convection heat transfer. As r_{in} increases, the

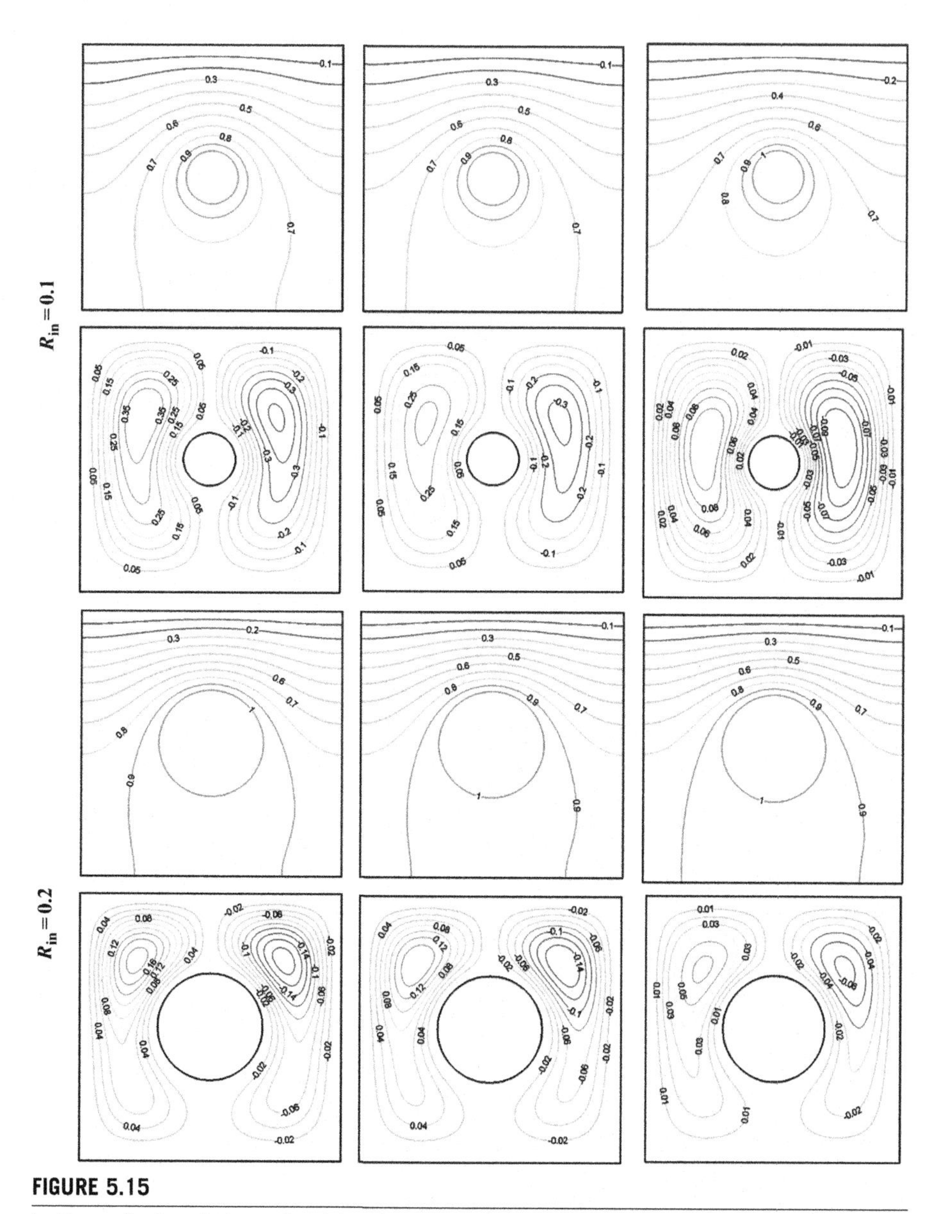

FIGURE 5.15

Isotherm (up) and streamline (down) contours for different values of Da and r_{in} at $\varepsilon = 0.5$ and $Ra = 10^4$.

Continued

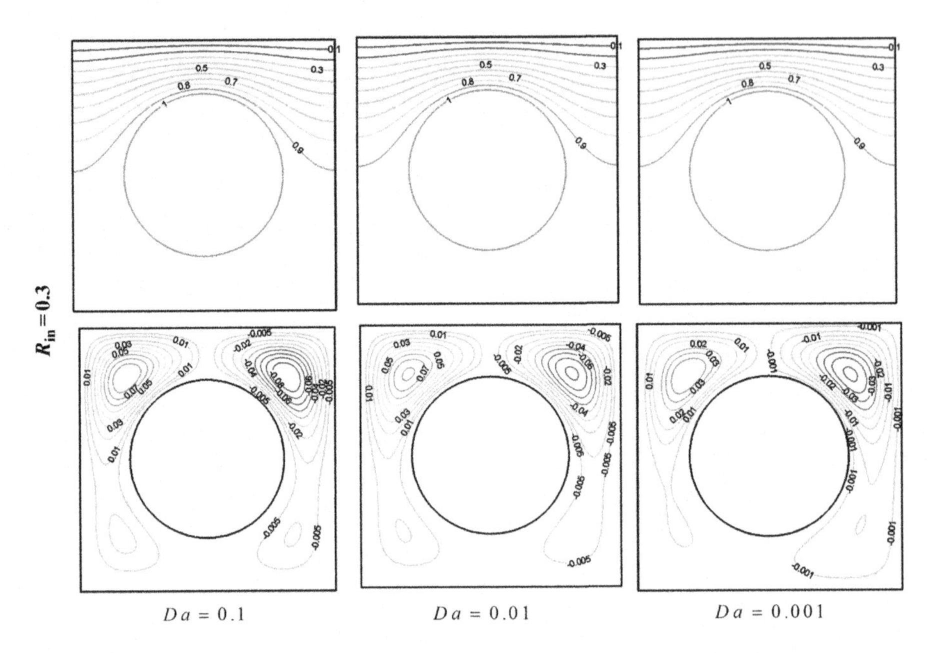

FIGURE 5.15, cont'd

thermal plume over the hot pipe gradually disappears; this occurs because of the decrease in the distance between the hot pipe and the cold wall of the enclosure; this results in the conduction heat transfer becoming more pronounced in the area above the pipe. Figure 5.16 also shows that the effects of Da and r_{in} on the streamlines and fluid flow is the same as their effect at lower Rayleigh numbers; as Da decreases the flow becomes weaker, and an increase in r_{in} suppresses fluid circulation because of the decrease in of the space available for fluid movement.

Effects of different governing parameters on the distribution of local Nusselt numbers on the hot pipe are shown in Fig. 5.17. Effect of the Da on the local Nusselt number over the wall is relatively small at low Rayleigh number over the pipe is relatively small at a low Rayleigh number (i.e., $Ra = 10^3$). At $Ra = 10^3$ the maximum value of the local Nusselt number is obtained at $\gamma = 90°$, where the distance between the cold and hot surfaces is small, whereas the minimum values correspond to $\gamma = 270°$. In addition, increasing r_{in} enhances the maximum values of local Nusselt numbers but diminishes their minimum values. Figure 5.17 indicates that the effect of Da a on the Nu_{loc} over the hot pipe is more pronounced at higher Rayleigh numbers ($Ra = 10^5$). Moreover, the figure shows that for a larger pipe radius ($r_{\text{in}} = 0.3$), the trend in local Nusselt numbers is similar to those of $Ra = 10^3$. Contrary to the $r_{\text{in}} = 0.3$, when the pipe radius increases to 0.3, the maximum value of the local Nusselt number is obtained at $\gamma = 270°$, where the thermal boundary layer is thinner; the maximum value of the local Nusselt number occurs at $\gamma = 90°$, which corresponds

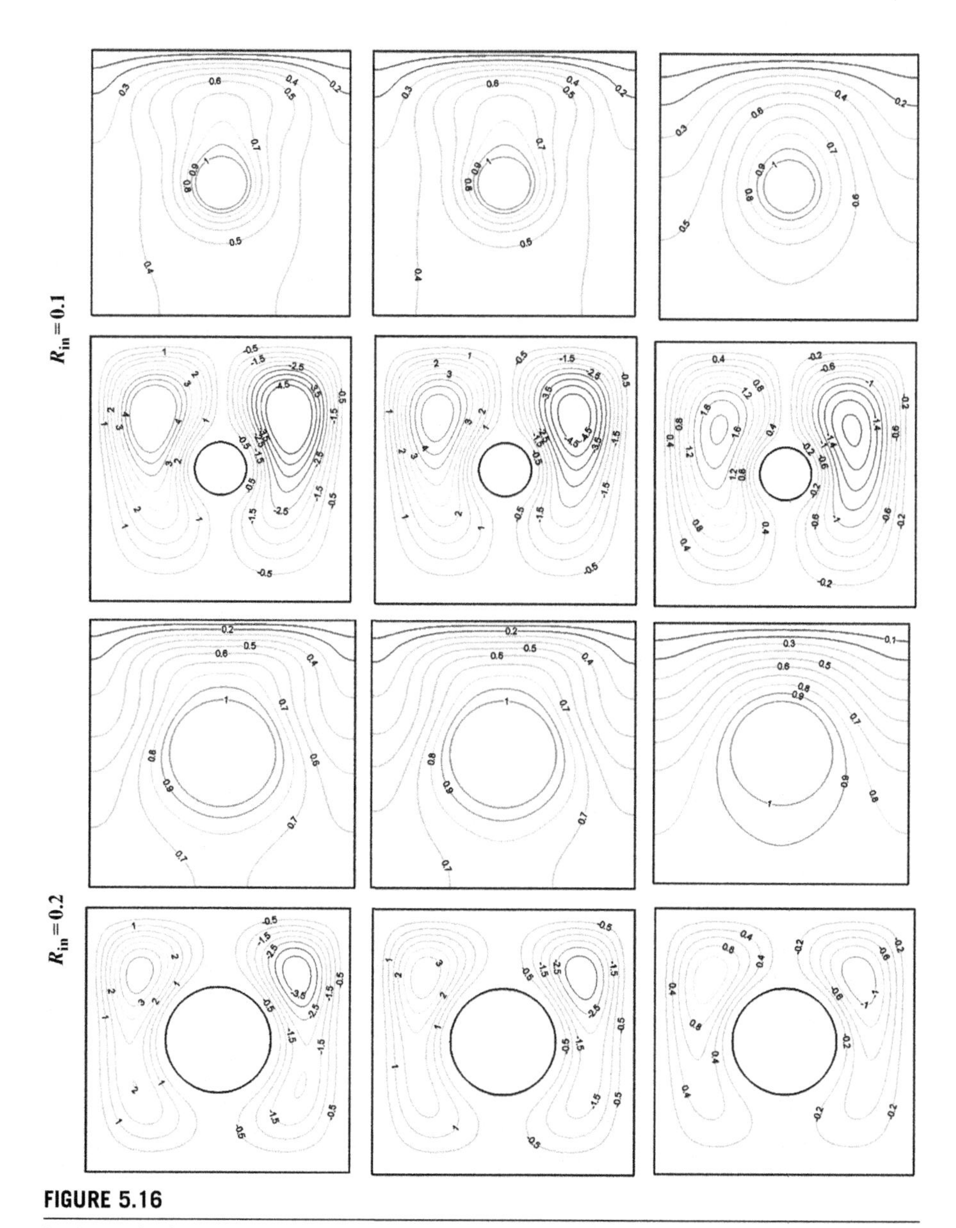

FIGURE 5.16

Isotherm (up) and streamline (down) contours for different values of Da and r_{in} at $\varepsilon=0.5$ and $Ra=10^5$.

Continued

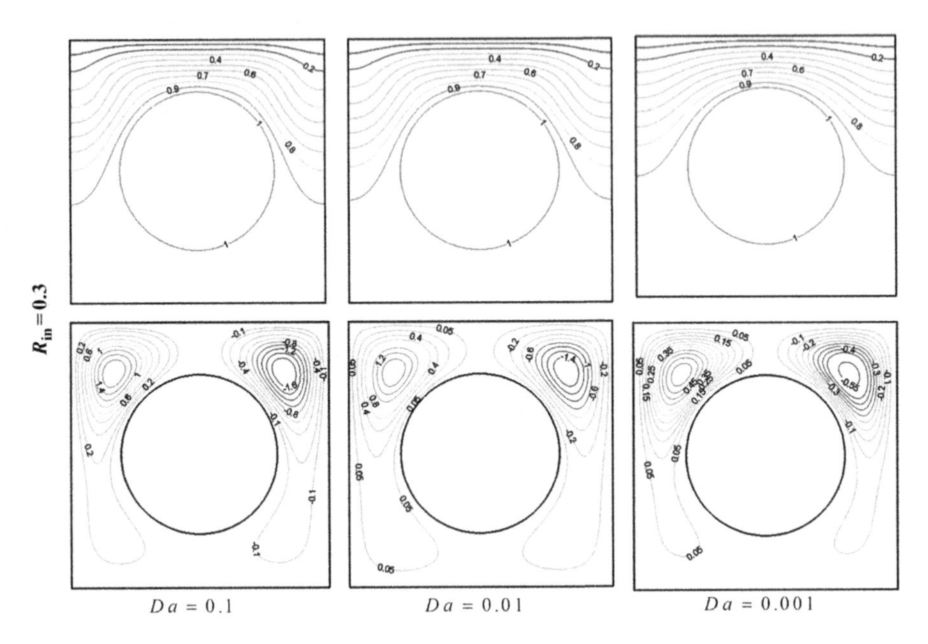

FIGURE 5.16, cont'd

to the existence of the thermal plumes. Figure 5.18 shows the effect of governing parameters on the average Nusselt number (Nu_{ave}) on the heated wall. For all cases, increase of Da and r_{in} decreases the value of Nu_{ave}.

5.3.3 NATURAL CONVECTION IN AN INCLINED, L-SHAPED, POROUS ENCLOSURE

5.3.3.1 Problem definition

The physical model along with important geometrical parameters and the mesh of the enclosure used in the present CVFEM program are shown in Fig. 5.19. The width and height of the enclosure are H. The right and top walls of the enclosure are maintained at constant cold temperatures T_c, whereas the inner circular wall is maintained at a constant hot temperature T_h and the two bottom and left walls (length $= H/2$) are thermally insulated. Under all cases $T_h > T_c$ condition is maintained. To assess the shape of the inner circular and outer rectangular boundaries, which consist of the right and top walls, a superelliptic function can be used:

$$\left(\frac{X}{a}\right)^{2n} + \left(\frac{Y}{b}\right)^{2n} = 1 \tag{5.12}$$

When $a = b$ and $n = 1$, the geometry becomes a circle. As n increases from 1, the geometry approaches a rectangle for $a \neq b$ and a square for $a = b$. The numerical

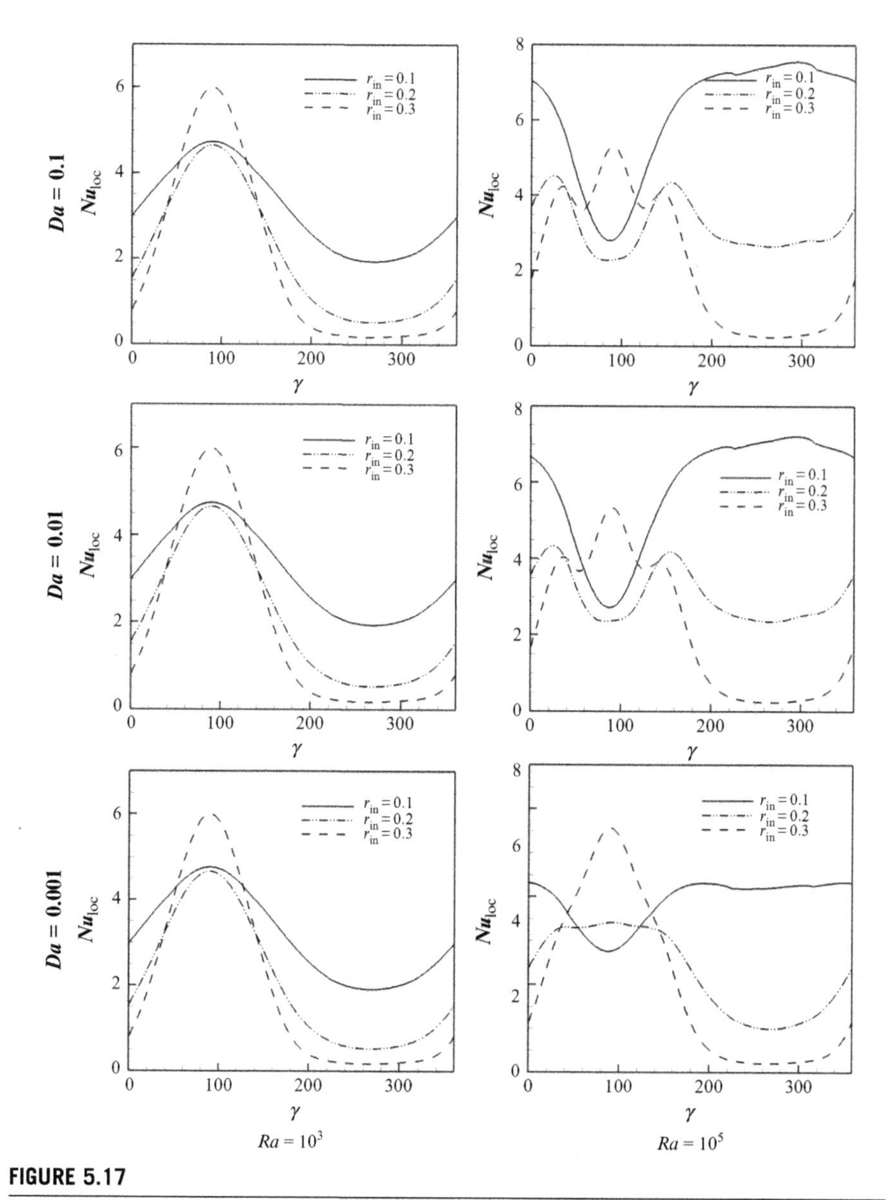

FIGURE 5.17

Effects of the dimensionless parameters on the local Nusselt number over the hot wall.

investigations are performed at a fixed Prandtl number ($Pr = 1$) equal to 1 and various values of nondimensional governing parameters, namely, the porosity ($\varepsilon = 0.1, 0.5, 0.9$), Darcy number ($Da = 10^{-1}, 10^{-3}, 10^{-5}$), and Rayleigh number ($Ra = 10^3, 10^4, 10^5$). The inclination angle ($\gamma = 0°, 30°, 45°$) is the geometric variable in this study.

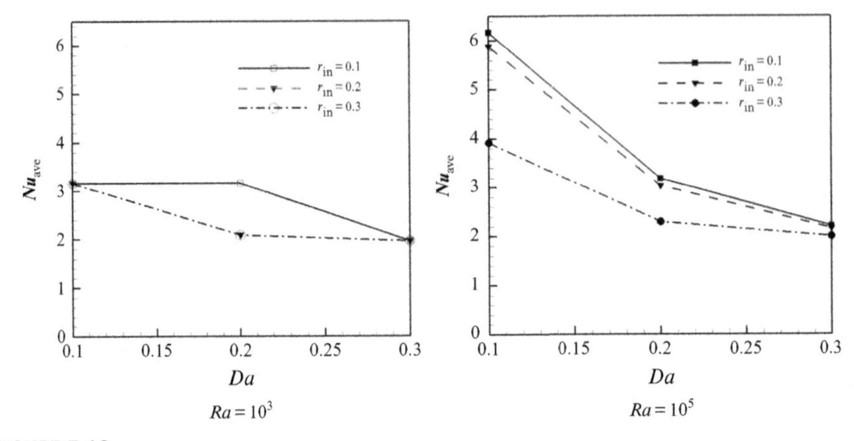

FIGURE 5.18

Effects of the dimensionless parameters on the average Nusselt number over the hot wall.

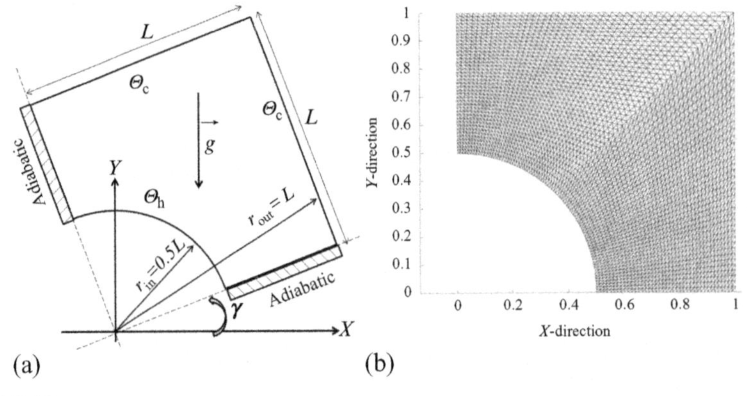

FIGURE 5.19

The geometry and boundary conditions (a) with the mesh of enclosure (b) considered in this work.

5.3.3.2 Effects of active parameters

Figure 5.20 illustrates the effect of different Rayleigh numbers on isotherms and streamlines at different inclination angles when $Da=0.1$ and $\varepsilon=0.5$. At $Ra=10^3$, the isotherms are parallel to each other and take the shape of the enclosure, which is the main characteristic of the conduction heat transfer mechanism. As the Rayleigh number increases, the isotherms become more distorted, and the stream function values increase; this increase is caused by the domination of the convective heat transfer mechanism at higher Rayleigh numbers. At Rayleigh numbers $Ra=10^3$ and 10^4, the maximum value of stream function occurs at $\gamma=45°$; this maximum

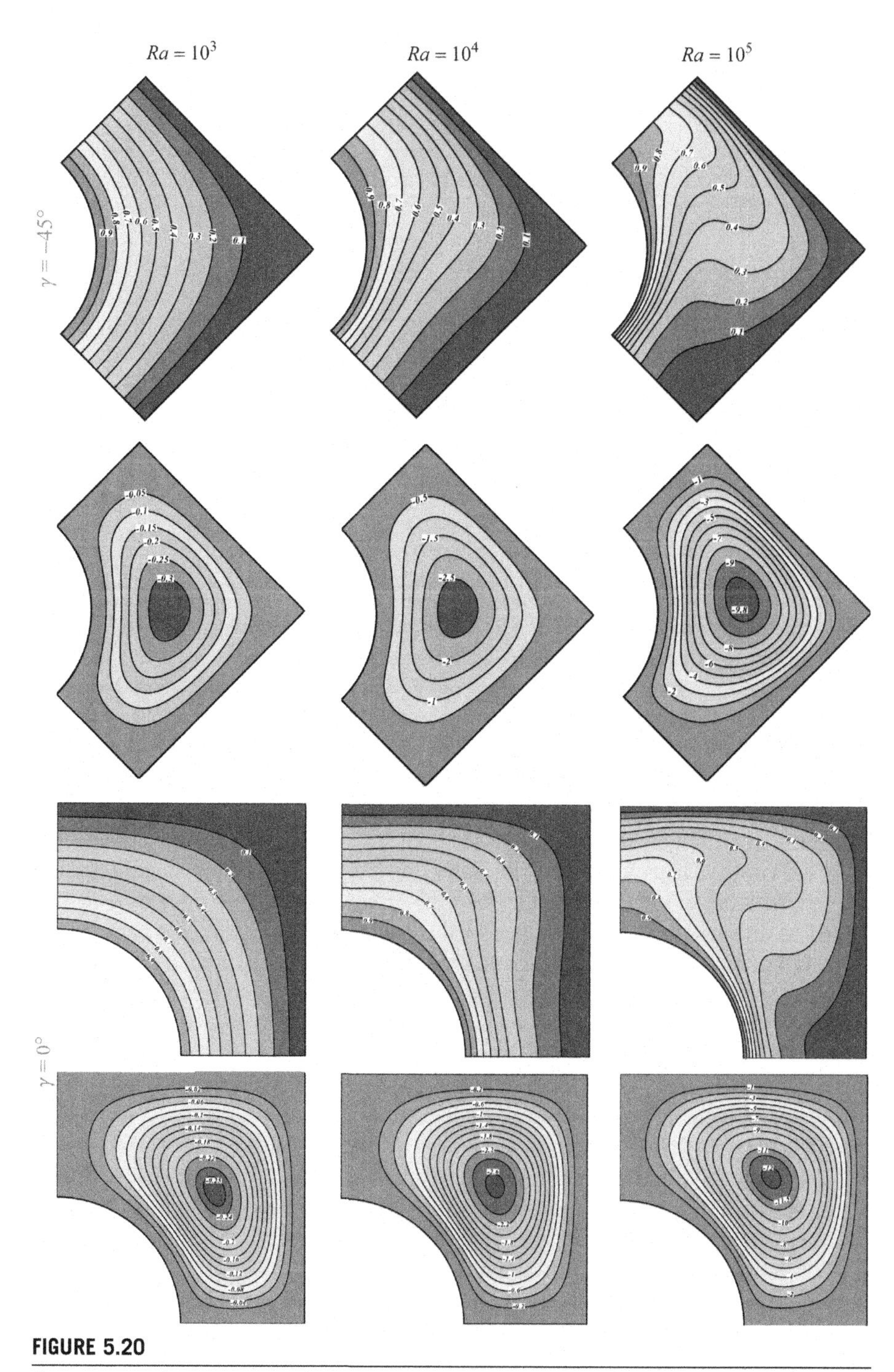

FIGURE 5.20

Comparison of the isotherms (up) and streamlines (down) contours for different values of Rayleigh numbers, γ and at $Da = 0.1$ and $\varepsilon = 0.5$.

Continued

FIGURE 5.20, cont'd

value corresponds to $\gamma = 0°$ at $Ra = 10^5$. At $\gamma = 45°$, the temperature counters and streamlines are symmetric with respect to the vertical lines that passes through the corner of the enclosure. It is due to the existence of a symmetric boundary condition and the geometry of the cavity with respect to these lines.

At $\gamma = 45°$ and 0°, an increase in the Rayleigh number causes the thermal boundary layer on the hot circular wall to decrease near the bottom of the enclosure; hence the local Nusselt number can be predicted to obtain its maximum value in this area. The isotherm shows that at $Ra = 10^4$, a thermal plume appears over the hot surface at $\gamma = 45°$. At this inclination angle, when Ra increases up to $Ra = 10^5$, the thermal plum growth impinges the hot fluid to the cold walls of the cavity.

Effects of the Darcy number on the fluid flow and heat transfer inside the enclosure are depicted in Fig. 5.21. In this case the effect of the Darcy number is investigated at different inclination angles of the enclosure when porosity $\varepsilon = 0.5$ and $Ra = 10^5$. The streamlines and isotherms are depicted for Darcy numbers $Da = 10^{-1}$, 10^{-3}, and 10^{-5}. Based on the definition of Darcy number, the direct proportionality of the Darcy number to permeability is clear. The permeability of a porous medium works as a measure the flow strength and can be considered as conductivity for fluid flow. Hence high permeability causes strong flow circulation in an enclosure, whereas low permeability inhibits flow circulation inside an enclosure and leads to weak flow. Comparison of isotherms shows that for higher Darcy numbers the convection heat transfer inside the enclosure increases, whereas conduction heat is the main heat transfer mechanism at low Darcy numbers. As a result, the isothermal contours are more distorted at high Darcy numbers, whereas they follow the form of the enclosure at low Darcy numbers.

The effects of ε along with the inclination angle of the enclosure are shown in Fig. 5.22 for $Da = 10^{-3}$ and $Ra = 10^5$. As a general observation, the effect of various porosity values on the streamlines and isotherms is less significant when compared with the effect of other governing parameters such as Darcy or Rayleigh numbers. From this figure it is clear that for a higher porosity, the drag terms are less significant, which leads to higher dimensionless flow velocities. Hence increasing ε results in higher flow circulation inside the enclosure and a greater magnitude of stream function. Figure 5.23a-c shows the effects of different governing parameters on the average Nusselt number over the circular hot wall of the enclosure when the other parameters are constant. Figure 5.23a shows the effect of Da on Nu_{ave}. As can be seen, the variation of Nu_{ave} with respect to Ra is negligible at $Da = 10^{-5}$. For $Da = 10^{-3}$, the variation of Nu_{ave} against Ra is small, whereas Nu_{ave} increases considerably as the Rayleigh number exceeds 10^4. At $Da = 10^{-5}$, the effect of the Rayleigh number on Nu_{ave} is noticeable, even at low Rayleigh numbers. Figure 5.23b indicates that the effect of increasing ε on Nu_{ave} is similar to the effect of increasing Da. The variation of Nu_{ave} from 10^3 to 10^4 is small, which is a characteristic of heat transfer at

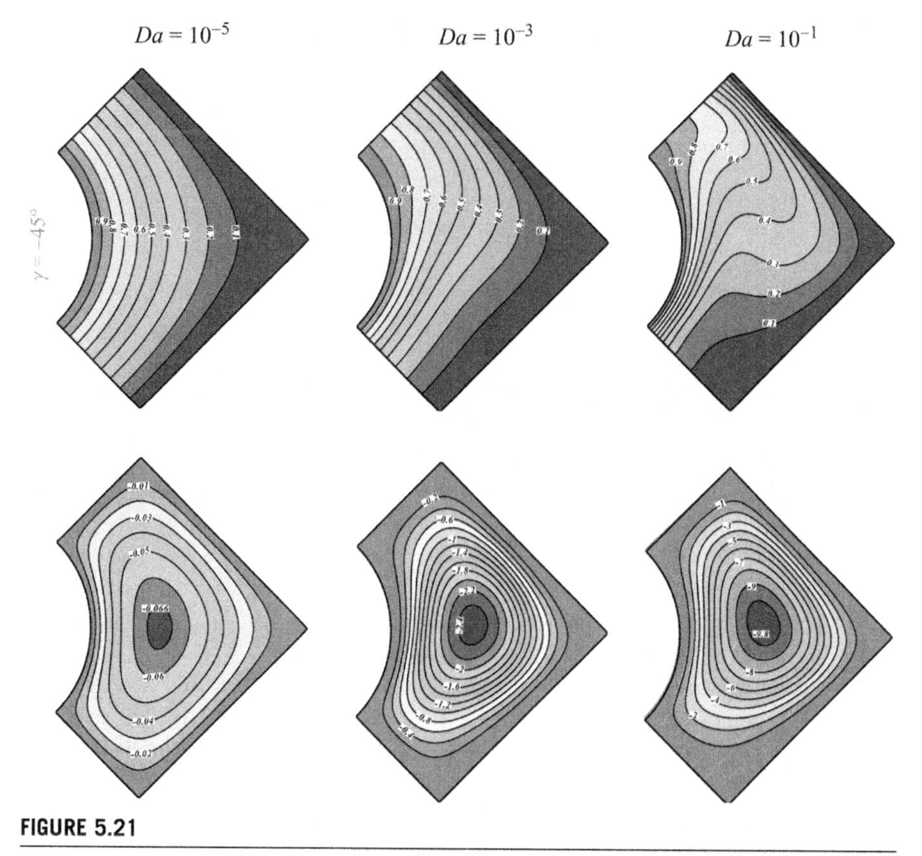

FIGURE 5.21

Comparison of the isotherms (up) and streamlines (down) contours for different values of Rayleigh numbers, γ and at $Ra = 10^5$ and $\varepsilon = 0.5$.

Continued

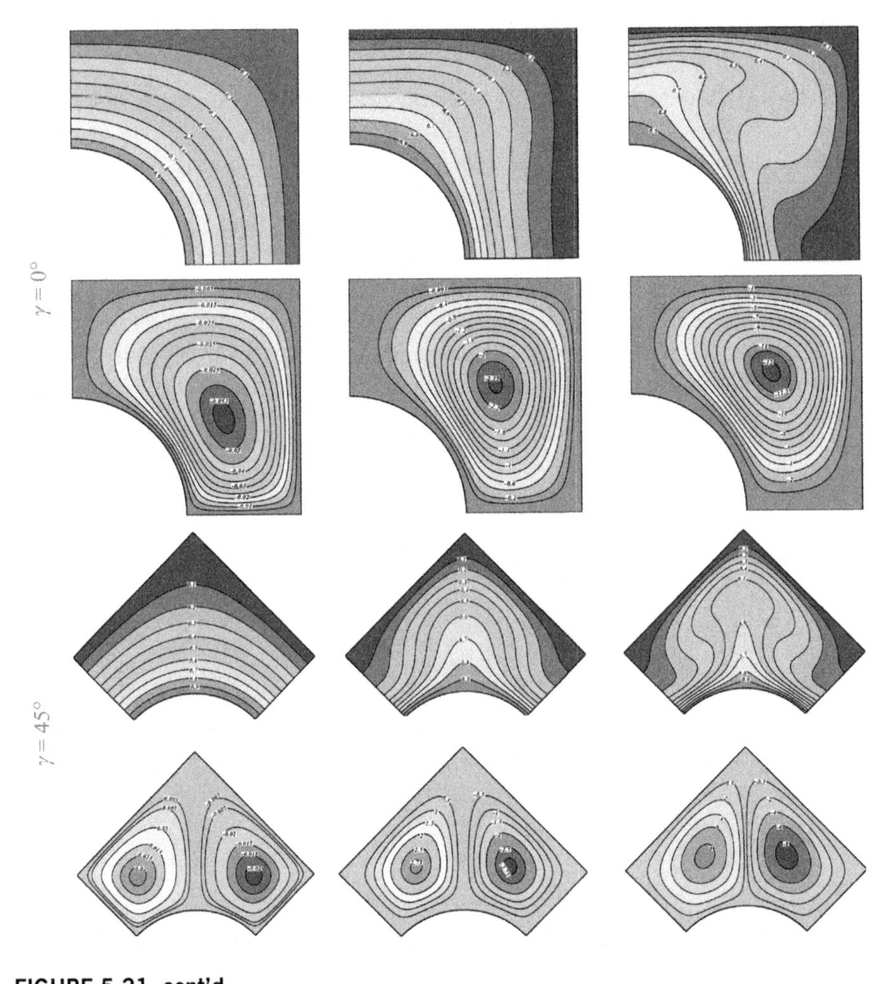

FIGURE 5.21, cont'd

$Da = 10^{-3}$. Figure 5.23c shows that the maximum value of Nu_{ave} is obtained at $\gamma = 45°$, whereas its minimum value corresponds to $\gamma = 0°$.

Effects of different governing parameters on the distribution of local Nusselt numbers on the hot wall are shown in Fig. 5.17. As was expected from the temperature contour, the effect of Da the on the local Nusselt number over the wall is relatively small at low Rayleigh numbers (i.e., $Ra = 10^3$). At $Ra = 10^3$ the maximum value of the local Nusselt number, is obtained at $\gamma = 90°$, where the distance between the cold and hot surfaces is small; the minimum values correspond to $\gamma = 270°$. In addition, increasing r_{in} enhances the maximum values of the local Nusselt numbers, while it diminishes the minimum value of that. Figure 5.17 indicates that the effect of Da a on the Nu_{loc} over the hot wall is more pronounced at higher Rayleigh numbers ($Ra = 10^5$). Moreover, the figure shows that for a larger wall radius ($r_{in} = 0.3$), the

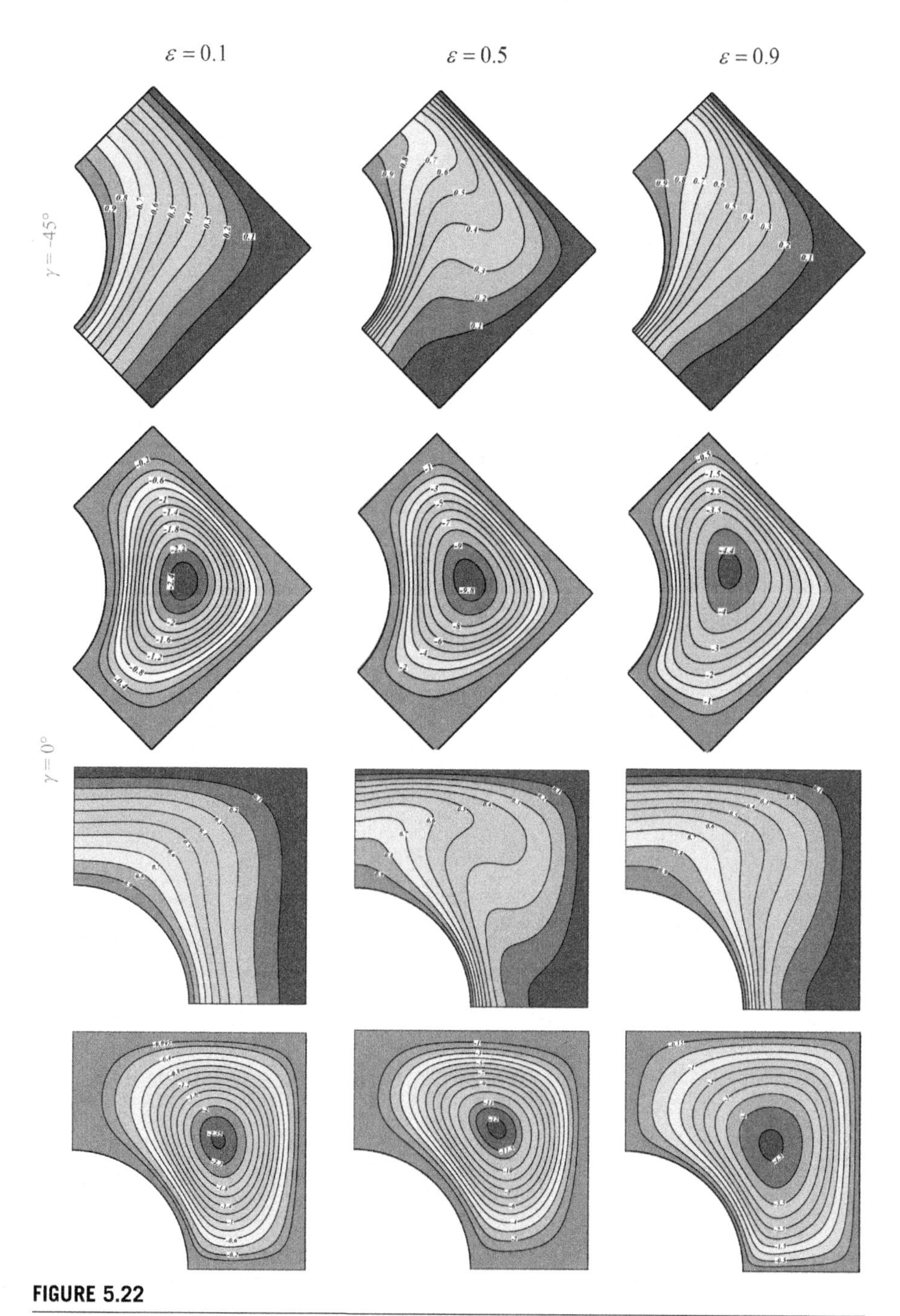

FIGURE 5.22

Comparison of the isotherms (up) and streamlines (down) contours for different values of Rayleigh numbers, γ and at $Ra = 10^5$ and $Da = 0.001$.

Continued

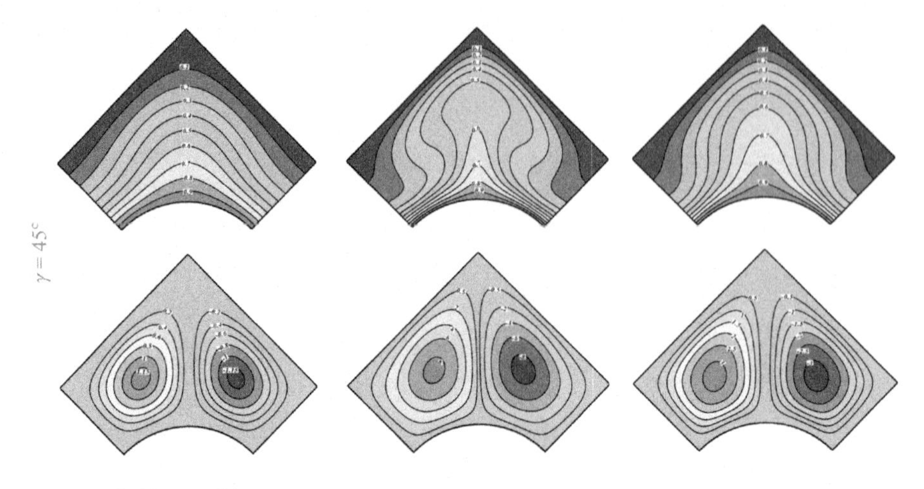

FIGURE 5.22, cont'd

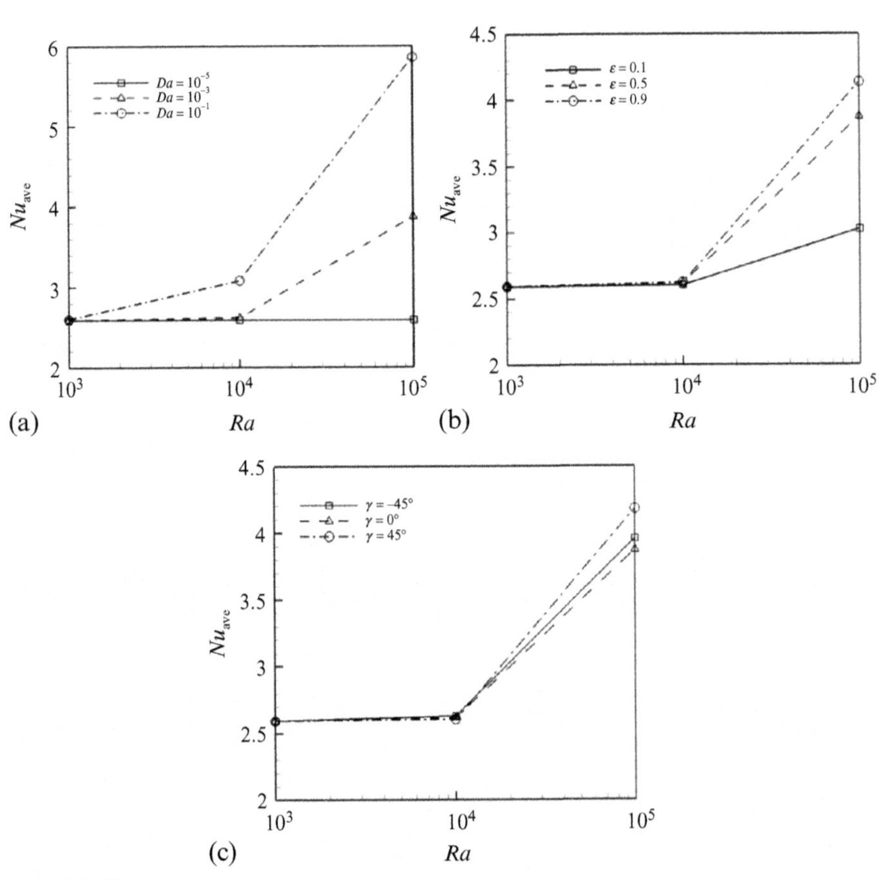

FIGURE 5.23

Effects of the Darcy number, porosity, Rayleigh number, and inclination angle on local Nusselt numbers: (a) $Da = 10^{-5}$ and $\gamma = 0°$, (b) $\varepsilon = 0.5$ and $\gamma = 0°$, and (c) $\varepsilon = 0.5$ and $Da = 10^{-3}$.

local Nusselt number trend is similar to that of $Ra=10^3$. Contrary to the $r_{in}=0.3$, when the wall radius increases up to 0.3, the maximum value of the local Nusselt number is obtained at $\gamma=270°$, where the thermal boundary layer is thinner; the maximum value of the local Nusselt number occurs at $\gamma=90°$, which corresponds to the existence of thermal plumes. Figure 5.18 shows the effect of governing parameters on the average Nusselt number (Nu_{ave}) on the heated wall. For all cases, an increase Da in and r_{in} decreases the value of Nu_{ave}.

REFERENCES

[1] P. Nithiarasu, K.N. Seetharamu, T. Sundararajan, Natural convective heat transfer in a fluid saturated variable porosity medium, Int. J. Heat Mass Transfer 40 (16) (1997) 3955–3967.

[2] W. Pakdee, P. Rattanadecho, Unsteady effects on natural convective heat transfer through porous media in cavity due to top surface partial convection, Appl. Therm. Eng. 26 (2006) 2316–2326.

[3] A.C. Baytas, I. Pop, Natural convection in a trapezoidal enclosure filled with a porous medium, Int. J. Eng. Sci. 39 (2001) 125–134.

[4] C. Saidi, F. Legay, B. Pruent, Laminar flow past a sinusoidal cavity, Int. J. Heat Mass Transfer 30 (1987) 649–660.

[5] P.K. Das, S. Mahmud, Numerical investigation of natural convection inside a wavy enclosure, Int. J. Therm. Sci. 42 (2003) 397–406.

[6] L. Adjlout, O. Imine, A. Azzi, M. Belkadi, Laminar natural convection in an inclined cavity with a wavy wall, Int. J. Heat Mass Transfer 45 (2002) 2141–2152.

[7] Z. Ziabakhsh, G. Domairry, M. Mozaffari, M. Mahbobifar, Analytical solution of heat transfer over an unsteady stretching permeable surface with prescribed wall temperature, J. Taiwan Inst. Chem. Eng. 41 (2) (2010) 169–177.

[8] S. Mahmud, P.K. Das, N. Hyder, A.K.M. Islam, Free convection in an enclosure with vertical wavy walls, Int. J. Therm. Sci. 41 (2002) 440–446.

[9] L.S. Yao, Natural convection along a vertical wavy surface, J. Heat Transfer 105 (1983) 465–468.

[10] A. Dalal, M.K. Das, Natural convection in a cavity with a wavy wall heated from below and uniformly cooled from the top and both sides, J. Heat Transfer 128 (2006) 717–725.

[11] J. Rostami, Unsteady natural convection in an enclosure with vertical wavy walls, Heat Mass Transfer 44 (2008) 1079–1087.

[12] Z. Ziabakhsh, G. Domairry, H.R. Ghazizadeh, Analytical solution of the stagnation-point flow in a porous medium by using the homotopy analysis method, J. Taiwan Inst. Chem. Eng. 40 (1) (2009) 91–97.

[13] B.V. Rathish Kumar, A study of free convection induced by a vertical wavy surface with heat flux in a porous enclosure, Numer. Heat Transfer A: Appl. 37 (2000) 493–510.

[14] X.B. Chen, P. Yu, S.H. Winoto, H.T. Low, Free convection in a porous cavity based on the Darcy-Brinkman-Forchheimer extended model, Numer. Heat Transfer A 52 (2007) 377–397.

[15] D.D. Ganji, M. Sheikholeslami, H.R. Ashorynejad, F. Shakeri, Two-dimensional viscous flow through permeable walls with expanding or contracting gaps, World Appl. Sci. J. 16 (Special Issue of Applied Math) (2012) 82–92.

[16] Z. Ziabakhsh, G. Domairry, H. Bararnia, H. Babazadeh, Analytical solution of flow and diffusion of chemically reactive species over a nonlinearly stretching sheet immersed in a porous medium, J. Taiwan Inst. Chem. Eng. 41 (2010) 22–28.
[17] H.C. Brinkmann, On the permeability of media consisting of closely packed porous particles, Appl. Sci. Res. 1 (1947) 81–86.
[18] G. Lauriat, V. Prasad, Natural convection in a vertical porous cavity: a numerical study for Brinkmann-extended Darcy formulation, Trans. ASME J. Heat Transfer 109 (1987) 295–320.
[19] A.M. Al-Amiri, Analysis of momentum and energy transfer in a lid driven cavity filled with a porous medium, Int. J. Heat Mass Transfer 43 (2000) 3513–3527.

A CVFEM code for lid driven cavity

```
implicit none
!!!!!!!!!!!!!===========Lid Driven Cavity=============!!!!!!!!!!!!
!!!!!!!!!!!!!=== Mohsen Sheikholeslami and Soheil Soleimani ===!!!!!!!!!!!!!!
integer rows,n,k1,k2,itert,nstep
integer cols,ii,iter,loopcount
integer i,j,k,n_seg,iterv,iters
parameter (cols=51,rows=51,n=rows*cols,n_seg=4)
integer n_s(1:n),n_b(1:n_seg),B(1:n_seg,1:max(rows,cols)),S(1:n,1:12)
real*8 dx,x(1:n),y(1:n),vor(1:n),sf(1:n),t(1:n)
real*8 dy,ulid,aps(1:n),apv(1:n),as(1:n,12),av(1:n,12)
real*8 V(1:n),xL(1:3),yL(1:3),RHS,BC(1:n),BB(1:n),TAX,QC(1:n),QB(1:n)
real*8 diff(1:3),Vele,Nx(1:3),Ny(1:3),diff_f,vxf,vyf,delx,dely,qf
real*8 dely2,delx2,dely1,delx1,Nu,SAX,SAY
real*8 Re,upper,lower
nstep=10000
Re=50.
ulid=1.0
dx=1./real(rows-1)
dy=1./real(cols-1)
do k=1 , rows
do j=1 , cols
i=j+(k-1)*cols
y(i)=real(k-1)/real(rows-1)
x(i)=real(j-1)/real(cols-1)
enddo
enddo
!support
do k=1,rows
do j=1,cols
i=j+(k-1)*cols
n_s(i)=0 !Nodes in support
do ii=1,12
S(i,ii)=0.0 !Global node number of ii_th node in support i
enddo
enddo
enddo
!boundary
```

```
n_b(1)=cols
n_b(2)=rows !outter radius
n_b(3)=cols
n_b(4)=rows !inner radius
do k=1,n_seg
do j=1,n_b(k)
B(k,j)=0.0 !Global node number of j_th node on i_th boundary
enddo
enddo
!SUPPORT DATA -----------------
! Support do internal points
do k=2,rows-1
do j=2,cols-1
i=j+(k-1)*cols
n_s(i)=6
S(i,1)=i+1
S(i,2)=i+1+cols
S(i,3)=i+cols
S(i,4)=i-1
S(i,5)=i-1-cols
S(i,6)=i-cols
S(i,7)=i+1 !last node is same as first node at internal nodes
enddo
enddo
!horizonal boundary (y=0) Non-corner points
do j=2,cols-1
i=j
n_s(i)=4
S(i,1)=i+1 ! arranged so that points are--
S(i,2)=i+1+cols ! --contiguous in counter clockwise count
S(i,3)=i+cols
S(i,4)=i-1
S(i,5)=0 ! last point zero: a boundary node flag
enddo
do k=2,rows-1
i=cols+(k-1)*cols
n_s(i)=4
S(i,1)=i+cols
S(i,2)=i-1
S(i,3)=i-1-cols
S(i,4)=i-cols
S(i,5)=0
enddo
do j=2,cols-1
```

```
i=j+(rows-1)*cols
n_s(i)=4
S(i,1)=i-1
S(i,2)=i-1-cols
S(i,3)=i-cols
S(i,4)=i+1
S(i,5)=0
enddo
do k=2,rows-1
i=1+(k-1)*cols
n_s(i)=4
S(i,1)=i-cols
S(i,2)=i+1
S(i,3)=i+1+cols
S(i,4)=i+cols
S(i,5)=0
enddo
!Corners
! x=0, y = rin
i=1
n_s(i)=3
S(i,1)=i+1
S(i,2)=i+1+cols
S(i,3)=i+cols
S(i,4)=0
! y = 0 x = rot
i=cols
n_s(i)=2
S(i,1)=i+cols
S(i,2)=i-1
S(i,3)=0
! x =0 y = rot
i=rows*cols
n_s(i)=3
S(i,1)=i-1
S(i,2)=i-1-cols
S(i,3)=i-cols
S(i,4)=0
! x=0 y = rin
i=1+(rows-1)*cols
n_s(i)=2
S(i,1)=i-cols
S(i,2)=i+1
S(i,3)=0
```

```
!END SUPPORT DATA -----------------
! BOUNDARY DATA ------------------
!NOTE counter-clockwise numbering and boundary segment index
do i=1,n_b(1)
B(1,i)=i
enddo
do i=1,n_b(2)
B(2,i)=i*cols
enddo
do i=1,n_b(3)
B(3,i)=rows*cols+1-i
enddo
do i=1,n_b(4)
B(4,i)=1+rows*cols-i*cols
enddo
!==========================================
!=================initial conditions=================
!==========================================
do i=1,n
vor(i)=0.0
sf(i)=0.0
enddo
do i=1,n
aps(i)=0.0
apv(i)=0.0
do j=1,n_s(i)+1
as(i,j)=0.0
av(i,j)=0.0
enddo
enddo
do iter=1,nstep
print*,iter
!==========================================
! ===============calculating the vorticity===============
!==========================================
do i=1,n
apv(i)=0.0
V(i)=0.0
do j=1,n_s(i)+1
av(i,j)=0.0
enddo
loopcount=n_s(i)
if (S(i,n_s(i)+1)==0) then
loopcount=n_s(i)-1
```

```
endif
do j=1,loopcount
k1=S(i,j) !The numbers of the nodes in the j_th elemnt are
k2=S(i,j+1) ! i, k1=S(i,J), and k2=S(i,j+1)
xL(1) = x(i) !xL=local x-coordinate of a node within a element
xL(2) = x(k1)
xL(3) = x(k2)
yL(1) = y(i) !yL=local y-coordinate of a node within a element
yL(2) = y(k1)
yL(3) = y(k2)
! nodal diffusivity (kappa)
diff(1)=1/Re
diff(2)=1/Re
diff(3)=1/Re
! Vele= element-volume
Vele=(xL(2)*yL(3)-xL(3)*yL(2)-xL(1)*yL(3)+xL(1)*yL(2)+yL(1)*xL(3)-yL(1)
*xL(2))/2.
V(i)=V(i)+Vele/3 !Contribution to control volume
Nx(1)= (yL(2)-yL(3))/(2.*Vele) !derivative of shape function wrt x
Nx(2)= (yL(3)-yL(1))/(2.*Vele)
Nx(3)= (yL(1)-yL(2))/(2.*Vele)
Ny(1)=-(xL(2)-xL(3))/(2.*Vele) !derivative of shape function wrt y
Ny(2)=-(xL(3)-xL(1))/(2.*Vele)
Ny(3)=-(xL(1)-xL(2))/(2.*Vele)
! Face 1–values of diffusivity and velocity
diff_f=(5.*diff(1)+5.*diff(2)+2.*diff(3))/12.
vxf= Ny(1)*sf(i)+Ny(2)*sf(k1)+Ny(3)*sf(k2)
vyf= -(Nx(1)*sf(i)+Nx(2)*sf(k1)+Nx(3)*sf(k2))
! Face 1–face area normal components
delx= ((xL(1)+xL(2)+xL(3))/3.)-((xL(1)+xL(2))/2.)
dely= ((yL(1)+yL(2)+yL(3))/3.)-((yL(1)+yL(2))/2.)
qf=vxf*dely-vyf*delx !Face 1 volume flux
! Face 1–Diffusion Coefficients
apv(i) = apv(i) + diff_f*(- Nx(1) * dely + Ny(1) * delx)
av(i,j) = av(i,j) + diff_f*( Nx(2) * dely - Ny(2) * delx)
av(i,j+1) = av(i,j+1) + diff_f*( Nx(3) * dely - Ny(3) * delx)
! Face 1–Advection Coefficients
apv(i)= apv(i) +max(qf,0.)
av(i,j)=av(i,j) +max(-qf,0.)
! Face 2–values of diffusivity and velocity
diff_f=(5*diff(1)+2*diff(2)+5*diff(3))/12.
vxf= Ny(1)*sf(i)+Ny(2)*sf(k1)+Ny(3)*sf(k2) !!!!ro face
vyf= -1*(Nx(1)*sf(i)+Nx(2)*sf(k1)+Nx(3)*sf(k2))
! Face 2–face area normal components
```

```
delx= ((xL(1)+xL(3))/2)-((xL(1)+xL(2)+xL(3))/3.)
dely= ((yL(1)+yL(3))/2)-((yL(1)+yL(2)+yL(3))/3.)
qf=vxf*dely-vyf*delx ! Face 1 volume flux
! Face 2–Diffusion Coefficients
apv(i) = apv(i) + diff_f*(- Nx(1) * dely + Ny(1) * delx)
av(i,j) = av(i,j) + diff_f*( Nx(2) * dely - Ny(2) * delx)
av(i,j+1) = av(i,j+1) + diff_f*( Nx(3) * dely - Ny(3) * delx)
! Face 2–Advection Coefficients
apv(i)= apv(i) + max(qf,0.)
av(i,j+1) = av(i,j+1) + max(-qf,0.)
enddo !end loop on nodes/elements in support
av(i,1)=av(i,1)+av(i,n_s(i)+1)
av(i,n_s(i)+1)=0
enddo !end loop on nodes in domain
!!!!!!!!!!!!!!!!!!!!!!!!!!!!!!!!!!!!!!!!!!!!!!!!!!!!!!!!!!!!!!!!!!!!!!!!!!!!!!!!!!!!!!!!!!
!==========SETTING THE VORTICITY BOUNDARY===========
!!!!!!!!!!!!!!!!!!!!!!!!!!!!!!!!!!!!!!!!!!!!!!!!!!!!!!!!!!!!!!!!!!!!!!!!!!!!!!!!!!!!!!!!!!
do k=1,n_b(1)
RHS=0.0
do j=1,n_s(B(1,k))
RHS=RHS+as(B(1,k),j)*sf(S(B(1,k),j))
enddo
vor(B(1,k))= (1/V(B(1,k)))*(aps(B(1,k))*sf(B(1,k))- RHS - 0.0 )
enddo
do k=1,n_b(2)
RHS=0.0
do j=1,n_s(B(2,k))
RHS=RHS+as(B(2,k),j)*sf(S(B(2,k),j))
enddo
vor(B(2,k))= (1/V(B(2,k)))*(aps(B(2,k))*sf(B(2,k))- RHS - 0.0)
enddo
Do k=1,n_b(3)
RHS=0.0
Do j=1,n_s(B(3,k))
RHS=RHS+as(B(3,k),j)*sf(S(B(3,k),j));
end do
if (k==1) then
upper=0
else
upper=0.5*sqrt((x(B(3,k))-x(B(3,k-1)))**2+(y(B(3,k))-y(B(3,k-1)))**2)
end if
if (k==n_b(3)) then
lower=0
else
```

```
lower=0.5*sqrt((x(B(3,k))-x(B(3,k+1)))**2+(y(B(3,k))-y(B(3,k+1)))**2)
end if
vor(B(3,k))= (1/V(B(3,k)))*( aps(B(3,k))*sf(B(3,k))- RHS - (lower+upper)*ulid )
end do
do k=1,n_b(4)
RHS=0.0
do j=1,n_s(B(4,k))
RHS=RHS+as(B(4,k),j)*sf(S(B(4,k),j))
enddo
vor(B(4,k))= (1/V(B(4,k)))*( aps(B(4,k))*sf(B(4,k))- RHS - 0.0 )
enddo
!========================================
do i=1,n
BC(i)=0.0
BB(i)=0.0
enddo
do k=1,n_b(1)
BC(B(1,k))=1.e18
BB(B(1,k))=vor(B(1,k))*1.e18
enddo
do k=1,n_b(2)
BC(B(2,k))=1.e18
BB(B(2,k))=vor(B(2,k))*1.e18
enddo
do k=1,n_b(3)
BC(B(3,k))=1.e18
BB(B(3,k))=vor(B(3,k))*1.e18
enddo
do k=1,n_b(4)
BC(B(4,k))=1.e18
BB(B(4,k))=vor(B(4,k))*1.e18
enddo
do i=1,n
QC(i)=0.0
if (S(i,n_s(i)+1).NE.0) then
QB(i)=0.0
else
QB(i)=0.0
endif
enddo
!========================================
do iterv=1,1
do i= 1,n
RHS=BB(i)+QB(i)
```

```
do j=1,n_s(i)
RHS=RHS+av(i,j)*vor(S(i,j))
enddo
vor(i)=RHS/(apv(i)+BC(i)+QC(i))
enddo
enddo
!==================================================
!==================================================
!=========== SOLVING THE STREAMFUNCTION===========
!==================================================
!==================================================
do i=1,n !loop on nodes
aps(i)=0.0
V(i)=0.0
do j=1,n_s(i)+1
as(i,j)=0.0
enddo
loopcount=n_s(i)
if (S(i,n_s(i)+1)==0) then
loopcount=n_s(i)-1
endif
do j=1,loopcount
k1=S(i,j)
k2=S(i,j+1)
xL(1) = x(i)
xL(2) = x(k1)
xL(3) = x(k2)
yL(1) = y(i)
yL(2) = y(k1)
yL(3) = y(k2)
diff(1)=1.0
diff(2)=1.0
diff(3)=1.0
Vele=(xL(2)*yL(3)-xL(3)*yL(2)-xL(1)*yL(3)+xL(1)*yL(2)+yL(1)*xL(3)-yL(1)
*xL(2))/2.
V(i)=V(i)+Vele/3.
Nx(1)= (yL(2)-yL(3))/(2.*Vele)
Nx(2)= (yL(3)-yL(1))/(2.*Vele)
Nx(3)= (yL(1)-yL(2))/(2.*Vele)
Ny(1)=-(xL(2)-xL(3))/(2.*Vele)
Ny(2)=-(xL(3)-xL(1))/(2.*Vele)
Ny(3)=-(xL(1)-xL(2))/(2.*Vele)
! Face 1–values of diffusivity and velocity
diff_f=(5.*diff(1)+5.*diff(2)+2.*diff(3))/12.
```

```
! Face 1--face area normal components
delx= ((xL(1)+xL(2)+xL(3))/3.)-((xL(1)+xL(2))/2.)
dely= ((yL(1)+yL(2)+yL(3))/3.)-((yL(1)+yL(2))/2.)
! Face 1--Diffusion Coefficients
aps(i) = aps(i) + diff_f*(- Nx(1) * dely + Ny(1) * delx)
as(i,j) = as(i,j) + diff_f*( Nx(2) * dely - Ny(2) * delx)
as(i,j+1) = as(i,j+1) + diff_f*( Nx(3) * dely - Ny(3) * delx)
! Face 2--values of diffusivity and velocity
diff_f=(5.*diff(1)+2.*diff(2)+5*diff(3))/12.
! Face 2--face area normal components
delx= ((xL(1)+xL(3))/2.)-((xL(1)+xL(2)+xL(3))/3.)
dely= ((yL(1)+yL(3))/2.)-((yL(1)+yL(2)+yL(3))/3.)
! Face 2--Diffusion Coefficients
aps(i) = aps(i) + diff_f*(- Nx(1) * dely + Ny(1) * delx)
as(i,j) = as(i,j) + diff_f*( Nx(2) * dely - Ny(2) * delx)
as(i,j+1) = as(i,j+1) + diff_f*( Nx(3) * dely - Ny(3) * delx)
enddo !end loop on nodes/elements in support
as(i,1)=as(i,1)+as(i,n_s(i)+1)
as(i,n_s(i)+1)=0
enddo !end loop on nodes in domain
!================ SET BOUNDARY================
do i=1,n
BC(i)=0.0
BB(i)=0.0
enddo
do k=1,n_b(1)
BC(B(1,k))=1.e18
enddo
do k=1,n_b(2)
BC(B(2,k))=1.e18
enddo
do k=1,n_b(3)
BC(B(3,k))=1.e18
enddo
do k=1,n_b(4)
BC(B(4,k))=1.e18
enddo
! ----------------------
do i=1,n
QC(i)=0
if (S(i,n_s(i)+1).NE.0) then
QB(i)=vor(i)*V(i)
else
QB(i)=0.0
```

```
endif
enddo
!------------------------
do iters=1,20
do i= 1,n
RHS=BB(i)+QB(i)
do j=1,n_s(i)
RHS=RHS+as(i,j)*sf(S(i,j))
enddo
sf(i)=RHS/(aps(i)+BC(i)+QC(i))
enddo
enddo
enddo
end
```

Index

Note: Page numbers followed by *f* indicate figures and *t* indicate tables.

D

P

R

S

T

V

Printed in the United States
By Bookmasters